WITHDRAWN
UTSA Libraries

Introduction to Microcontrollers

Architecture, Programming, and Interfacing of the Motorola 68HC12

Academic Press Series in Engineering

Series Editor
J. David Irwin
Auburn University

Designed to bring together interdependent topics in electrical engineering, mechanical engineering, computer engineering, and manufacturing, the Academic Press Series in Engineering provides state-of-the-art handbooks, textbooks, and professional reference books for researchers, students, and engineers. This series provides readers with a comprehensive group of books essential for success in modern industry. A particular emphasis is given to the applications of cutting-edge research. Engineers, researchers, and students alike will find the Academic Press Series in Engineering to be an indispensable part of their design toolkit.

Published books in the series:
Industrial Controls and Manufacturing, 1999, E. Kamen
DSP Integrated Circuits, 1999, L. Wanhammar
Time Domain Electromagnetics, 1999, S. M. Rao
Single- and Multi-Chip Microcontroller Interfacing for the Motorola 68HC12, 1999, G. J. Lipovski
Control in Robotics and Automation, 1999, B. K. Ghosh, N. Xi, T. J. Tarn
Soft Computing and Intelligent Systems, 1999, N. K. Sinha, M. M. Gupta

Introduction to Microcontrollers

Architecture, Programming, and Interfacing of the Motorola 68HC12

G. Jack Lipovski
Department of Electrical and Computer Engineering
University of Texas
Austin, Texas

A Harcourt Science and Technology Company

San Diego London Boston
New York Sydney Tokyo Toronto

This book is printed on acid-free paper. ♾

Copyright © 1999 by Academic Press

All rights reserved.
No part of this publication may be reproduced or
transmitted in any form or by any means, electronic
or mechanical, including photocopy, recording, or
any information storage and retrieval system, without
permission in writing from the publisher.

Academic Press
A Harcourt Science and Technology Company
525 B. St., Suite 1900, San Diego, California 92101-4495, USA
http://www.apnet.com

Academic Press
24–28 Oval Road, London NW1 7DX, UK
http://www.hbuk.co.uk/ap/

Library of Congress Catalog Card Number: 99-65099
ISBN: 0-12-451831-1

Printed in the United States of America

99 00 01 02 03 DS 9 8 7 6 5 4 3 2 1

Library
University of Texas
at San Antonio

Dedicated to my father,
Joseph Lipovski

LIMITED WARRANTY AND DISCLAIMER OF LIABILITY

ACADEMIC PRESS ("AP") AND ANYONE ELSE WHO HAS BEEN INVOLVED IN THE CREATION OR PRODUCTION OF THE ACCOMPANYING CODE ("THE PRODUCT") CANNOT AND DO NOT WARRANT THE PERFORMANCE OR RESULTS THAT MAY BE OBTAINED BY USING THE PRODUCT. THE PRODUCT IS SOLD "AS IS" WITHOUT WARRANTY OF ANY KIND (EXCEPT AS HEREAFTER DESCRIBED), EITHER EXPRESSED OR IMPLIED, INCLUDING, BUT NOT LIMITED TO, ANY WARRANTY OF PERFORMANCE OR ANY IMPLIED WARRANTY OF MERCHANTABILITY OR FITNESS FOR ANY PARTICULAR PURPOSE. AP WARRANTS ONLY THAT THE MAGNETIC DISK(S) ON WHICH THE CODE IS RECORDED IS FREE FROM DEFECTS IN MATERIAL AND FAULTY WORKMANSHIP UNDER THE NORMAL USE AND SERVICE FOR A PERIOD OF NINETY (90) DAYS FROM THE DATE THE PRODUCT IS DELIVERED. THE PURCHASER'S SOLE AND EXCLUSIVE REMEDY IN THE EVENT OF A DEFECT IS EXPRESSLY LIMITED TO EITHER REPLACEMENT OF THE DISK(S) OR REFUND OF THE PURCHASE PRICE, AT AP'S SOLE DISCRETION.

IN NO EVENT, WHETHER AS A RESULT OF BREACH OF CONTRACT, WARRANTY OR TORT (INCLUDING NEGLIGENCE), WILL AP OR ANYONE WHO HAS BEEN INVOLVED IN THE CREATION OR PRODUCTION OF THE PRODUCT BE LIABLE TO PURCHASER FOR ANY DAMAGES, INCLUDING ANY LOST PROFITS, LOST SAVINGS OR OTHER INCIDENTAL OR CONSEQUENTIAL DAMAGES ARISING OUT OF THE USE OR INABILITY TO USE THE PRODUCT OR ANY MODIFICATIONS THEREOF, OR DUE TO THE CONTENTS OF THE CODE, EVEN IF AP HAS BEEN ADVISED OF THE POSSIBILITY OF SUCH DAMAGES, OR FOR ANY CLAIM BY ANY OTHER PARTY.

ANY REQUEST FOR REPLACEMENT OF A DEFECTIVE DISK MUST BE POSTAGE PREPAID AND MUST BE ACCOMPANIED BY THE ORIGINAL DEFECTIVE DISK, YOUR MAILING ADDRESS AND TELEPHONE NUMBER, AND PROOF OF DATE OF PURCHASE AND PURCHASE PRICE. SEND SUCH REQUESTS, STATING THE NATURE OF THE PROBLEM, TO ACADEMIC PRESS CUSTOMER SERVICE, 6277 SEA HARBOR DRIVE, ORLANDO, FL 32887, 1-800-321-5068. AP SHALL HAVE NO OBLIGATION TO REFUND THE PURCHASE PRICE OR TO REPLACE A DISK BASED ON CLAIMS OF DEFECTS IN THE NATURE OR OPERATION OF THE PRODUCT.

SOME STATES DO NOT ALLOW LIMITATION ON HOW LONG AN IMPLIED WARRANTY LASTS, NOR EXCLUSIONS OR LIMITATIONS OF INCIDENTAL OR CONSEQUENTIAL DAMAGE, SO THE ABOVE LIMITATIONS AND EXCLUSIONS MAY NOT APPLY TO YOU. THIS WARRANTY GIVES YOU SPECIFIC LEGAL RIGHTS, AND YOU MAY ALSO HAVE OTHER RIGHTS WHICH VARY FROM JURISDICTION TO JURISDICTION.

THE RE-EXPORT OF UNITED STATES ORIGIN SOFTWARE IS SUBJECT TO THE UNITED STATES LAWS UNDER THE EXPORT ADMINISTRATION ACT OF 1969 AS AMENDED. ANY FURTHER SALE OF THE PRODUCT SHALL BE IN COMPLIANCE WITH THE UNITED STATES DEPARTMENT OF COMMERCE ADMINISTRATION REGULATIONS. COMPLIANCE WITH SUCH REGULATIONS IS YOUR RESPONSIBILITY AND NOT THE RESPONSIBILITY OF AP.

Contents

Appendices

PREFACE

Programming is an essential engineering skill. To almost any engineer, it is as important as circuit design to an electrical engineer, as statistics to a civil engineer, and as heat transfer to a chemical engineer. The engineer has to program in high-level languages to solve problems. He or she also should be able to read assembly-language programs to understand what a high-level language does. Finally, he or she should understand the capabilities of a microcontroller because they are components in many systems designed, marketed, and maintained by engineers. The first goal of this book then is to teach engineers how a computer executes instructions. The second goal is to teach the engineer how a high-level language statement converts to assembler language. A final goal is to teach the engineer what can be done on a small computer and how the microcomputer is interfaced to the outside world. Even the nonprogramming engineer should understand these issues. Although this book is written for engineers, it will serve equally well for anyone, even hobbyists, interested in these goals.

The reader is taught the principles of assembly-language programming by being shown how to program a particular microcomputer, the Motorola 6812. The important thing about the 6812 is that it has a straightforward yet powerful instruction set, midway between smaller and more powerful microcontrollers; from it the reader can adjust to these smaller or more powerful microcontrollers. The best way to learn these principles is to write a lot of programs, debug them, and see them work on a real microcontroller. This hands-on experience can be inexpensively obtained on the 6812. Several 6812 boards, which do everything described in this book, are available for under $100. (This price doesn't include the personal computer that hosts the 6812 target system.)

The following discussion outlines the book and explains several decisions that were made when we wrote the book. Optional chapters are available for readers having various interests. The main skills taught in each chapter are summarized.

Chapters 1 to 3 discuss programming using hand-translated machine code, and the implementation of machine instructions in an idealized microcontroller. The assembler is not introduced until Chapter 4. This gives the engineering student a fine feeling for the machine and how it works, and helps him or her resolve problems encountered later with timing in input/output programming or with the use of addressing modes in managing data structures. Chapter 1 explains how a microprocessor interacts with the memory and how it executes the instruction cycle. The explanation focuses on a microcomputer and is simplified to provide just enough background for the remainder of the text. Simple instructions and elementary programs are introduced next. Pointing out that there is no best program to solve a problem, Chapter 1 observes what makes a good program and encourages the reader to appreciate good programming style. A discussion of the available organizations of 6812 microcontrollers concludes this chapter.

In Chapter 2, the main concept is the alternative forms of the same kind of instruction on the 6812. Rather than listing the instructions alphabetically, as is desirable in a reference book, we group together instructions that perform the same type of function. Our groups are the classical ones, namely, the move, arithmetic, logical, edit, control, and input/output groups. Although other groupings are also useful, this one seems to encourage the student to try alternative instructions as a way of looking for the best instruction for his or her purpose. The 6812 has an extensive set of addressing modes that can be used with most instructions; these are covered in Chapter 3. The different addressing modes are introduced with a goal of explaining why these modes are useful as well as how they work. Examples at the end of the chapter illustrate the use of these modes with the instructions introduced in Chapter 2.

The end of Chapter 3 shows the use of program-relative addressing for position independence and the use of stack addressing for recursion and reentrancy.

Chapters 4 to 6 show how a program can be more easily written in assembler and the high-level C language and translated into machine code by an assembler. Chapter 4 introduces the assembler, explains assembler directives and symbolic addresses, and introduces limitations of forward referencing in a two-pass assembler. Assembly language examples that build on the examples from previous chapters conclude Chapter 4. Chapter 5, which may be omitted if the reader is not going to write assembler language programs, provides insights needed by programmers who write large assembler language programs. A general discussion of related programs, including macro, conditional, and relocatable assemblers and linkers, is given.

Chapter 6 develops assembler language subroutines. It illustrates techniques used in assembler language at an implementation level (such as passing arguments in registers). The tradeoffs between macros and subroutines as a means to insert the same program segment into different places in a program are discussed.

Chapter 7 covers arithmetic routines. An implementation is shown for unsigned and signed multiplication and division. Conversion between different bases is discussed and examples are given illustrating conversion from ASCII decimal to binary and vice versa. Stack operation and Polish notation are shown to be useful in realizing arithmetic routines of any complexity. Multiple-precision integer arithmetic is discussed by means of examples of 32-bit operations including multiplication and division. Floating-point representations are introduced, and the mechanics of common floating- point operations are discussed. Finally, a 6812-oriented introduction of fuzzy logic is presented.

Chapter 8, which may be omitted if the reader is already familiar with C, discusses compilers and interpreters, and briefly introduces C programming to provide a background for later chapters.

Chapter 9 introduces the implementation of C procedures. Several constructs in C, such as switch statements, are shown implemented in assembler language. The techniques C uses to hold local variables and to pass arguments to a subroutine on the stack are shown implemented in assembler language.

Chapter 10 covers elementary data structures. The simplest, including the character string used in earlier chapters, and the more involved deque and linked list structures are related to the addressing modes available in the 6812. The main theme of this chapter is that the correct storage of data can significantly improve the efficiency of a program.

Chapter 11 introduces input/output programming. Input and output devices are characterized. Then the 6812's parallel ports are described. Input and output software is illustrated with some examples. We then show 6812 synchronization hardware, to introduce gadfly and interrupt synchronization. Finally we show how D-to-A and A-to-D conversion is done.

Chapter 12 shows how the assembly language of a different microcontroller might be learned once that of the 6812 has been learned. Although we would like to discuss other popular microcontrollers, we believe that we could fill another book doing that. To illustrate the idea, we look at the near relatives less costly than the 6812, in particular, the 6805, 6808, and 6811. We also discuss briefly more powerful microcontrollers such as the 68300, 500, and M·CORE series. The main theme is that once you understand the instruction set of one microcontroller, you can quickly learn to program efficiently on other microcontrollers.

This book systematically develops the concepts of programming of a microcontroller in high-level language and assembly language. It also covers the principles of good programming practice through top-down design and the use of data structures. It is suitable as an introductory text for a core course in an engineering curriculum on assembly language programming or as the first course on microcomputers that demonstrates what a small computer can do. It may also be used by those who want to delve more deeply into assembly language principles and practices. You should find, as we have, that programming is a skill that magnifies the knowledge and control of the programmer, and you should find that programming, though very much an important engineering skill, is fun and challenging. This book is dedicated to show you that.

Problems are a major part of a good textbook. We have developed over twenty problems for each chapter, and for each section we generally have at least two problems, one that can be assigned for homework, while the other can be used in a quiz or exam. Some of these problems are "brain teasers" that are designed to teach the student that even simple programs should be tested, generally at their extreme values, to be sure they work. Often the obvious and simple solutions succumb to truncation or overflow errors. Also, problems in Chapter 11, including the keyless entry design and the experiment that plays "The Eyes of Texas" on a pair of earphones, are absolutely great motivators for sophomores, when they get them to work on a real microcontroller board. They see how exciting computer engineering is. This is having a significant impact on retention. An instructor's manual, available from the publisher, includes solutions to all the problems given at the end of each chapter.

This book was developed largely from a book by the author and T. J. Wagner on the 6809. The author expresses his gratitude for the contributions made by Dr. Wagner through his writing of much of the earlier book.

List of Figures

Figure	Title	Page

List of Tables

Acknowledgments

The author would like to express his deepest gratitude to everyone who contributed to the development of this book. The students of EE 319K at the University of Texas at Austin during fall 1998 significantly helped correct this book; special thanks are due to Chao Tan, Brett Wisdom, Alex Winbow, Eric Wood, John Prochnow, and Nathan Madino for finding the most errors. This text was prepared and run off using a Macintosh and LaserWriter, running WriteNow. I am pleased to write this description of the Motorola 6812, which is an incredibly powerful component and a vehicle for teaching a considerable range of concepts.

G. J. L.

About the Author

G. Jack Lipovski has taught electrical engineering and computer science at the University of Texas since 1976. He is a computer architect internationally recognized for his design of the pioneering database computer, CASSM, and the parallel computer, TRAC. His expertise in microcomputers is also internationally recognized by his being a past director of Euromicro and an editor of *IEEE Micro*. Dr. Lipovski has published more than 70 papers, largely in the proceedings of the annual symposium on computer architecture, the IEEE transactions on computers, and the national computer conference. He holds eight patents, generally in the design of logic-in-memory integrated circuits for database and graphics geometry processing. He has authored seven books and edited three. He has served as chairman of the IEEE Computer Society Technical Committee on Computer Architecture, member of the Computer Society Governing Board, and chairman of the Special Interest Group on Computer Architecture of the Association for Computer Machinery. He has been elected Fellow of the IEEE and a Golden Core Member of the IEEE Computer Society. He received his Ph.D. from the University of Illinois, 1969, and has taught at the University of Florida and at the Naval Postgraduate School, where he held the Grace Hopper chair in Computer Science. He has consulted for Harris Semiconductor, designing a microcomputer, and for the Microelectronics and Computer Corporation, studying parallel computers. He founded the company Linden Technology Ltd. and is the chairman of its board. His current interests include parallel computing, database computer architectures, artificial intelligence computer architectures, and microcomputers.

1

Basic Computer Structure and the 6812

Computers, and microcomputers in particular, are among the most useful tools that humans have developed. They are not the news media's mysterious half-human forces implied by "The computer will decide . . ." or "It was a computer error!" No, computers are actually like levers; as a lever amplifies what the human arm can do, so the computer amplifies what the human brain can do. Good commands are amplified, and the computer is a great tool, but bad commands are likewise amplified, and good commands incorrectly programmed are also amplified. "To err is human, but to really foul things up, you need a computer." You have to study and exercise this tool to make it useful; that is the purpose of this book. The computer also has to be used with insight and consideration for its effects on society, but that will not be studied in this book.

We shall study the computer as an engineer studies any tool—we begin by finding out just how it ticks. We make our discussion concrete using the well-designed Motorola 6812 microcomputer, as a means of teaching the operations of computers in general. In this chapter we introduce basic computer structure. We discuss memory, how memory words are read to tell the microcomputer what to do, and how these words are written and read to save the microcomputer's data. Finally, we describe a small but useful subset of 6812 instructions to show how a computer reads and carries out an instruction and a program, to introduce the idea of programming.

After reading this chapter, you should be able to approach a typical instruction, to be introduced in the next two chapters, with an understanding about what the mnemonic, the machine code, and a sequence of memory reads and writes may mean for that instruction. This chapter then provides background for the discussion of instructions that we will present in the next two chapters.

1.1 Basic Computer Structure

What is a microcomputer, and how does it execute the instructions that a programmer writes for it? This question is explored now at a level of abstraction that will be adequate for this text. We do know that many readers will object to one aspect of the following discussion, and we want to answer that objection a priori, so that those readers will not miss the point. We will introduce a seemingly large number of terms. Don't miss the objective: We are really introducing concepts. The reader should think about the concepts rather than memorize definitions. Like your physics text, this text has to use terms in a fairly precise way to avoid ambiguity. Your physics text, you may recall, used the word "work" in a very precise way, as the product of force times distance, which is a bit different from the conversational use of the word "work" as used in the expression, "He's doing a lot of work." We will use terms such as "read" and "fetch" in a similar way. When defined, they will be written in *italics* and will be listed in the index. We ask you to learn the term and its meaning even though you do not have to memorize the wording of the definition. But take heart, because although we do have a few concepts that have to be learned, and we have to learn the terms for those concepts, we do not have many formulas or equations to deal with. Accept our challenge to understand these terms; then you will enjoy the latter discussions even more than if you muddle through this section without thinking about the terminology.

You probably know what microcomputers and computers are, to some degree, but let us discuss the term "computer" so that if you get into an argument about whether a hand calculator is a computer, you can respond knowledgeably.

A microcomputer is a kind of computer or, more accurately, a kind of von Neumann computer, named after the scientific giant of our century who invented it. All *von Neumann computers* have four components: *memory, controller, data operator* (sometimes called an arithmetic-logic unit), and *input-output* (*I/O*), which are connected by an address and data bus. A simplified diagram of a computer is shown in Figure 1.1. Briefly, the memory stores both the data and the program, and the input-output provides the communication with the outside world. In a conventional personal computer system input-output is done through peripherals such as CRTs, keyboards, scanners, printers, modems, and so on. In typical microcontroller applications the input-output system provides the necessary connections, or interfacing, to the device, of which the microcontroller is a component, such as an automobile, kitchen appliance, toy, or laboratory instrument. The data operator performs arithmetic and logical operations on data, such as addition, ANDing, and so on. The controller controls the flow of information between the components by means of control lines (which are not shown in Figure 1.1), directing what computation is to be done. The input/output, controller, and data operator may themselves contain a small amount of memory in *registers*.

A microcomputer is a computer that is implemented using low-cost integrated circuits (ICs) and is therefore cheap enough to be used in an incredible range of applications where a large computer would be infeasible. For the purposes of this book, if the data operator and the controller are together on a single IC, but other ICs are needed to complete the computer, that IC is called a *microprocessor*; the computer that uses a

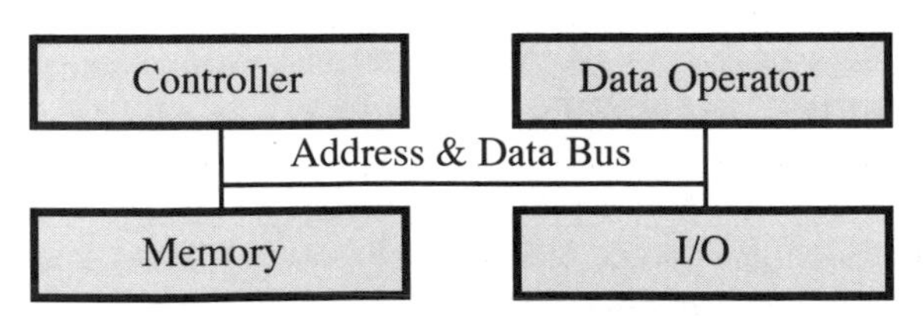

Figure 1.1. Simplified Computer Structure

microprocessor is called a *microcomputer*; and, if a complete computer is on a single integrated circuit, that integrated circuit is called a *single-chip microcontroller*.

Some aspects of microcomputers apply to all computers. We will often discuss an aspect of the computer and, of course, that aspect applies to microcontrollers, on which we are concentrating. The microcomputer's, or microcontroller's, controller and data operator is abbreviated *MPU* (microprocessor unit). The abbreviation CPU (central processor unit) is often used to denote the controller and data operator, but that term leads subtly to the idea that the CPU is central and most important; but this is misleading, especially when a computer system has many MPUs, none of which is "central."

We now look more closely at the memory and the MPU. We can think of the memory as a large number of cells, each able to store a 0 or a 1—that is, a binary digit or one bit of memory. The bits are grouped together in units called *bytes*, which consist of 8 bits. Within a particular byte the bits are labeled b7, . . ., b0 as shown.

b7	b6	b5	b4	b3	b2	b1	b0

Byte

The right-hand bits are called lower-order or least significant, and the left-hand bits are called higher-order or most significant. There is nothing sacred about this labeling, and, in fact, many computer manufacturers reverse it. A *word* in memory is the number of bits that are typically read or written as a whole. In small microcomputers, a word is one byte, so that the terms "word" and "byte" are often used interchangeably. In this text, the 6812 can read an 8-bit or a 16-bit word, which is two bytes. In a 16-bit word, bits are numbered from 15 (on the left) to 0 (on the right). In the memory, each byte has an address between 0 and $2^N - 1$, where N is the total number of address bits. In the 6812, N is essentially 16, so each address between 0 and 65,535 is described by its 16-bit binary representation (see Appendix 1), although some 6812 versions can extend this range.

The MPU controls the memory by means of a clock and a read/write line, and communicates to it by an address bus and a data bus, shown in Figure 1.1. A *line* or *wire* carries one bit of information at each instance of time by the level of voltage on it. Each line can carry a true (1) or a false (0) value. A *bus* is a collection of lines that can move a word or an address in parallel, so that one bit of a word or address is on one line of the bus. The data bus moves an 8-bit or 16-bit word to or from memory and from or to the MPU, and the address bus moves a 16-bit address from the MPU to the memory. A *clock* is a signal that is alternately a 0 and 1 (a square wave). A *clock cycle* is the time interval from when the clock just becomes 0 until the next time it just becomes 0, and the *clock rate*, or *clock frequency*, is the reciprocal of the clock cycle time. Contemporary 6812 microcontrollers essentially use an 8 MHz clock.

In one clock cycle, the MPU can *read* a word from the memory by putting an address on the address bus and making the *read/write line* 1 for the cycle and then picking up the word on the data bus at the end of the cycle. It can also *write* a word into the memory in one clock cycle by putting the address on the address bus and putting the word on the data bus and making the read/write line 0 for the cycle. A read or a write to memory is also called an *access* to memory.

We can enlarge our description of how the memory works. Assume that we want to get the contents of a particular word or byte from memory, that is, read a word from memory. The MPU first puts a 1 for read on the read/write line and then puts the address of the desired word or byte on the address bus throughout the duration of a clock cycle. The memory is designed so that, at the end of the clock cycle, the desired word is put on the data bus. The MPU then places a copy of the contents of the word or byte on the data bus into some register inside the MPU as required by the instruction that it is executing. This is done without changing the contents in memory of the word or byte addressed.

To write a word into memory, the address of the word is put on the address bus, the word is put on the data bus, and the read/write line has 0 (to indicate a write) for a full clock cycle. The memory is designed to store the word at that address at the end of the clock cycle. After the word is stored at the given address, the MPU may still retain in one of its registers a copy of the word that has just been written.

The MPU can read or write an 8-bit or a 16-bit word in memory in one clock cycle. Such a memory is usually called *random access memory* or *RAM* because each byte is equally accessible or can be selected at random without changing the time of the access. With microcomputer applications, it is not unusual to have part of the memory bytes in *ROM* (*read only memory*). A ROM is capable of a read operation but not a write operation; its words are written when it is made at the factory and are retained even when the power is turned off. If the data in a RAM are lost when power is turned off, the RAM is termed *volatile;* otherwise, it is termed *nonvolatile*. RAM memories are essentially volatile. The term RAM is also used almost universally to imply memories that you can read and write in, even though ROM memories can be randomly accessed with a read operation. The part of memory that is in ROM is typically used to store a *program* for a microcomputer that only executes one program. For example, the microcontroller used in an automobile would be running the same program every time it is used, so that the part of the memory that is used for storing the program is in ROM.

1.2 The Instruction

We now examine the notion of an *instruction,* which is an indivisible operation performed by the MPU. It can be described statically as a collection of bits stored in memory or as a line of a program or, dynamically, as a sequence of actions by the controller. In this discussion we begin with a simplified dynamic view of the instruction and then develop a static view. Examples are offered to combine these views to explain the static aspects of the operation code, addressing mode, machine code, and mnemonics. We conclude with an expanded view of the dynamic aspects of the instruction cycle.

The controller will send commands to memory to read or write and will send commands to all other parts of the computer to effectively carry out the intentions of the programmer. The specification of what the control unit is to do is contained in a program, a sequence of instructions stored, for the most part, in consecutive bytes of memory. To execute the program, the MPU controller repeatedly executes the *instruction cycle* (or *fetch/execute cycle*):

1. Fetch (read) the next instruction from memory.
2. Execute the instruction.

As we shall see with the 6812 MPU, reading an instruction from memory will require that one or more bytes be read from memory. To execute the instruction, some bytes may also be read or written. These two activities, read and execute, seem to be read or write operations with the memory but are quite different to the MPU, and we use different terms for them. To *fetch* means to read a word from memory to be used as an instruction in the controller. The first step in the cycle shown previously is the *fetch phase*. To *recall* means to read a word into the MPU that is not part of the instruction. The recall and write operations are done in the second step of the instruction, which is called the *execute phase*. Thus, when we talk about fetching a word, you can be sure that we are talking about reading the instruction, or part of the instruction. We will not use "fetch" to describe an operation of reading data to be input to the data operator.

The 6812's registers are shown in Figure 1.2, where the top five registers hold 16 bits and the condition code register holds 8 bits. The 16-bit D register is composed of two 8-bit registers A and B; D, A, and B are called *accumulators* because arithmetic operations can be done with their contents with the results placed back in the registers to accumulate the result. This accumulating aspect of registers D, A, and B will be assumed to be understood so that we often refer to (register) "D," "A," or "B" rather than "accumulator D," "accumulator A," or "accumulator B." The registers A and B are always the left and right halves of register D; if you put $12 in register A and $34 in register B then read register D, it has $1234. Similarly, if you put $5678 in register D, then reading register A gives $56 and reading register B gives $78. Registers X and Y are

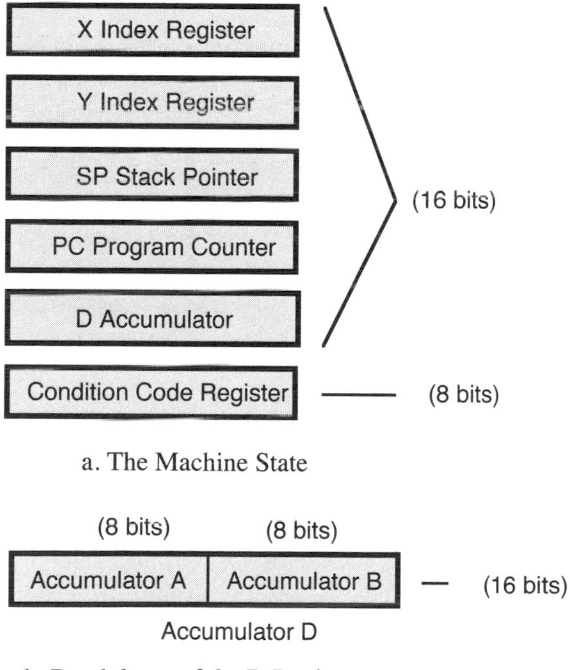

Figure 1.2. Registers in the 6812

index registers, and SP is a stack pointer; they are used in address calculations. The program counter, PC, is used to fetch the instruction. It is called a counter because it normally increments each time it is used. The condition code register (CC) has bits that are used to indicate results of tests that can be used in conditional branch instructions.

At the beginning of the instruction cycle it is assumed that the program counter contains the address of the first byte of the instruction. As each byte of the instruction is fetched, the PC is incremented by 1, so that the PC always has the address of the next byte of the instruction to be read from memory. When the instruction cycle is completed, the PC then automatically contains the address of the first byte of the next instruction.

We now look at the instruction statically as one or more memory bytes or as a line of a program. This discussion will introduce new concepts, but we have tried to keep the number down so that the examples can be discussed without your having too many terms to deal with. The examples will help to clarify the concepts that we introduce below.

Each instruction in a microcomputer carries out an operation. The types of operations provided by a von Neumann computer can be categorized as follows:

1. Move.
2. Arithmetic.
3. Logical.
4. Edit.
5. Control.
6. Input/output.

At this time, we are interested in how instructions are stored in memory as part of a program and how they are executed by the 6812.

After the instruction is fetched, the execute phase of the fetch execute cycle will often use an address in memory for the input data of the operation or for the location of the result of the operation. This location is called the *effective address*. The Motorola 6812, like most microcomputers, is a *one-address computer* because each instruction can specify at most one effective address in memory. For instance, if an instruction were to move a word from location 100 in memory into register A, then 100 is the effective address. This effective address is generally determined by some bits in the instruction. The *addressing mode* specifies how the effective address is to be determined, and there are binary numbers in the instruction that are used to determine the address. The effective address is calculated at the beginning of the execute phase, just after the instruction is fetched and before any of the operations actually take place to execute the instruction.

An instruction in the 6812 is stored in memory as one or more bytes. The first, and possibly only, byte of the instruction is generally the operation code byte. The *operation code byte* contains the *operation code (opcode*, for short), which specifies the operation to be carried out and the specification of the addressing mode. The remaining bytes of the instruction, if any, specify the effective address according to the given addressing mode. The bytes representing the instruction can be represented as a sequence of ones and zeros, that is, a binary number. The trouble with this is that it is hard to remember and to check an 8-bit or longer string of ones and zeros. To make it easier, we can represent the bit pattern as a *hexadecimal number*. A hexadecimal number will be prefixed by a dollar sign ($) to distinguish it from a decimal number. (For example, 193 = $C1. If you are unfamiliar with hexadecimal numbers, see Appendix 1.) When the opcode, addressing modes, and constants used to determine the address are represented either as binary or hexadecimal numbers, we call this representation the *machine code* because it is the actual format used to store the instruction in the *machine* (microcomputer), and this format is used by the machine to determine what is to be done.

Machine code is quite useful for making small changes in a program that is being run and corrected or *debugged*. However, writing even a moderately long program in machine code is a punishment that should be reserved for the fifth level of Dante's inferno. In Chapter 4 we discuss how text produced by an editor is converted by an *assembler* to the machine code stored in the computer's memory. The text input to the assembler is called *source code*. In a line of source code, to make remembering the instructions easier, a three- or four-character *mnemonic* is used to describe the operation, and its addressing information may be given as a hexadecimal or a decimal number. A line of source code, consisting of mnemonics and addressing information, can be converted by hand into their hexadecimal machine code equivalents using Motorola's CPU12 Reference Guide (you can order it from Motorola by using reference number CPU12RG/D). In the first three chapters, we want to avoid using the assembler, to see clearly just how the computer ticks. We will hand-convert mnemonics and addressing information to hexadecimal machine code and work with hexadecimal machine code.

We now look at a load immediate instruction in detail, to introduce concepts about instructions in general. The load instruction will move a byte into an accumulator, either A or B. Its simplest addressing mode is called *immediate*. For instance, to put a specific number, say $2F, in accumulator A, execute the instruction whose source code line is

```
LDAA #$2F
```

where the symbol "#" denotes immediate addressing and the symbol "$" is used to indicate that the number that follows is in hexadecimal. This instruction is stored in memory as the two consecutive bytes:

$86
$2F

(Look in CPU12 Reference Guide, Instruction Set Summary, for the mnemonic `LDAA` and, under it, find $86 in under the Machine Coding column, in the row beginning LDAA #opr8i, which also has the addressing mode IMM for immediate addressing.)

Looking dynamically at an instruction, an operation (e.g., add, subtract, load, clear, etc.) may be carried out with inputs (or *operands)* and may produce a result. The instruction is executed in the instruction cycle as follows.

1. Fetch the first byte of the instruction from memory.
2. Increment the PC by one.
3. Decode the opcode that was fetched in step 1.
4. Repeat steps 1 and 2 to fetch all bytes of the instruction.
5. Calculate the effective address to access memory, if needed.
6. Recall the operand from memory, if needed.
7. Execute the instruction, which may include writing the result into memory.

The controller fetches the first byte, $86. The program counter is incremented. The controller decodes $86. The controller fetches the second byte, $2F, putting it into accumulator A. The program counter is incremented. After this instruction is executed, another instruction is fetched and executed.

1.3 A Few Instructions and Some Simple Programs

Now that we have examined the instruction from static and dynamic points of view, we will look at some simple programs. The machine code for these programs will be described explicitly so that you can try out these programs on a real 6812 and see how they work, or at least so that you can clearly visualize this experience. The art of determining which instructions have to be put where is introduced together with a discussion of the bits in the condition code register. We will discuss what we consider to be a good program versus a bad program, and we will discuss what is going to be in Chapters 2 and 3. We will also introduce an alternative to the immediate addressing mode using the load instruction. Then we bring in the store, add, software interrupt, and add with carry instructions to make the programs more interesting as we explain the notions of programming in general.

We first consider some variations of the load instruction to illustrate different addressing modes and representations of instructions. We may want to put another number into register A. Had we wanted to put \$3E into A rather than \$2F, only the second byte of the instruction would be changed, with \$3E replacing \$2F. The same instruction as

```
LDAA    #$3E
```

could also be written using a decimal number as the immediate operand: for example,

```
LDAA #62
```

Either line of source code would be converted to machine code as follows:

$86
$3E

We can load register B using a different opcode byte. Had we wanted to put \$2F into accumulator B, the first byte would be changed from \$86 to \$C6 and the instruction mnemonic would be written

```
LDAB #$2F
```

We now introduce the direct addressing mode. Although the immediate mode is useful for initializing registers when a program is started, the immediate mode would not be able to do much work by itself. We would like to load words that are at a known location but whose actual value is not known at the time the program is written. One could load accumulator B with the contents of memory location \$0840. This is called *direct addressing*, as opposed to immediate addressing. The addressing mode, direct, uses no pound sign "#" and a 2-byte address value as the effective address; it loads the word at this address into the accumulator. The instruction mnemonic for this is

```
LDAA    $0840
```

and the instruction appears in memory as the three consecutive bytes.

$B6
$08
$40

Notice that the "#" is missing in this mnemonic because we are using direct addressing instead of immediate addressing. Also, the second two bytes of the instruction give the address of the operand, high byte of the address first.

The store instruction is like the load instruction described earlier except that it works in the reverse manner (and a STAA or STAB with the immediate addressing mode is neither sensible nor available). It moves a word from a register in the MPU to a memory location specified by the effective address. The mnemonic, for store from A, is STAA; the instruction

```
STAA 2090
```

will store the byte in A into location 2090 (decimal). Its machine code is

$7A
$08
$2A

where we note that the number 2090 is stored in hexadecimal as $082A. With direct addressing, two bytes are always used to specify the address even though the first byte may be zero.

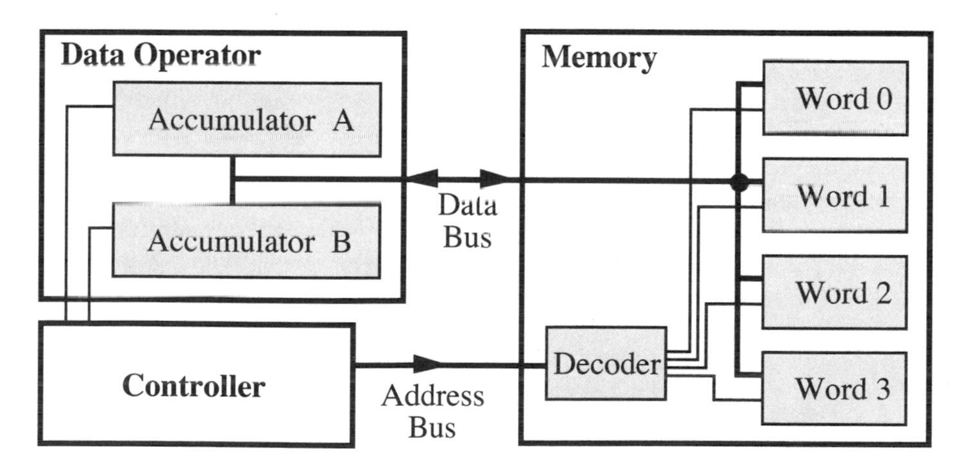

Figure 1.3. Registers and Memory

Figure 1.3 illustrates the implementation of the load and store instruction in a simplified computer, which has two accumulators and four words of memory. The data operator has accumulators A and B, and a memory has four words and a decoder. For the instruction `LDAA 2`, which loads accumulator A with the contents of memory word 2, the controller sends the address 2 on the address bus to the memory decoder; this decoder enables word 2 to be read onto the data bus, and the controller asserts a control signal to accumulator A to store the data on the data bus. The data in word 2 is not lost or destroyed. For the instruction `STAB 1`, which stores accumulator B into the memory word 1, the controller asserts a control signal to accumulator B to output its data on the data bus, and the controller sends the address 1 on the address bus to the memory decoder; this decoder enables word 1 to store the data on the data bus. The data in accumulator B is not lost or destroyed.

The `ADD` instruction is used to add a number from memory to the number in an accumulator, putting the result into the same accumulator. To add the contents of memory location $0BAC to accumulator A, the instruction

```
ADDA $0BAC
```

appears in memory as

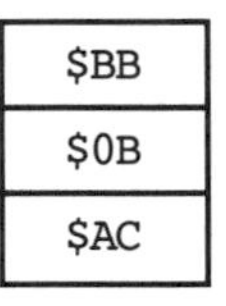

The addition of a number on the bus to an accumulator such as accumulator A is illustrated by a simplified computer having a data bus and an accumulator (Figure 1.4).

The data operator performs arithmetic operations using an adder (see Figure 1.4). Each 1-bit adder, shown as a square in Figure 1.4b, implements the truth table shown in Figure 1.4a. Registers A, B, and S may be any of the registers shown in Figure 1.2 or instead may be data from a bus. The two words to be added are put in registers A and B, Cin is 0, and the adder computes the sum, which is stored in register S. Figure 1.4c shows the symbol for an adder. Figure 1.4d illustrates addition of a memory word to accumulator A. The word from accumulator A is input to the adder while the word on the data bus is fed into the other input. The adder's output is written into accumulator A.

When executing a program, we need an instruction to end it. When using the debugger MCUez or HiWave with state-of-the-art hardware, background (mnemonic `BGND`) halts the microcontroller, when using the debugger DBUG_12 with less-expensive hardware, software interrupt (mnemonic `SWI`) serves as a halt instruction. In either case, the `BRA` instruction discussed in the next chapter can be used to stop. When you see the instruction `SWI` or `BGND` in a program, think "halt the program."

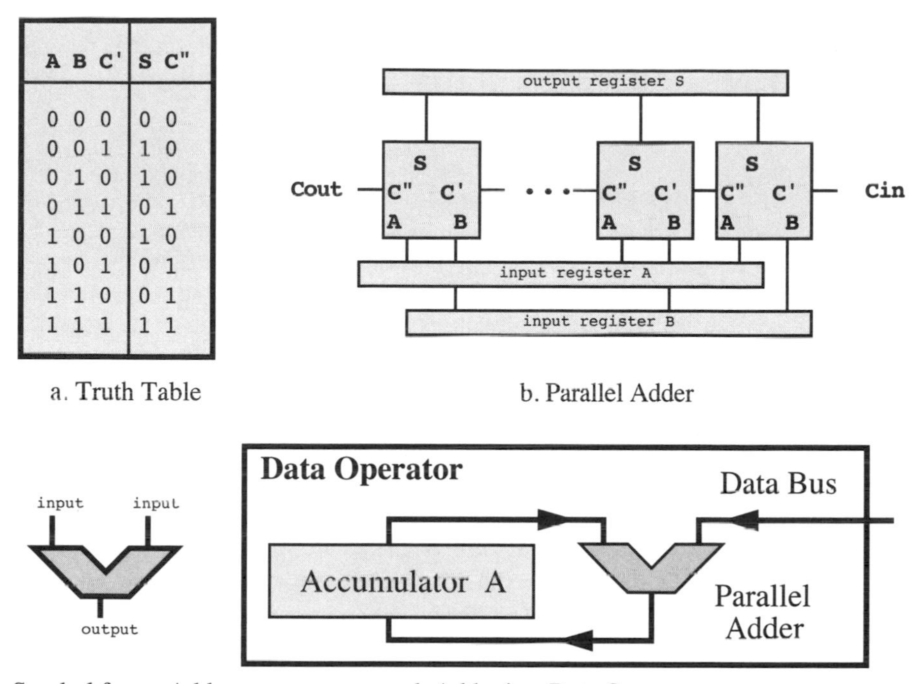

Figure 1.4. Data Operator Arithmetic

The four instructions in Figure 1.5 can be stored in locations 2061 through 2070, or \$80D through \$816 in hexadecimal; its execution adds two numbers in locations \$840 and \$841, putting the result into location \$842.

Location	Contents	Mnemonic	Comment
080D	B6	LDAA $840	; get 1st operand
080E	08		
080F	40		
0810	BB	ADDA $841	; add 2nd operand
0811	08		
0812	41		
0813	7A	STAA $842	; store sum
0814	08		
0815	42		
0816	00	BGND	; halt (use SWI in DEBUG-12)

Figure 1.5. Program for 8-Bit Addition

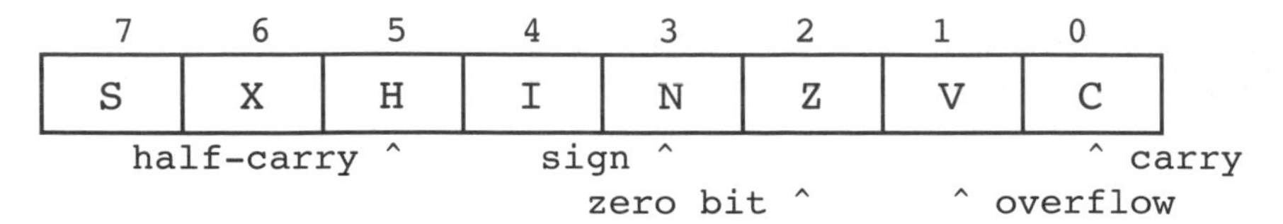

Figure 1.6. Bits in the Condition Code Register

We will now look at condition code bits in general and the carry bit in particular. The carry bit is really pretty obvious. When you add by hand, you may write the carry bits above the bits that you are adding so that you will remember to add it to the next bit. When the microcomputer has to add more than eight bits, it has to save the carry output from one byte to the next, which is Cout in Figure 1.4b, just as you remembered the carry bits in adding by hand. This carry bit is stored in bit C of the condition code register shown in Figure 1.6. The microcomputer can input this bit into the next addition as Cin in Figure 1.4b. For example, when adding the 2-byte numbers $349E and $2570, we can add $9E and $70 to get $0E, the low byte of the result, and then add $34, $25, and the carry bit to get $5A, the high byte of the result. See Figure 1.7. In this figure, C is the carry bit obtained from adding the contents of locations 11 and 13; (m) is used to denote the contents of memory location m, where m may be 10, 11, 12, etc. The *carry bit* (or *carry* for short) in the *condition code register* (Figure 1.6) is used in exactly this way.

The carry bit is also an error indicator after the addition of the most significant bytes is completed. As such, it may be tested by conditional branch instructions, to be introduced later. Other characteristics of the result are similarly saved in the controller's condition code register. These are, in addition to the carry bit C, N (negative or sign bit), Z (zero bit), V (two's-complement overflow bit or signed overflow bit), and H (*half-carry* bit) (see Figure 1.6). How 6812 instructions affect each of these bits is shown in the CPU12 Reference Guide, Instruction Set Summary, in the rightmost columns.

We now look at a simple example that uses the carry bit C. Figures 1.8 and 1.9 list two equally good programs to show that there is no way of having exactly one correct answer to a programming problem. After the example, we consider some ways to know if one program is better than another. Suppose that we want to add the two 16-bit numbers in locations $810, $811 and $812, $813, putting the sum in locations $814, $815. For all numbers, the higher-order byte will be in the smaller-numbered location. One possibility for doing this is the following instruction sequence, where, for compactness, we give only the memory location of the first byte of the instruction.

<- C

($810)	($811)
($812)	($813)
($814)	($815)

Figure 1.7. Addition of Two-Byte Numbers

Location	Contents	Mnemonic	Comment
820	F6 08 11	LDAB $811	; get low byte of 1st
823	B6 08 10	LDAA $810	; get high byte of 1st
826	FB 08 13	ADDB $813	; add low byte of 2nd
829	B9 08 12	ADCA $812	; add high byte of 2nd
82C	7B 08 15	STAB $815	; store low sum byte
82F	7A 08 14	STAA $814	; store high sum byte
832	00	BGND	; halt

Figure 1.8. Program for 16-Bit Addition

In the program segment above, the instruction `ADCA $812` adds the contents of A with the contents of location $812 and the C condition code bit, putting the result in A. At that point in the sequence, this instruction adds the two higher-order bytes of the two numbers together with the carry generated from adding the two lower-order bytes previously. This is, of course, exactly what we would do by hand, as seen in Figure 1.7. Note that we can put this sequence in any 19 consecutive bytes of memory as long as the 19 bytes do not overlap with data locations $810 through $815. Finally, the notation A:B is used for putting the accumulator A in tandem with B or concatenating A with B. This concatenation is just the double accumulator D. We could also have used just one accumulator with the following instruction sequence.

In this new sequence, the load and store instructions do not affect the carry bit C. (See the CPU12RG/D manual Instruction Set Summary. We will understand why instructions do not affect C as we look at more examples.) Thus, when the instruction `ADCA $812` is performed, C has been determined by the `ADDB $813` instruction.

The two programs above were equally acceptable. However, we want to discuss guidelines to writing good programs early in the book, so that you can be aware of them to know what we are expecting for answers to problems and so that you can develop a good programming style. A good program is shorter and faster and is generally clearer than a bad program that solves the same problem. Unfortunately, the fastest program is

Location	Contents	Mnemonic	Comment
820	B6 08 11	LDAA $811	; get low byte of 1st
823	BB 08 13	ADDA $813	; add low byte of 2nd
826	7A 08 15	STAA $815	; store low sum byte
829	B6 08 10	LDAA $810	; get high byte of 1st
82C	B9 08 12	ADCA $812	; add high byte of 2nd
82F	7A 08 14	STAA $814	; store high sum byte
832	00	BGND	; halt

Figure 1.9. Alternative Program for 16-Bit Addition

almost never the shortest or the clearest. The measure of a program has to be made on one of the qualities, or on one of the qualities based on reasonable limits on the other qualities, according to the application. Also, the quality of clarity is difficult to measure but is often the most important quality of a good program. Nevertheless, we discuss the shortness, speed, and clarity of programs to help you develop good programming style.

The number of bytes in a program (its length) and its execution time are something we can measure. A short program is desired in applications where program size affects the cost of the end product. Consider two manufacturers of computer games. These products feature high sales volume and low cost, of which the microcomputer and its memory are a significant part. If one company uses the shorter program, its product may need fewer ROMs to store the program, may be substantially cheaper, and so may sell in larger volume. A good program in this environment is a short program. Among all programs doing a specific computation will be one that is the shortest. The quality of one of these programs is the ratio of the number of bytes of the shortest program to the number of bytes in the particular program. Although we never compute this *static efficiency* of a program, we will say that one program is more statically efficient than another to emphasize that it takes fewer bytes than the other program.

The CPU12RG/D manual Instruction Set Summary gives the length of each instruction by showing its format. For instance, the `LDAA #$2F` instruction is shown alphabetically under `LDAA` in the line IMM. The pattern 86 ii, means that the opcode is $86 and there is a one-byte immediate operand ii, so the instruction is two bytes long.

The speed or execution time of a program is prized in applications where the microcomputer has to keep up with a fast environment, such as in some communication switching systems, or where the income is related to how much computing can be done. A faster computer can do more computing and thus make more money. However, speed is often overemphasized: "My computer is faster than your computer." To show you that this may be irrelevant, we like to tell this little story about a computer manufacturer. This is a true story, but we will not use the manufacturer's real name for obvious reasons. How do you make a faster version of a computer that executes the same instruction? The proper answer is that you run a lot of programs, and find instructions that are used most often. Then you find ways to speed up the execution of those often-used instructions. Our company did just that. It found one instruction that was used very, very often: It found a way to really speed up that instruction. The machine should have been quite a bit faster, but it wasn't! The most common instruction was used in a routine that waited for input-output devices to finish their work. The new machine just waited faster than the old machine that it was to replace. The moral of the story is that many computers spend a lot of time waiting for input-output work to be done. A faster computer will just wait more. Sometimes speed is not as much a measure of quality as it is cracked up to be. But then in other environments, it is the most realistic measure of a program. As we shall see in later chapters, the speed of a particular program can depend on the input data to the program. Among all the programs doing the same computation with specific input data, there will be a program that takes the fewest number of clock cycles. The ratio of this number of clock cycles to the number of clock cycles in any other program doing the same computation with the same input data is called the *dynamic efficiency* of that program. Notice that dynamic efficiency does depend on the input data but not on the clock rate of the microprocessor. Although we never calculate

Location	Contents	Mnemonic	Comment
820	FC 08 10	LDD $810	; get 1st 16-bit word
823	F3 08 12	ADDD $812	; add 2nd 16-bit word
826	7C 08 14	STD $814	; store 16-bit word
829	00	BGND	; halt

Figure 1.10. Most Efficient Program for 16-Bit Addition

dynamic efficiency explicitly, we do say that one program is more dynamically efficient than another to indicate that the first program performs the same computation more quickly than the other one over some range of input data.

The CPU12RG/D manual Instruction Set Summary gives the instruction timing. For instance, the `LDAA #$2F` instruction is shown alphabetically under `LDAA` for the mode IMM. The Access Detail column indicates that this instruction takes one memory cycle of type P, which is a program word fetch. Generally, a memory cycle is 125 ns.

The *clarity* of a program is hard to evaluate but has the greatest significance in large programs that have to be written by many programmers and that have to be corrected and maintained for a long period. Clarity is improved if you use good documentation techniques, such as comments on each instruction that explain what you want them to do, and flowcharts and precise definitions of the inputs, outputs, and the state of each program, as explained in texts on software engineering. Some of these issues are discussed in Chapter 5. Clarity is also improved if you know the instruction set thoroughly and use the correct instruction, as developed in the next two chapters.

While there are often two or more equally good programs, the instruction set may provide significantly better ways to execute the same operation, as illustrated by Figure 1.10. The 6812 has an instruction `LDD` to load accumulator D, an instruction `ADDD` to add to accumulator D, and an instruction `STD` to store accumulator D, which, for accumulator D, are analogous to the instructions `LDAA`, `ADDA`, and `STAA` for accumulator A. The following program performs the same operations as the programs given above but is much more dynamically and statically efficient and is clearer.

If you wish to write programs in assembly language, full knowledge of the computer's instruction set is needed to write the most efficient, or the clearest, program. The normal way to introduce an instruction set is to discuss operations first and then addressing modes. We will devote Chapter 2 to the discussion of instructions and Chapter 3 to the survey of addressing modes.

In summary, you should aim to write good programs. As we saw with the example above, there are equally good programs, and generally there are no best programs. Short, fast, clear programs are better than the opposite kind. Yet the shortest program is rarely the fastest or the clearest. The decision as to which quality to optimize is dependent on the application. Whichever quality you choose, you should have as a goal the writing of clear, efficient programs. You should fight the tendency to write sloppy programs that just barely work or that work for one combination of inputs but fail for others. Therefore, we will arbitrarily pick one of these qualities to optimize in the problems at the end of the chapters. We want you to optimize static efficiency in your solutions, unless we state otherwise in the problem. Learning to work toward a goal should help you write better programs for any application when you train yourself to try to understand what goal you are working toward.

1.4 MC68HC812A4 and MC68HC912B32 Organizations

The 6812 is currently available in two implementations, which are designated the MC68HC812A4 (abbreviated the 'A4) and MC68HC912B32 (abbreviated the 'B32). These are discussed herein.

The 'A4 can operate in the single-chip mode or the expanded bus mode. In the *single-chip mode*, the 'A4 can be the only chip in a system, for it is self-sufficient. The processor, memory, controller, and I/O are all in the chip. (See Figure 1.11.) The memory consists of 1K words of RAM and 4K words of *electrically erasable programmable memory* (*EEPROM*). The I/O devices include a dozen parallel I/O registers, a *serial peripheral interface* (*SPI*), a *serial communication interface* (*SCI*), a *timer,* and an *A/D converter*.

The 'A4's expanded bus mode removes three or four of the parallel ports, using their pins to send the address and data buses to other chips. RAM, ROM, *erasable programmable read-only memory* (*EPROM*), and *programmable read-only memory* (*PROM*) can be added to an expanded bus. In a narrow expanded mode, ports A and B are removed for address lines, and port C is an 8-bit data bus. Port D is available for parallel I/O. In a wide expanded mode (see Figure 1.12), ports A and B are removed, their pins are used for the address bus, and ports C and D are a 16-bit data bus. Port D is unavailable for parallel I/O. In both modes, ports E, F, and G are available for bus control, chip selects, and memory control, or else for parallel I/O.

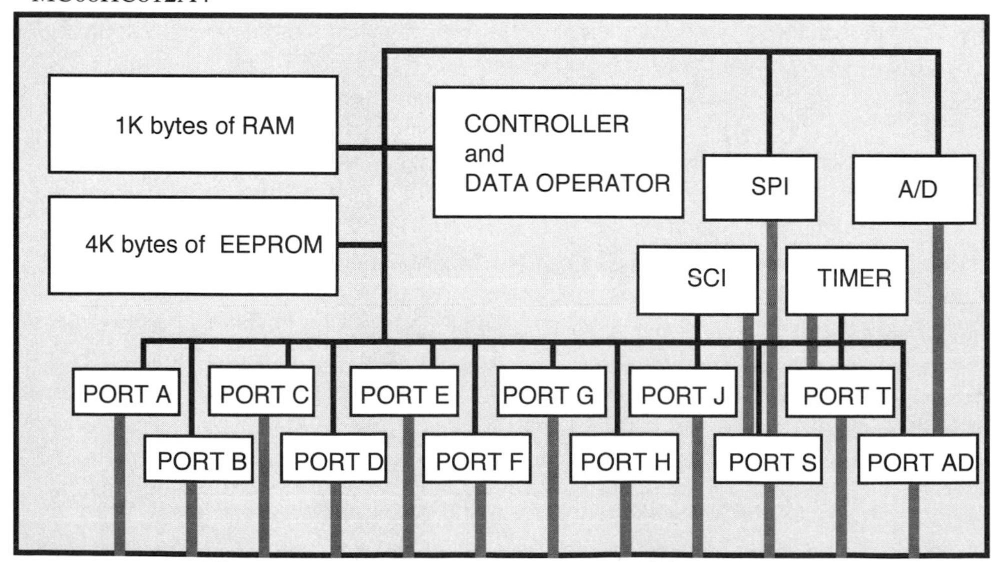

Figure 1.11. Single-Chip Mode of the MC68HC812A4

MC68HC812A4

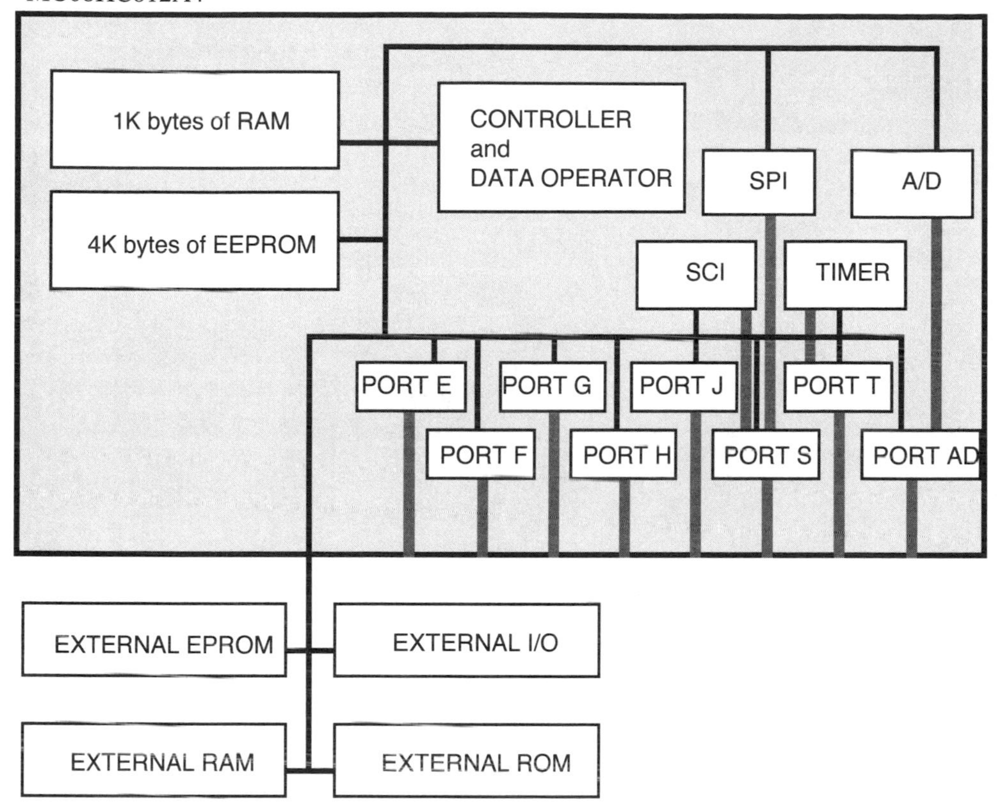

Figure 1.12. Expanded Wide Multiplexed Bus Mode of the MC68HC812A4

A *memory map* is a description of the memory showing what range of addresses is used to access each part of memory or each I/O device. Figure 1.13a presents a memory

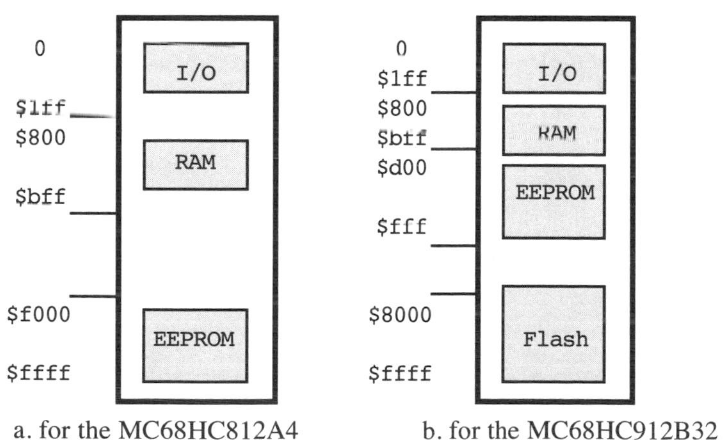

a. for the MC68HC812A4 b. for the MC68HC912B32

Figure 1.13. Memory Maps for the 6812

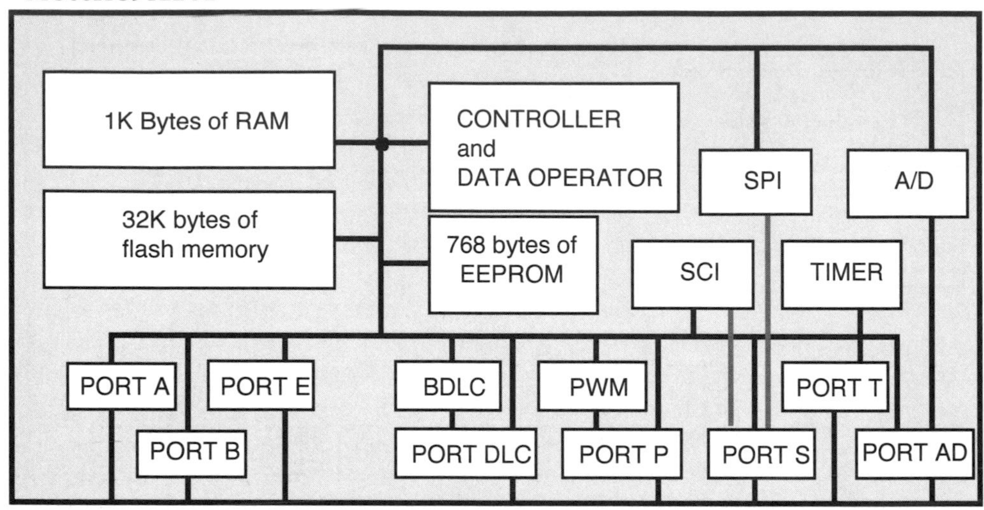

Figure 1.14. Single-Chip Mode of the MC68HC912B32

map for the 'A4. Actually, EEPROM, RAM, and I/O may be put anywhere in memory (on a 2K or 4K boundary), but we will use them in the locations shown in Figure 1.13a throughout this text. I/O is at lowest address, and RAM is at $800 to $bff. The EEPROM at $F000 has a monitor. Usually your data is put in RAM, and your program may be put in RAM or in EEPROM.

The 'B32 can also operate in the single-chip mode or the expanded bus mode, but in the latter mode, address and data are time-multiplexed on the same pins. In the single-chip mode, the 'B32 can be the only chip in a system. The processor, memory, controller, and I/O are all in the chip. (See Figure 1.14.) The controller and data operator execute the 6812 instruction set discussed earlier. The memory consists of 1K words of RAM, 768 bytes of EEPROM, and 32K words of *flash memory,* which is like EEPROM. The I/O devices include eight parallel I/O registers, a serial peripheral interface (SPI), a serial communication interface (SCI), a timer, a *pulse-width modulator* (*PWM*), a *Byte Data Link Communication Module* (*BDLC*), and an A/D converter.

The expanded bus mode of the 'B32 removes two of the parallel ports, ports A and B, using their pins to send the time-multiplexed address and data buses to other chips. The address and data buses are time-multiplexed; in the first part of each memory cycle, the 16-bit address is output on the pins, and in the second part, data is output or input on the indicated pins. In a narrow expanded mode, port A is used for an eight-bit data bus. In a wide expanded mode (see Figure 1.15), ports A and B pins are used for a 16-bit data bus. In both modes, port E can be used for bus control or else for parallel I/O. RAM, ROM, EPROM, and PROM can be added to the expanded bus.

Figure 1.13b presents a memory map for the 'B32. I/O is at lowest address, and RAM is at $800 to $bff. A small EEPROM is at $d00 to $fff. Flash memory at $8000 to $ffff has a monitor. Usually your data is put in RAM, and your program may be put in RAM or in flash memory.

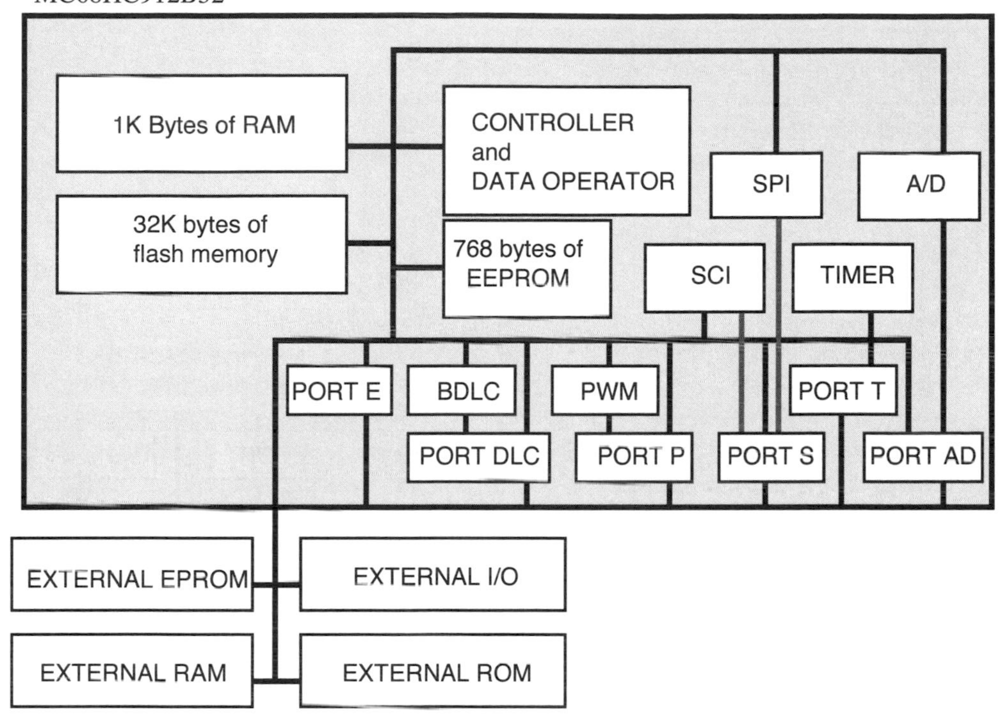

Figure 1.15. Expanded Wide Multiplexed Bus Mode of the MC68HC912B32

A significant advantage of the 'A4 or 'B32 is that either can be used in either the single chip or either narrow or wide expanded multiplexed bus mode. The former mode is obviously useful when the resources within the microcontroller are enough for the application—that is, when there are enough memory and I/O devices for the application. The latter mode is required when more memory is needed, when a program is in an EPROM and has to be used with the 'A4, or when more or different I/O devices are needed than are in the 'A4.

1.5 Variable Word Width

We have glibly stated that a 6812 can have either 8-bit or 16-bit word widths. The fastidious reader might wonder how this takes place. This optional section provides details on how a 6812's word widths can be either 8 bits or 16 bits wide, as discussed.

The word width is a function of the instruction and of the bus mode discussed in the last section. As noted in the last section, external memory can utilize a narrow or a wide data bus. We first consider an external memory using the wide data bus and comment on such a memory using a narrow bus at the end of our discussion. Figure 1.16 illustrates a 6812 system with an internal memory at locations \$800 to \$bff and an external memory at locations \$7000 to \$7fff, each composed of two 8-bit-wide memory "banks."

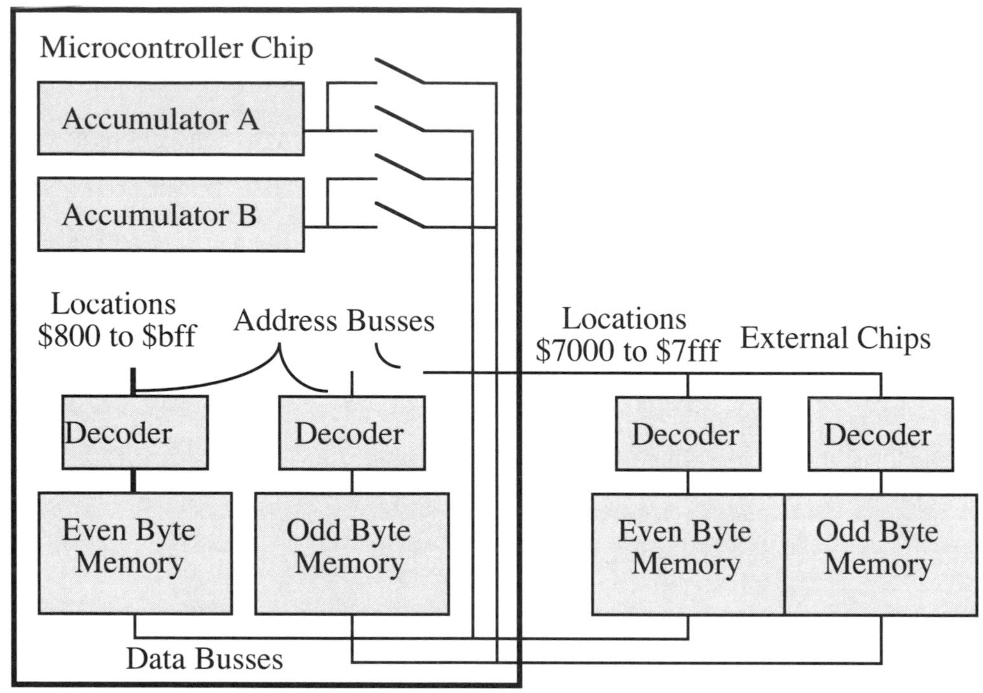

Figure 1.16. Variable Word Width Implementation

Each internal 8-bit wide memory bank's decoder can decode addresses on different buses, while both external 8-bit wide memory banks' decoders decode the same address. The even memory bank stores even bytes, such as 0, 2, 4, . . . while the odd memory bank stores odd bytes, such as 1, 3, 5, ... The effective address computed by the instruction is generally shifted one bit right to derive the address that is presented to the memories (memory address). The original effective address's least significant bit indicates which 8-bit memory is accessed.

We will consider reading an 8-bit word first, which is the simplest operation. Then we discuss writing an 8-bit word, reading a 16-bit word, and writing a 16-bit word.

Consider reading an 8-bit word into accumulator A. A whole 16-bit word can be read from both banks of either the internal memory at locations $800 to $bff or external memory at locations $7000 to $7fff, using the same memory address; and the switches to the right of accumulator A can feed the byte from an even or odd memory bank to it, depending on whether the byte address is even or odd. Writing accumulator A into memory can be effected by putting that register's data on both even and odd byte data buses and giving the command to write to an even or odd bank depending on whether the effective address is even or odd. The bank that does not get the command to write will see the byte on its data input but won't write that byte into its memory.

Consider reading a 16-bit word into accumulator D, which is accumulator A concatenated with accumulator B. If the address is even, then a 16-bit word is simply read from both banks, using the same memory address, through both data buses and written into both accumulators A and B. Reading a 16-bit word from an odd effective address in the internal memory can be done in one memory cycle; the even-byte memory bank can decode a memory address one above the memory address of the odd bank, and the switches next to the accumulators can reroute the bytes into the correct accumulators. However, reading a 16-bit word from an odd address in the external memory is done in two memory cycles; the byte at the lower address is read first, as discussed in the previous paragraph, then the byte at the higher address is read.

Consider writing a 16-bit word from accumulator D. If the address is even, simply write a byte from both accumulators A and B through both data buses into both banks. If the address is odd, writing 16 bits is easily done in the internal memory; each byte is written, but at different addresses in the even and odd byte memory banks. But if the address is odd, writing 16 bits into the external memory is done by writing the lower addressed byte first in one memory cycle, then writing the other byte in the next cycle.

Each access to read or write a byte through an 8-bit narrow bus takes one memory cycle. Each access to read or write a 16-bit word uses two memory cycles.

Cycle counts in the CPU12RG/D manual Instruction Set Summary are for instructions and data read and written in internal RAM. These counts may be higher when instructions and data are read from or written into external RAM.

At the beginning of this chapter, we proposed to show you how an instruction is executed in a microcontroller. But from this discussion of 6812 memory operations, you see that a simple operation becomes significantly more complex when it is implemented in a state-of-the-art microcontroller. Just reading a 16-bit word from memory might be done several ways depending on where the word is located, inside or outside the microcontroller, or using an even or an odd address. But from the point of view of how an instruction is executed in the 6812, a simple model that explains the concept fully is better than the fully accurate model that accounts for all the techniques used to implement the operation; we simply state that a 16-bit word is read from memory. The reader should understand how instructions are executed, but from now on in this book, we will use a simplified model of the hardware to explain how an instruction is actually implemented in hardware (in an idealized microcontroller, rather than the real 6812).

1.6 Summary and Further Reading

In this chapter we examined the computer and the instruction in some detail. You should be prepared to study each of the instructions in the 6812 in the following two chapters with respect to the details that we introduced for the load instruction in this chapter. We will expand the ideas of programming, introduced at the end of this chapter, as we progress through the book. Many questions may remain unanswered, though, after reading this chapter. We want you to continue reading the following chapters as we discuss the way to use this marvelous tool.

In this book, we use Motorola's CPU12RG/D manual to provide essential information needed to write and read machine code for the 6812 in compact and neat form. This manual is a summary of key tables and figures in the manual CPU12RM/AD, which contains complete information on the execution of each instruction. Additionally, there is a manual for the 'B32, M68HC912B32TS/D, and a manual for the 'A4, M68HC812A4TS/D, which describe their I/O systems. Motorola is generous with these manuals and maintains them better than we can in an appendix in this book, so we recommend that you order the manuals from Motorola that you need to accompany this book. Finally, if you are already an accomplished assembly language programmer on another computer or microcomputer, you might find this book too simple and spread out. We might offer Chapter 1 of the text *Single- and Multiple-Chip Microcontroller Interfacing* (G. J. Lipovski), as a condensed summary of much of the material in this text.

Do You Know These Terms?

This is a list of all italicized words in this chapter. You should check these terms to be sure that you recognize their meaning before going on to the next chapter. These terms also appear in the index, with page numbers for reference, so you can look up those that you do not understand.

von Neumann computer
memory
controller
data operator
input-output (I/O)
register
microprocessor
microcomputer
single-chip microcontroller
MPU
byte
word
line
wire
bus
clock
clock cycle
clock rate
clock frequency
read
read/write line
write
access
random access memory (RAM)
read only memory (ROM)
volatile
nonvolatile
program
instruction
instruction cycle
fetch/execute cycle
fetch
fetch phase
recall
execute phase
accumulator
effective address
one-address computer
addressing mode
operation code byte
operation code (op code)
hexadecimal number
machine code
machine
debug
assembler
source code
mnemonic
immediate addressing
operand
direct addressing
carry bit (carry)
condition code register
half-carry
static efficiency
dynamic efficiency
clarity
single-chip mode
electrically erasable programmable memory (EEPROM)
serial peripheral interface (SPI)
serial communication interface (SCI)
timer
A/D converter
erasable programmable read-only memory (EPROM)
programmable read-only memory (PROM)
memory map
flash memory
pulse-width modulator (PWM)
Byte Data Link Communication Module (BDLC)

PROBLEMS

1. Is a Hewlett-Packard handheld calculator, model 21 (or any programmable calculator that you may select), a (von Neumann) computer, and why?

2. What do the following terms mean: memory, controller, data operator, input/output?

3. What are a microcomputer, a microprocessor, and a single-chip microcontroller?

4. Describe the terms clock, data bus, address bus, and read/write line. Discuss the operation of reading a word from memory using these terms.

5. How many memory read cycles are needed for the following instructions, using the CPU12 reference manual? How many are fetch operations, and how many are recall operations? How many memorize cycles are used?

(a) `LDAA #19`
(b) `LDAB #18`
(c) `ADDA $3FB2`
(d) `ADDA 23` (Use 16-bit direct addressing)
(e) `STAA 199` (Use 16-bit direct addressing)

6. While executing a particular program, (PC) = 2088, (A) = 7, and (B) = 213 before the following sequence is executed:

```
LDAA  #10
ADDA  2142
STAA  2139
```

Location	Contents
2139	8
2140	7
2141	16
2142	251
2143	19

If the contents of memory locations 2139 through 2143 are as shown on the right before the sequence is executed, what will be the contents of A, B, and PC after the sequence is executed? (<u>Location</u> is the memory location, or address, and <u>Contents</u> is the memory contents. All numbers given are in decimal.) What will the C bit be equal to after the sequence is executed?

7. Write a program to add two 3-byte numbers in the same manner as the last program segment of this chapter.

8. Select goals for good programs in the following applications, and give a reason for the goals. The goals should be: static or dynamic efficiency or clarity.

(a) A 75,000-instruction program
(b) A program for guidance of a space satellite
(c) A controller for a drill press
(d) An automobile engine controller
(e) Programs for sale to a large number of users (like a Basic interpreter)

9. What is the effective address in the following instructions, assuming the opcode byte is at $802?

(a) `LDAA 122`
(b) `LDAA #122`
(c) `ADDA $3452`
(d) `ADDA #125`

10. Rewrite Figure 1.5 to subtract the 8-bit number in location $840 from the 8-bit number in location $841, putting the result into location $842. Use `SUBA`.

11. How many clock cycles does it take to execute the program in Figure 1.5? (See the operation code bytes table in the CPU12RG/D manual Instruction Set Summary.) If a memory clock cycle is 125 nsec, how long does this program take, in real time?

12. Rewrite Figure 1.8 to subtract the 16-bit number in location $812 from the 16-bit number in location $810, putting the result into location $814. Use `SUBB` and `SBCA`.

13. Rewrite Figure 1.9 to subtract the 16-bit number in location $812 from the 16-bit number in location $810, putting the result into location $814. Use `SUBA` and `SBCA`.

14. How many clock cycles does it take to execute the program in Figure 1.8 and the program in Figure 1.9? (See the operation code bytes table in the CPU12RG/D manual.) If a memory clock cycle is 125 nsec, how long does this program take, in real time?

15. Rewrite Figure 1.10 to subtract the 16-bit number in location $812 from the 16-bit number in location $810, putting the result into location $814. Use `SUBD`.

16. How many clock cycles does it take to execute the program in Figure 1.10? (See the operation code bytes table in the CPU12RG/D manual.) If a memory clock cycle is 125 nsec, how long does this program take to execute in real time?

17. How many parallel ports of the narrow expanded mode of the MC68HC812A4, which are already on the microcontroller chip, can be used? How many parallel ports of the wide expanded mode of the MC68HC812A4, which are already on the microcontroller chip, can be used? In the latter case, how many can be used for output?

18. How many parallel ports of the narrow expanded mode of the MC68HC912B32, which are already on the microcontroller chip, can be used? How many parallel ports of the wide expanded mode of the MC68HC912B32, which are already on the microcontroller chip, can be used? In the latter case, how many can be used for output?

19. In Figure 1.16, an external memory is 16 bits wide, but most static random access memories (SRAMs) are 8 bits wide. If a 16 Kbyte external memory is to be used as shown in this figure, what kind of SRAM chips should be ordered? How should the microcontroller address bus be attached to the SRAM chip address pins?

20. In Figure 1.16, an internal memory is 16 bits wide, while the older 6811 microcontroller's internal memory is 8 bits wide. A word of EEPROM can be programmed each 10 milliseconds in either case. How long does it take to write a 1 Kbyte program in EEPROM in each microcontroller?

The MC68HC912B32 die.

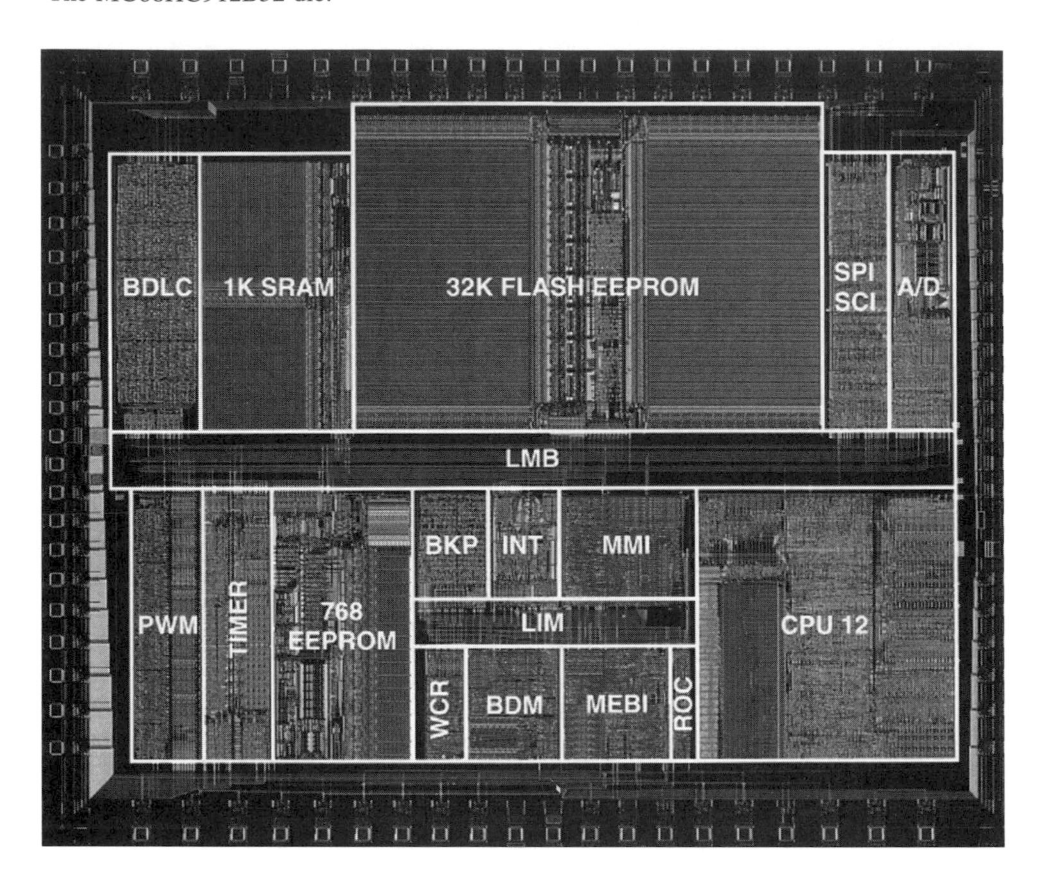

2

The Instruction Set

In our study of how the computer ticks, we think that you will be motivated to read this chapter because it will describe the actions the computer can do. It will supply a key ingredient that you need to write programs, so that the computer can magnify your ideas as a lever can magnify your physical capabilities. The next chapter completes the study of the instruction set by describing the addressing modes used with these instructions.

In order to learn the possible actions or operations that a computer may execute, you need to keep a perspective. There is a lot of detail. You do need to learn this detail to be able to program the 6812. But learning about that microcomputer must be viewed as a means to an end, that is, to understand the operations of any computer in general. While you learn the details about programming the 6812, get the feel of programming by constantly relating one detail to another and questioning the reason for each instruction. When you do this, you will learn much more than the instruction set of a particular computer—you will learn about computing.

We have organized this chapter to facilitate your endeavor to compare and to associate details about different instructions and to offer some answers to questions that you might raise about these instructions. This is done by grouping similar instructions together and studying the groups one at a time, as opposed to listing instructions alphabetically or by presenting a series of examples and introducing new instructions as needed by each example as we did in Chapter 1. We group similar instructions together into a class and present each class one at a time. As mentioned in Chapter 1, the instructions for the 6812, as well as any other computer, may be classified as follows:

1. Move instructions
2. Arithmetic instructions
3. Logic instructions
4. Edit instructions
5. Control instructions
6. Input output instructions
7. Special instructions

We have added, as a separate section, the special instructions that are generally arithmetic instructions usually not used by compilers but that provide the 6812 with some unique capabilities. We now examine each instruction class for the 6812. This discussion of classes, with sections for examples and remarks, is this chapter's outline.

At the conclusion of the chapter, you will have all the tools needed to write useful programs for the 6812, in machine code. You should be able to write programs on the order of 20 instructions long, and you should be able to write the machine code for these programs. If you have a laboratory parallel to a course that uses this book, you should be able to enter these programs, execute them, debug them, and, using this hands-on experience, you should begin to understand computing.

2.1 Move Instructions

Behold the humble move instructions, for they labor to carry the data for the most elegant instructions. You might get excited when you find that this computer has a fairly fancy instruction like multiply, or you might be disappointed that it does not have floating-point instructions like the ones most big machines sport. Studies have shown that, depending on the kind of application program examined, between 25 and 40% of the instructions were move instructions, while only 0.1% of the instructions were multiplies. As you begin to understand computing, you will learn to highly regard these humble move instructions and to use them well.

Move instructions essentially move one or two bytes from memory to a register (or vice versa) or between registers or memory locations. The two aspects of these instructions that give most readers some problems are the setting of condition codes and the allowable addressing modes. We shall take some care with the setting of condition codes in this chapter and the allowable addressing modes in the next chapter.

The two simplest instructions from the move class are the load and store instructions. These have already been examined for accumulators A, B, and D; they also may be used with index registers like X. For example, in the load instruction

```
LDX   2062
```

the high byte of X is taken from location 2062 while the low byte of X is taken from location 2063. An exactly parallel situation holds for the store instruction

```
STX   2337
```

where the high byte of X is put into location 2337 while the low byte of X is put into location 2338. In addition to X, there are load and store instructions for index register Y and stack pointer S. They work exactly as described for D.

The `TST` and `CLR` instructions are two more examples in the move class of instructions of the 6812. The clear instruction `CLR` is used to initialize the accumulators or memory locations with the value zero. As such, `CLRA` can replace instructions such as `LDAA #0` in the sequence in Figure 2.1. Further, the two instructions in Figure 2.1 can be replaced by the single instruction `CLR 2090.`

Location	Contents	Mnemonic	Comment
820	86 00	LDAA #0	; generate constant 0
822	7A 08 2A	STAA 2090	; move to $82A

Figure 2.1. A Program Segment to Clear a Byte

Table 2.1. Move Instructions Using an Effective Address

LDAA opr	(E) -> A; Load Accumulator A
LDAB opr	(E) -> B; Load Accumulator B
LDD opr	(E):(E+1) -> D; Load Accumulator D
LDS opr	(E):(E+1) -> SP; Load Stack Pointer
LDX opr	(E):(E+1) -> X; Load Index Register X
LDY opr	(E):(E+1) -> Y; Load Index Register Y
STAA opr	A -> (E); Store Accumulator A to Memory
STAB opr	B -> (E); Store Accumulator B to Memory
STD opr	D -> (E):(E+1); Store Accumulator D
STS opr	SP -> (E):(E+1); Store Stack Pointer
STX opr	X -> (E):(E+1); Store Index Register X
STY opr	Y -> (E):(E+1); Store Index Register Y
CLR opr, CLRA, CLRB	0 -> (E); Clear Memory, or A or B
TST opr, TSTA, TSTB	(E) - 0; Test Memory, or A or B
LEAS opr	E -> SP; Load Effective Address into SP
LEAX opr	E -> X; Load Effective Address into X
LEAY opr	E -> Y; Load Effective Address into Y

Notice that although `CLRA` and `LDAA #0` make the same move, `CLRA` clears C, whereas `LDAA #0` does not affect C. Your program may need to use C later. The test instruction TST, sometimes called a "half a load " instruction, adjusts the N and Z bits in the condition code register exactly as a load instruction does but without actually loading the byte into an accumulator. The versatile "load effective address" instructions, `LEAX`, `LEAY`, and `LEAS`, load one of the index registers or stack pointer with the effective address computed in an indirect index address calculation, which will be discussed in Chapter 3. These instructions do not affect the condition code register bits.

Table 2.1 lists the move instructions that use addressing modes. The expressions such as (E) -> A; accurately describe the instruction's principle effect. E is the effective address, (E) is the word in memory at location E, A is accumulator A, and -> is a data transfer, so (E) -> A; means that the word in memory at the location determined by the instruction's effective address is put into accumulator A. The CPU12RG/D manual Instruction Set Summary further gives the opcode bytes, allowable addressing modes for each instruction, and condition code modifications that result from executing these instructions. These same conventions are used with the tables that follow in this chapter.

Table 2.2 lists move instructions that *push* on or *pull* from the stack. The *stack* pointed to by register SP is called the *hardware stack*. A program can have other stacks as well. A stack is an abstraction of a stack of letters on your desk. Received letters are put on top of the stack; when read, they are removed from the top of the stack.

In the computer, letters become bytes and the memory that stores the stack becomes a *buffer* with a stack pointer as follows. One decides to put this buffer, say, from \$b80 to \$bff. See Figure 2.2. The amount of storage allocated to the buffer should be the worst case number of bytes saved on the stack. Usually, we allow a little extra to prevent a stack overflow. SP points to the top byte that is on the stack. The SP register is generally initialized once to the high end of the buffer, at the beginning of the program,

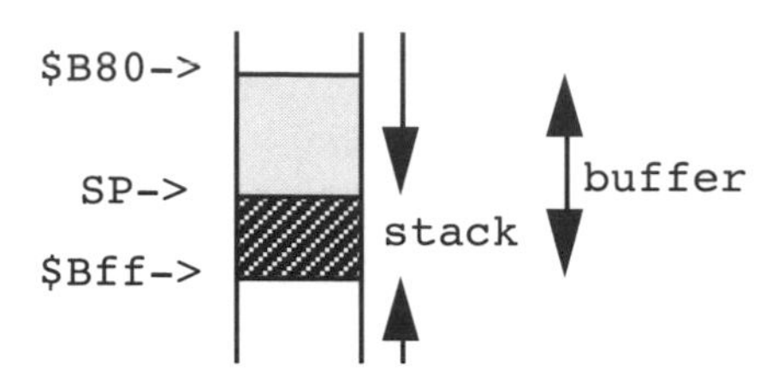

Figure 2.2. A Stack

and is thereafter adjusted only by push and pull instructions and, perhaps, the `LEAS` instruction to move it. For example, at the beginning of the program, the instruction `LDS #$c00` initializes the stack so that the first push is into location $bff.

If a byte is pushed onto the stack, SP is decremented by 1, and a byte, from one of the 8-bit registers, is put into location (SP). If one byte is removed or pulled from the stack, the byte is transferred to one of the 8-bit registers, and SP is incremented by 1. If two bytes are pushed onto the stack, SP is decremented by 2, and two bytes, from one of the 16-bit registers, are put into location (SP) and (SP+1). If two bytes are pulled from the stack, two bytes from location (SP) and (SP+1) are put into one of the 16-bit registers and SP is incremented by 2. Any of the registers, except SP, may be pushed or pulled from the stack for which SP is the stack pointer. `PSHB` will push B onto the stack, `PULD` will pull two words from the stack, putting the combined word in accumulator D. The order for 16-bit words is always that the low byte is pushed before the high byte and the high byte pulled before low byte.

The stack will later be used for saving and restoring the program counter when executing a subroutine and saving and restoring all the registers when executing an interrupt. It will later also be used to store procedure arguments and local variables.

The hardware stack and the stack pointer SP must be used with some care in computers like the 6812. There may be a temptation to use it as an index register, to move it around to different data locations. This is very dangerous. Interrupts, which may occur randomly, save data on the hardware stack, and programs used to aid in the testing and debugging of your program generally use interrupts. Such a program may be very

Table 2.2. Stack Move Instructions

Instruction	Operation
PSHA	`SP - 1 -> SP; A -> (SP); Push Accumulator A onto Stack`
PSHB	`SP -1 -> SP; B -> (SP); Push Accumulator B onto Stack`
PSHC	`SP - 1 -> SP; (CCR) -> (SP); Push CCR onto Stack`
PSHD	`SP - 2 -> SP; D -> (SP):(SP+1); Push Accumulator D`
PSHX	`SP - 2 -> SP; X -> (SP):(SP+1); Push Index Register X`
PSHY	`SP - 2 -> SP; Y -> (SP):(SP+1); Push Index Register Y`
PULA	`(SP) -> A; SP + 1 -> SP; Pull Accumulator A from Stack`
PULB	`(SP) -> B; SP + 1 -> SP; Pull Accumulator B from Stack`
PULC	`(SP) -> CCR; SP + 1 -> SP; Pull CCR from Stack`
PULD	`(SP):(SP+1) -> D; SP + 2 -> SP; Pull D from Stack`
PULX	`(SP):(SP+1) -> X; SP + 2 -> SP; Pull Index Register X`
PULY	`(SP):(SP+1) -> Y; SP + 2 -> SP; Pull Index Register Y`

Table 2.3. Special Move Instructions

EXG rl,r2	rl <-> r2; 8-bit or 16-bit Register Exchange
TFR rl,r2	8-bit or 16-bit Register to Register Transfer
SEX rl,r2	Alternate for some TFR rl,r2 instructions
TAB	A -> B; Transfer A to B
TBA	B -> A; Transfer B to A
TAP	A -> CCR; Translates to TFR A,CCR
TPA	CCR -> A; Translates to TFR CCR,A
TSX	SP -> X; Translates to TFR SP,X
TSY	SP -> Y; Translates to TFR SP,Y
TXS	X -> SP; Translates to TFR X,SP
TYS	Y -> SP; Translates to TFR Y,SP
XGDX	Translates to EXG D,X
XGDY	Translates to EXG D,Y
MOVB oprl,opr2	(E) -> (e); 8-bit Memory to Memory Move
MOVW oprl,opr2	(E):(E+1) -> (e):(e+1); 16-bit Move

difficult to test and debug because some data in your program may be overwritten in your attempt to test and debug it. On the other hand, this same stack is the best place to save data used by a subroutine, which is not used by other subroutines, as we explain later. Incidentally, the word "pop" is used instead of "pull" in many textbooks.

The transfer and exchange instructions in Table 2.3, `TFR` and `EXG`, allow the transfer of register Rl to R2 or the exchange of Rl and R2, respectively, where Rl and R2 are any pair of 8 or 16 bit registers. You can move data from an 8-bit register to a 16-bit one or vice versa. As an example, the instruction `TFR D,Y` puts the contents of D into Y, and `EXG D,X` exchanged the contents of accumulator D and index register X.

The `TFR` or `EXG` machine code consists of an operation code byte and a *post byte*. The opcode byte is obtained from the CPU12RG/D manual Instruction Set Summary, and the post byte (see Table 3 therein) can be obtained as follows: The source is the left nibble and the destination is the right nibble; their values are: 0, accumulator A; 1, accumulator B; 2, condition code register; 4, accumulator D; 5, index register X; 6, index register Y; 7, stack pointer SP. As an example, the instruction `TFR D,Y` is stored in memory as the two bytes:

$B7
$46

The post byte's left four bits indicate the source of the transfer, which is D, and the right four bits indicate the destination of the transfer, which is Y. When transferring from an 8-bit to a 16-bit register, the sign bit is extended so that positive numbers remain positive and negative numbers remain negative; the sign extend mnemonic `SEX` can be used as an alternative to the `TFR` mnemonic in these cases. Figure 2.3a illustrates sign extension when transferring the data from an 8-bit register, like A, shown on the top, into a 16-bit register like X, shown on the bottom. The low-order byte is moved bit-by-bit from the flip-flops in the 8-bit to the flip-flops in the 16-bit register. The high-order byte's flip-flops are loaded with the low-order byte's sign bit. The `EXG` instruction similarly permits exchanging the contents of two registers, and the post byte coding is the same, but when moving from an 8-bit to a 16-bit register, instead of

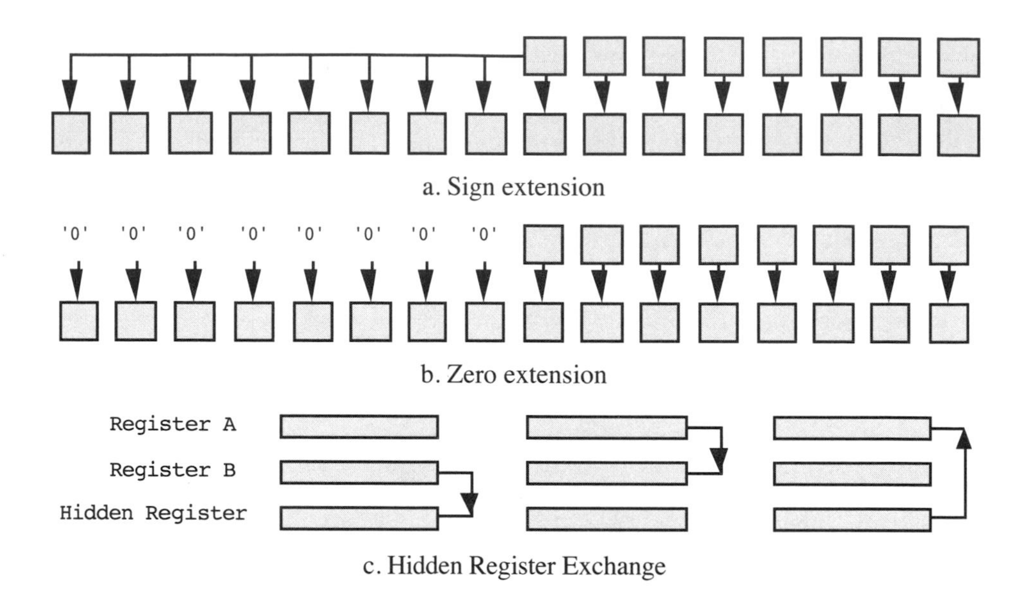

a. Sign extension

b. Zero extension

c. Hidden Register Exchange

Figure 2.3. Transfers between Registers

extending the sign, it merely fills the high byte with zeros (see Figure 2.3b). Exchanges are accomplished by means of a *hidden register* (see Figure 2.3c). The instruction `EXG A,B` first copies register B into the hidden register. Then it copies A into B. Then it copies the hidden register into A. Such hidden registers are not in the description of the 6812's register set (Figure 1.2) but are additional registers within the data operator.

The `MOVB` and `MOVW` instructions implement a constant-to-memory or a memory-to-memory move. The instruction below puts the constant $04 into location $803.

```
MOVB   #4,$803
```

This instruction is coded as shown below; the *prefix byte* $18 precedes the opcode byte $0B. See the CPU12RG/D manual. In effect, the opcode is a 16-bit opcode $180B. The prefix byte $18 is used in the 6812 to encode a number of instructions. It is as if, when the 6812 fetches the prefix $18, it thinks: Oh, this is one of the 2-byte op codes, so fetch another byte to get the complete opcode. The third byte is the immediate operand $04, and the last two bytes are the direct address $803.

$18
$0B
$04
$08
$03

This move instruction moves the immediate operand \$04 into a hidden register and then moves the data from the hidden register into location \$803. From the CPU12RG/D manual, we observe that its execution takes four memory cycles. The alternative to this instruction is the program segment:

```
LDAA    #4
STAA    $803
```

which is encoded as follows:

\$86
\$04
\$7A
\$08
\$03

This pair of instructions takes the same number of bytes as the `MOVB` instruction. Further, the `LDAA` instruction executes in one memory cycle, and the `STAA` instruction executes in three memory cycles. The `MOVB` instruction is neither statically nor dynamically more efficient than the pair of instructions, `LDAA` and `STAA`. However, it is clearer. We recommend using the `MOVB` instruction to write a constant into memory.

But if the same constant is written into two places in memory, as in

```
LDAA    #4
STAA    $803
STAA    $807
```

then a program sequence using the `MOVB` instruction is less efficient:

```
MOVB #4,$803
MOVB #4,$807
```

This program sequence takes ten bytes and executes in eight memory cycles, while the program sequence above it takes eight bytes and executes in seven memory cycles.

The `MOVB` instruction also moves data from any memory location to any memory location. `MOVB $801,$803` moves a byte from location \$801 to location \$803. The `MOVW` instruction similarly moves 16 bits of data.

Missing move instructions can often be implemented by combinations of other move instructions. Because there is no instruction to "load" the condition code register, it can be loaded through accumulator A or B with the `TFR` instruction. For example, to put 3 into the condition code, execute the code shown in Figure 2.4.

Location	Contents	Mnemonic	Comment
820	86 03	LDAA #3	; generate constant 3
822	B7 02	TFR A,CCR	; move to cc register

Figure 2.4. Program Segment to Initialize the Condition Code Register

2.2 Arithmetic Instructions

The computer is often used to compute numerical data, as the name implies, or to keep books or control machinery. These operations need arithmetic instructions, which we now study. However, you must recall that computers are designed and programs are written to enhance static or dynamic efficiency. Rather than have the four basic arithmetic instructions that you learned in grade school—add, subtract, multiply, and divide—computers have the instructions that occur most often in programs. Rather than having the sophisticated square root as an instruction, for instance, we will see the often-used increment instruction in a computer. Let us look at them. See Table 2.4.

We have already discussed the add instructions: ADCA, ADCB, ADDA, ADDB, and ADDD. The corresponding subtraction instructions, SBCA, SBCB, SUBA, SUBB, and SUBD, are the obvious counterparts of add instructions, where the carry condition code bit holds the borrow. However, 16-bit add and subtract instructions with carry, ADCD and SBCD, are missing; multiple-byte arithmetic must be done one byte at a time rather than two bytes at a time. Comparisons are normally made by subtracting two numbers and checking if the result is zero, negative, positive, or a combination of these. But using the subtract instruction to compare a fixed number against many numbers requires that the fixed number has to be reloaded in the register each time the subtraction is performed. To streamline this process, compare instructions are included that do not change the contents of the register used. These compare instructions are used to compare the contents of registers A, B, D, X, Y, and SP with the contents of memory locations in order to give values to the condition code bits C, V, N, and Z. Finally, note that DEC, INC, and NEG are provided for often-used special cases of add and subtract instructions, to improve efficiency.

Table 2.4. Add Instructions Using an Effective Address

ADCA opr	A + (E) + C -> A Add with Carry to A
ADCB opr	B + (E) + C -> B Add with Carry to B
ADDA opr	A + (E) -> A Add without Carry to A
ADDB opr	B + (E) -> B Add without Carry to B
ADDD opr	D + (E):(E+1) -> D; ADD to D without Carry
CMPA opr	A - (E); Compare Accumulator A with Memory
CMPB opr	B - (E); Compare Accumulator B with Memory
CPD opr	D - (E):(E+1); Compare D to Memory (16-Bit)
CPS opr	SP - (E):(E+1); Compare SP to Memory (16-Bit)
CPX opr	X - (E: E+1); Compare X to Memory (16-Bit)
CPY opr	Y - (E: E+1); Compare Y to Memory (16-Bit)
SBCA opr	A - (E) - C -> A; Subtract with Borrow from A
SBCB opr	B - (E) - C -> B; Subtract with Borrow from B
SUBA opr	A - (E) -> A; Subtract from Accumulator A
SUBB opr	B - (E) -> B; Subtract from Accumulator B
SUBD opr	D - (E):(E+1) -> D; Subtract from Acc. D
DEC opr, DECA, DECB	(E) - 1 -> (E); Decrement Memory Byte/Reg.
INC opr, INCA, INCB	(E) + 1 -> (E); Increment Memory Byte/Reg.
NEG opr, NEGA, NEGB	0 - (E) -> (E); two's Complement Negate

Location	Contents	Mnemonic	Comment
800	FC 08 52	LDD $852	; get two low bytes of first
803	F3 08 56	ADDD $856	; add two low bytes of second
806	7C 08 56	STD $856	; save two low bytes of second
809	FC 08 50	LDD $850	; get two high bytes of first
80C	F9 08 55	ADCB $855	; add third byte of second
80F	B9 08 54	ADCA $854	; add high byte of second
812	7C 08 54	STD $854	; save two high bytes of scnd.

Figure 2.5. Program Segment for 32-Bit Addition

Figure 2.5 illustrates a simple example of an arithmetic operation: adding a 4-byte number at $850 to a 4-byte number at $854. ADDD can be used to add the two low-order bytes, but ADCB and ADCA are needed to add the high-order bytes.

Arithmetic instructions are really very simple and intuitively obvious, except for the condition code bits. Addition or subtraction uses the same instruction for unsigned as for two's-complement numbers, but the test for overflow is different (see Appendix 1). The programmer has to use the correct condition code test after instruction completion; for example, SUBA $876 sets C = 1 if, and only if, there has been an unsigned overflow; that is, A – ($876) produces a borrow or, when each number is treated as an unsigned number, A < ($876). (Here A and ($876) denote the contents of A and the contents of location $876.) Similarly, V = 1 if, and only if, a two's-complement (signed) overflow occurs, when A and ($876) are treated as two's-complement numbers; i.e., A – ($876) is not in the 8-bit two's-complement range. Note again that subtraction is performed with the same instruction, such as SUBA, regardless of whether the numbers are two's-complement or unsigned numbers.

Table 2.5 shows special instructions used to improve efficiency for commonly used operations. ABX, ABY, ABA, CBA, and SBA use accumulator B, and DES, DEX, DEY, INS, INX, and INY increment or decrement index registers and the stack pointer.

Multiply instructions MUL and EMUL multiply unsigned numbers in specific registers. EMULS similarly multiplies signed numbers. One may also multiply a signed or unsigned number by two with the arithmetic shift-left instructions discussed with the edit class, such as ASLA, ASLB, ASLD, and ASL 527. Divide instructions IDIV, FDIV, and EDIV divide unsigned numbers in specific registers. IDIVS similarly divides signed numbers. One may divide a two's-complement number by two with corresponding arithmetic shift-right instructions, e.g., ASRA, ASRB, and ASR 327.

0110	0110	0110	0110
1010	1010	1010	1010
0110	01100	011110	0111100
add top number to 0	shift it left	shift left, add top	shift left

Figure 2.6. Multiplication

Table 2.5. Arithmetic Instructions That Do Not Use an Effective Address

ABX	B + X -> X; Translates to LEAX B,X
ABY	B + Y -> Y; Translates to LEAY B,Y
ABA	A + B -> A; Add Accumulators A and B
CBA	A - B; Compare 8-Bit Accumulator
DAA	Adjust Sum to BCD; Decimal Adjust Accumulator A
DES	SP - 1 -> SP; Translates to LEAS -1,SP
DEX	X - 1 -> X; Decrement Index Register X
DEY	Y - 1 -> Y; Decrement Index Regisfer Y
EDIV	Y:D / X -> Y; Divide (unsigned), Remainder -> D;
EDIVS	Y:D / X -> Y; Divide (signed), Remainder -> D;
EMUL	D * Y -> Y:D; 16 x 16 to 32-bit Multiply (unsigned)
EMULS	D * Y -> Y:D; 16 x 16 to 32-bit Multiply (signed)
FDIV	D / X -> X; r -> D; 16 x 16 Fractional Divide (unsigned)
IDIV	D / X -> X; r -> D; 16 x 16 Integer Divide (unsigned)
IDIVS	D / X -> X; r -> D; 16 x 16 Integer Divide (signed)
INS	SP + 1 -> SP; Translates to LEAS 1,SP
INX	X + 1 -> X; Increment Index Register X
INY	Y + 1 -> Y; Increment Index Register Y
MUL	A * B -> D; 8 x 8 to 16-bit Multiply (unsigned)
SBA	A - B -> A; Subtract B from A

Multiplication can be done by addition and shifting almost as multiplication is done by hand, but to save hardware, the product is shifted rather than shifting the multiplier to be added to it. Figure 2.6 multiplies a 4-bit unsigned number 0110 by another 4-bit number 1010, to get an 8-bit product. First, because the most significant bit of the multiplier 1010 is one, add the number 0110 into the initial product 0, then shift the product one bit left, twice, and then add the number 0110 into the product, and shift the product one bit left. The answer is 00111100.

Actually, modern microcontrollers execute several shift-and-add operations in one clock cycle. Thus, `EMUL`, which multiplies a 16-bit by a 16-bit unsigned number, takes only three clock cycles. Signed multiplication sign extends its multiplier rather than filling with zero, and it treats the sign bit differently. Therefore use the `EMULS` instruction for signed numbers. These remarks apply analogously to the division instructions `EDIV` and `EDIVS`. The other instructions: `EDIV`, `EDIVS`, `EMULS`, `FDIV`, `IDIV`, `IDIVS`, and `MUL`, are similarly used. Note that after division, the remainder is in

Location	Contents	Mnemonic	Comment
840	FC 08 52	LDD $852	; load first number
843	FD 08 54	LDY $854	; load second number
846	13	EMUL	; multiply
847	7C 08 56	STD $856	; save result

Figure 2.7. Program Segment for 16-Bit Unsigned Multiplication

Location	Contents	Mnemonic	Comment
830	B6 08 74	LDAA $874	; get low byte of 1st
833	BB 08 63	ADDA $863	; add low byte of 2nd
836	18 07	DAA	; correct for decimal
838	7A 08 63	STAA $863	; store low sum byte
83B	B6 08 73	LDAA $873	; get high byte of 1st
83E	B9 08 62	ADCA $862	; add high byte of 2nd
841	18 07	DAA	; correct for decimal
843	7A 08 62	STAA $862	; store high sum byte

Figure 2.8. Program Segment for BCD Addition

accumulator D, and the quotient is in an index register. As an example of a multiplication of two 16-bit unsigned numbers at $852 and $854 to get a 16-bit product into $856, see Figure 2.7.

The special instruction, DAA (decimal adjust accumulator A), adds binary-coded decimal numbers. Briefly, two decimal digits per byte are represented with binary-coded decimal, the most significant four bits for the most significant decimal digit and the least significant four bits for the least significant decimal digit. Each decimal digit is represented by its usual 4-bit binary code, so the 4-bit sequences representing 10 through 15 are not used. To see how the decimal adjust works, consider adding a four-digit binary coded decimal number in the two bytes at $873 to a similar number at $862, as shown in Figure 2.8. DAA "corrects" ADDA's result. The DAA instruction may be used after ADDA or ADCA but can't be used with any other instructions such as ADDB, DECA, or SUBA.

Our next example illustrates the use of arithmetic instructions, with a move instruction to put the desired intermediate result in the correct register for the next operation. This example involves conversion of temperature from Celsius to Fahrenheit. If temperature T is measured in degrees Celsius, then the temperature in Fahrenheit is ((T * 9) / 5) + 32. Suppose T, a signed 16-bit number representing degrees Celsius, is in accumulator D. The program in Figure 2.9 evaluates the formula and leaves the temperature, in Fahrenheit, in accumulator D.

Location	Contents	Mnemonic	Comment
812	CD 00 09	LDY #9	; get multiplier
815	18 13	EMULS	; multiply by D
817	CE 00 05	LDX #5	; get divisor
81A	18 14	EDIVS	; divide to convert to F
81C	B7 64	TFR Y,D	; move quotient to D
81E	C3 00 20	ADDD #32	; correct for freezing point

Figure 2.9. Program Segment for Conversion from Celsius to Fahrenheit

2.3 Logic Instructions

Logic instructions (see Table 2.6) are used to set and clear individual bits in A, B, and CCR. They are used by compilers, programs that translate high-level languages to machine code, to manipulate bits to generate machine code. They are used by controllers of machinery because bits are used to turn things on and off. They are used by operating systems to control input-output (I/O) devices and to control the allocation of time and memory on a computer. Logic instructions are missing in calculators. That makes it hard to write compilers and operating systems for calculators, no matter how much memory they have. Returning to a problem at the end of Chapter 1, we now say a programmable calculator is not a von Neumann computer because it does not have logic instructions or any efficient replacements for these instructions with combinations of other instructions. (This differentiation may be pedagogically satisfying, but unfortunately, von Neumann's original computer is not a von Neumann computer by this definition. Because we are engineers and not historians, we say that programmable calculators, and von Neumann's original computer, are not von Neumann computers in the strictest sense because they cannot support compilers and operating systems efficiently.)

Consider now the logic instructions that make a computer a computer and not a calculator. The most important logic instructions carry out bit-by-bit logic operations on accumulators A or B with a memory location or an immediate value. (See Figure 2.10 for a summary of the common logic operations.) For example, the instruction ANDB $817 carries out a bit-by-bit AND with the contents of B and the contents of location $817, putting the result in B (see Figure 2.11b). The ANDing is implemented by AND gates in the MPU, shown in Figure 2.11a. Compare to Figure 1.4d. The OR and AND instructions are used, among other things, to set or clear control bits in registers used in input-output operations.

The two instructions ANDA and ANDCC do the same thing as ANDB except that ANDCC uses only immediate addressing and the condition code register CCR. As an example, ANDCC #$FE clears the carry bit in the condition code register, that is, puts C = 0 leaving the other bits unchanged. This instruction is used only to clear condition code bits and is not used to modify other data bits. The same remarks hold for the OR instructions, ORAA, ORAB, and ORCC, and for the exclusive-OR instructions, EORA and EORB (see Figure 2.11a again: Exchange the AND gates with OR or exclusive-OR gates). The mnemonics CLC, CLI, SEC, SEI, SEV, CLV are merely special cases of the ANDCC and ORCC instructions; they are given these special mnemonic names to permit assembly-language programs written for the 6811 to be used without modification in the 6812. While the ANDCC instruction is used to clear bits in the CCR register, the ORCC instruction is used to set bits in that register. There is no EORCC instruction.

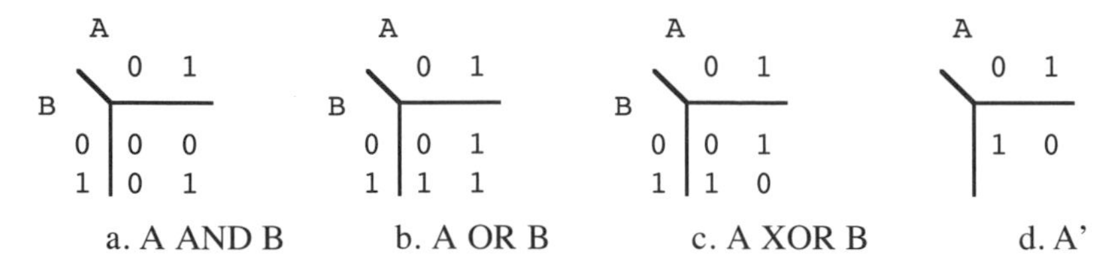

a. A AND B b. A OR B c. A XOR B d. A'

Figure 2.10. Common Logic Operations

Table 2.6. Logic Instructions

ANDA opr	A AND (E) -> A; Logical And A with Memory
ANDB opr	B AND (E) -> B; Logical And B with Memory
BITA opr	A AND (E); Logical And A with Memory
BITB opr	B AND (E); Logical And B with Memory
EORA opr	A XOR (E) -> A; Exclusive-OR A with Memory
EORB opr	B XOR (E) -> B; Exclusive-OR B with Memory
ORAA opr	A OR (E) -> A; Logical OR A with Memory
ORAB opr	B OR (E) -> B; Logical OR B with Memory
BCLR opr,msk	(E) AND ~ mask -> (E); Clear Bit(s) in Memory
BSET opr,msk	(E) OR mask -> (E); Set Bit(s) in Memory
COM opr	~ (E) ->(E); Equivalent to $FF - (E) -> (E)
ANDCC opr	CCR AND (E) -> CCR; Logical And CCR with Memory
CLC	0 -> C; Is same as ANDCC #$FE
CLI	0 -> I; Is same as ANDCC #$EF (enables int.)
CLV	0 -> V; Is same as ANDCC #$FD
ORCC opr	CCR OR (E) -> CCR; Logical OR CCR with Memory
SEC	1 -> C; Is same as ORCC #1
SEI	1 -> I; Is same as ORCC #$10 (inhibits int.)
SEV	1 -> V; Is same as ORCC #2
COMA	~ A -> A; Complement Accumulator A
COMB	~ B -> B; Complement Accumulator B

Consider this example. Suppose that we need to clear bits 0 and 4; set bits 5, 6, and 7; and leave bits 1, 2, and 3 unmodified in accumulator A. The following instructions carry out these modifications.

```
ORAA   #$E0   Set bits 5, 6, and 7, leaving others unchanged
ANDA   #$EE   Clear bits 0 and 4, leaving others unchanged
```

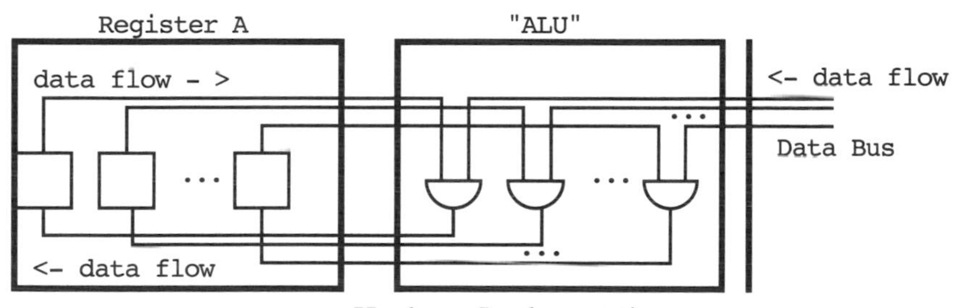

a. Hardware Implementation

```
        1 0 1 1 0 1 0 1   <- contents of $817
    AND 0 0 0 1 1 1 0 0   <- initial value of B
        0 0 0 1 0 1 0 0   <- final value of B
```

b. Logical Operation

Figure 2.11. Bit-by-Bit AND

Table 2.7. Edit Instructions

ASL opr	Arithmetic Shift Left
ASR opr	Arithmetic Shift Right
LSL opr	Logical Shift Left; same as ASL
LSR opr	Logical Shift Right
ROL opr	Rotate Memory Left through Carry
ROR opr	Rotate Memory Right through Carry
ASLA	Arithmetic Shift Left Accumulator A
ASLB	Arithmetic Shift Left Accumulator B
ASLD	Arithmetic Shift Left Accumulator D
ASRA	Arithmetic Shift Right Accumulator A
ASRB	Arithmetic Shift Right Accumulator B
LSLA	Logical Shift Accumulator A to Left
LSLB	Logical Shift Accumulator B to Left
LSLD	Same as ASLD
LSRA	Logical Shift Accumulator A to Right
LSRB	Logical Shift Accumulator B to Right
LSRD	Logical Shift Right Accumulator D
ROLA	Rotate A Left through Carry
ROLB	Rotate B Left through Carry
RORA	Rotate A Right through Carry
RORB	Rotate B Right through Carry

The complement instruction `COM` takes the complement of the bit-by-bit contents of A, B, or a memory location, putting the result in the same place. Finally, the `BIT` instruction, for bit test, determines the bits as though the `AND` instruction had been performed with A or B and the contents of a byte from memory. With the `BIT` instruction, however, the contents of A and B are unchanged. It is to the `AND` instruction what the `CMP` instruction is to the `SUB` instruction; it is used to avert the need to reload the register after the condition code bits are set as in the `AND` instruction.

Logic instructions are used primarily to set and clear and to test and change (logically invert) bits in a word. These instructions are used to build operating systems, compilers, and other programs that control resources and format data. These are the instructions that make a computer so much more useful than a programmable calculator.

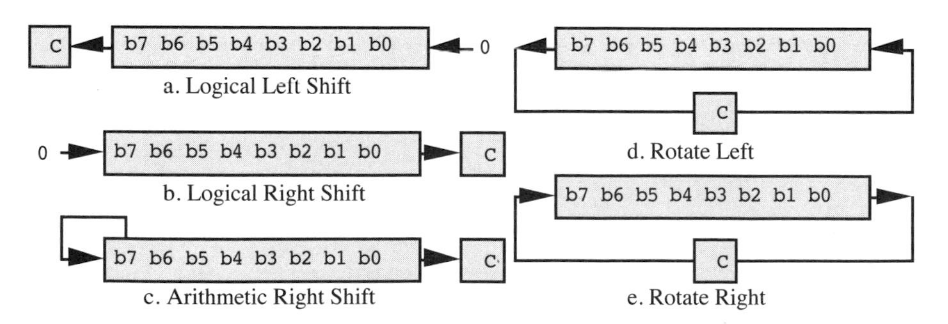

Figure 2.12. Shifts and Rotates

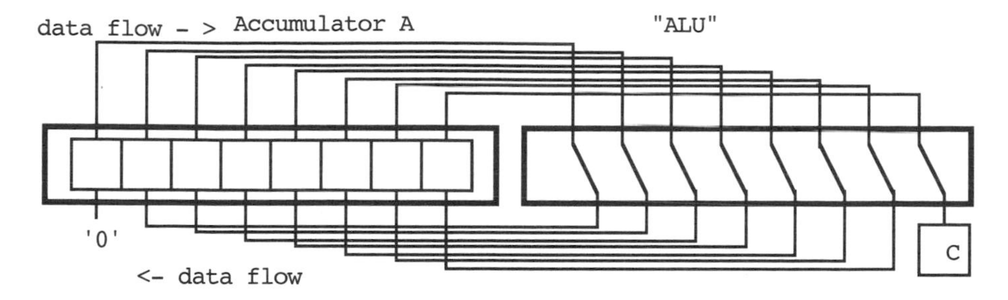

Figure 2.13. Shift Hardware

2.4 Edit Instructions

Edit instructions (see Table 2.7) rearrange bits of data without generating new bits as an ADD does. Large machines have complex edit instructions, but microcomputers have simple ones. For example, the arithmetic shift-left instructions shift all the bits left, putting the most significant bit into the carry bit of the condition code register and putting a zero in on the right (the same as LSLA) (see Figure 2.12a). This, except for overflow, doubles the unsigned or signed number contained in A. The ASR instruction keeps the sign bit unchanged and shifts all other bits to the right, putting the least significant bit into the carry bit (see Figure 2.12c). As mentioned in the discussion of the arithmetic class of instructions, ASR divides the original two's-complement number contained in an accumulator or memory location by two (rounding down).

The shift operation is generally done in the MPU, and is shown in Figure 2.13 for LSRA. The MPU shifts the data because these data paths follow the same data paths as addition and logical operations (see Figures 1.4d and 2.11a). Each flip-flop's output feeds through the multiplexer in the MPU to the input of the next bit to the right; a zero is fed into the leftmost flip-flop and the rightmost flip-flop's output is put in the carry C.

Location	Contents	Mnemonic	Comment
823	18 0E	TAB	; copy byte
825	58	LSLB	; shift left byte
826	58	LSLB	; shift left byte
827	58	LSLB	; shift left byte
828	58	LSLB	; shift left byte
829	44	LSRA	; shift right byte
82A	44	LSRA	; shift right byte
82B	44	LSRA	; shift right byte
82C	44	LSRA	; shift right byte
82D	18 06	ABA	; combine nibbles

Figure 2.14. Program Segment to Swap Nibbles

Location	Contents	Mnemonic	Comment
812	FC 08 56	LDD $856	; get bits to be inserted
815	84 FC	ANDA #$FC	; remove bits to be inserted
817	C4 7F	ANDB #$7F	; in both bytes
819	7C 08 56	STD $856	; temporarily save value
81C	B6 08 54	LDAA $854	; get bits to be inserted
81F	C7	CLRB	; AND low byte with all 0's
820	49	LSRD	; shift 16-bit data right
821	84 03	ANDA #3	; remove extraneous bits
823	BA 08 56	ORAA $856	; combine first part with
826	FA 08 57	ORAB $857	; second part, in both bytes
829	7C 08 56	STD $856	; save the result

Figure 2.15. Program Segment for Insertion of Some Bits

The remaining shifts and rotates (e.g., LSR, LSL, ROR, ROL) are easily understood by looking at Figure 2.12. The rotate instructions are used with multiple-byte arithmetic operations such as division and multiplication. Edit instructions are generally used to rearrange bits. For example, Figure 2.14 swaps the nibbles in accumulator A.

In a slightly more interesting problem, we insert the three low-order bits of the byte in $854 into bits 9 to 7 of the 16-bit word at location $856. This program in Figure 2.15 illustrates the use of logical instructions to remove unwanted bits and to combine bits and edit instructions to move bits into the desired bit positions. A program segment like this is used in C structs using bit fields and is commonly used in inserting bits into I/O ports that do not line up with whole bytes or whole 16-bit words.

In this example, observe that all logical and many edit instructions are performed on 8-bit operands in each instruction. However, instructions are so designed that pairs of instructions, on accumulator A and accumulator B, effectively work on accumulator D.

2.5 Control Instructions

The next class of instructions, the control instructions, are those that affect the program counter PC. After the MOVE class, this class composes the most-often-used instructions. Control instructions are divided into conditional branching instructions and other control instructions. We discuss conditional branching first and then the others.

The BRA instruction loads the PC with a new value, using relative addressing discussed in §3.3. It adds the last byte of its instruction, called the *offset,* to the PC. Branch statements have "long" branch counterparts where each mnemonic is prefaced with an L, such as LBRA, and the offset is two bytes, enabling the programmer to add larger values to the PC, to branch to locations further from the instruction.

Table 2.8. Conditional Branch Instructions

(L)BCC rel	(Long) Branch if Carry Clear (if C = 0)
(L)BCS rel	(Long) Branch if Carry Set (if C = 1)
(L)BEQ rel	(Long) Branch if Equal (if Z = 1)
(L)BGE rel	(Long) Branch if Signed Greater Than or Equal
(L)BGT rel	(Long) Branch if Signed Greater Than
(L)BHI rel	(Long) Branch if Unsigned Higher
(L)BHS rel	(Long) Branch if Higher or Same; same as BCC
(L)BLE rel	(Long) Branch if Signed Less Than or Equal
(L)BLO rel	(Long) Branch if Lower; same as BCS
(L)BLS rel	(Long) Branch if Unsigned Lower or Same
(L)BLT rel	(Long) Branch if Signed Less Than
(L)BMI rel	(Long) Branch if Minus (if N = 1)
(L)BNE rel	(Long) Branch if Not Equal (if Z = 0)
(L)BPL rel	(Long) Branch if Plus (if N = 0)
(L)BRA rel	(Long) Branch Always
(L)BRN rel	(Long) Branch Never
(L)BVS rel	(Long) Branch if Overflow Bit Set (if V = 1)
(L)BVC rel	(Long) Branch if Overflow Bit Clear (if V = 0)

Conditional branch instructions test the condition code bits. As noted earlier, these bits have to be carefully watched, for they make a program look so correct that you want to believe that the hardware is at fault. The hardware is rarely at fault. The condition code bits are often the source of the fault because the programmer mistakes where they are set and which ones to test in a conditional branch. The instructions should now be reviewed with regard to how they affect the condition code bits. See the right columns of the CPU12RG/D manual Instruction Set Summary. Note that move instructions generally either change the N and Z bits or change no bits; arithmetic instructions generally change all bits; logic instructions change the N and Z bits; and edit instructions change all bits. However, there are many exceptions, and these exceptions are precisely the ones that cause mystifying errors. There is sound rationale for which bits are set and the way they are set. Some of that is discussed in this chapter. But most of it is simply learned by experience. We conclude by reminding you that when your program does not work and you have checked every angle, carefully examine the setting and testing of the condition code bits. Now we look at the testing of these bits in detail.

Eight simple branching instructions test only a single condition code register bit: `BNE`, `BEQ`, `BPL`, `BMI`, `BVC`, `BVS`, `BCC`, and `BCS`. The letters S and C are used for "set" and "clear" (to 1 and 0, respectively) in branching instruction mnemonics.

Frequently, two numbers are compared, as in a compare instruction or a subtraction. One would like to make a branch based on whether the result is positive, negative, and so forth. Table 2.8 shows the test and the branching statement to make depending on whether the numbers are interpreted as signed numbers or unsigned numbers. The branch mnemonics for the two's-complement numbers, or signed, numbers case are the ones usually described in mathematical "greater or less" prose. For example, `BLT` for "branch if less than," `BLE` for "branch if less than or equal to," and so forth. The mnemonics for unsigned numbers are described in mathematical "high or low" prose, offbeat enough to

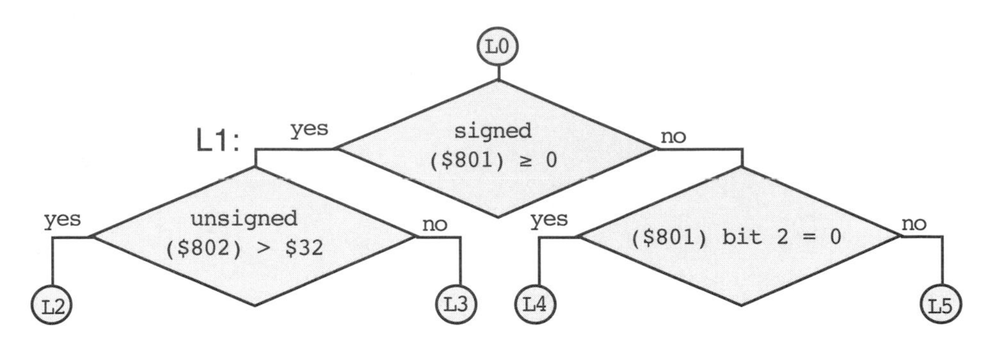

Figure 2.16. Decision Tree

keep you from confusing them with the signed ones, for example, BLO for "branch if lower," BLS for "branch if lower or the same," BHI for "branch if higher," and BHS for "branch if higher or the same." Notice that BLO is the same instruction as BCS, and BHS is the same instruction as BCC. Here then is an example of two different mnemonics describing the same instruction, something that is sometimes warranted when the programmer will be using the same instruction with two distinct meanings.

Figure 2.16 illustrates a flow chart of several tests that form a decision tree. In Figure 2.17 and subsequent figures, labels are written left justified and end in a colon. Upon entry at label L0, location $801 is tested; if it is positive, go to L2 if location 802's unsigned value is greater than $32, otherwise go to L3. If location $801 is negative, go to L4 if location $801 bit 2 is zero, otherwise go to L5. The program segment in Figure 2.17 implements this decision tree.

Do not be concerned about calculating the relative branch offsets; §3.3 will discuss this calculation. One should consult Table 2.8 for a while to make sure that the correct branch is being chosen. For example, to test a register value greater than or equal to a memory value, you might be tempted to use the simple branch BPL for signed numbers instead of BGE. The problem is that you want the branch test to work even when subtraction or comparison generates a signed overflow. But this is just exactly when the

Location	Contents	Mnemonic	Comment
82A	B6 08 01	L0: LDAA $801	; get byte, also set N and Z
82D	2A 06	BPL L1	; if N is zero, goto L1
82F	85 04	BITA #4	; check bit 2
831	27 0D	BEQ L4	; if zero, go to L4
833	20 0D	BRA L5	; otherwise go to L5
835	B6 08 02	L1: LDAA $802	; get byte
838	81 32	CMPA #$32	; compare to $32
83A	22 02	BHI L2	; if higher, go to L2
83C	20 06	BRA L3	; otherwise go to L3

Figure 2.17. Program Segment for a Decison Tree

Table 2.9. Other Control Instructions

BRCLR opr,msk,rel	Branch if (E) AND (msk) = 0
BRSET opr,msk,rel	Branch if (E)' AND (msk) = 0
DBEQ cntr,rel	cntr-1 -> cntr; if cntr = 0, Branch.
DBNE cntr,rel	cntr-1 -> cntr; if cntr != 0, Branch.
IBEQ cntr,rel	cntr+1 -> cntr; If cntr = 0, Branch.
IBNE cntr,rel	cntr+1 -> cntr; If cntr != 0, Branch.
TBEQ cntr,rel	If cntr = 0 then Branch.
TBNE cntr,rel	If cntr != 0 then Branch.
JMP opr	Jump; Address -> PC
JSR opr	Jump to Subroutine
BSR rel	Branch to Subroutine
RTS	Return from Subroutine.
CALL opr,page	Call subroutine in extended memory
RTC	Return from Call
BGND	Place CPU in Background Mode
NOP	No Operation
RTI	Return from Interrupt
WAI	WAIT for interrupt
STOP	STOP All Clocks
SWI	Software Interrupt
TRAP	Unimplemented opcode trap

sign is incorrect; then using BPL cannot be used to replace BGE. Thus, after a compare or subtract between signed numbers, use BGE rather than BPL. You might also be tempted to use BPL for the unsigned test. However, if accumulator A has \$80 and the immediate operand is \$32, then N = 0 after performing the test. Thus BPL takes the branch, even though it should not because \$32 is not higher than \$80. Thus after an unsigned number comparison or subtraction, use BHS rather than BPL.

A rather amusing instruction, BRN L, which "branches never" regardless of the location L, is the opposite to the "branch always" instruction. It is useful because any branching instruction can be changed to a BRA or BRN instruction just by changing an opcode byte. This allows a programmer to choose manually whether a particular branch is taken while he or she is debugging a program.

Location	Contents	Mnemonic	Comment
820	1C 08 02 40	L1: BSET \$0802,#64	; set bit 6 in byte \$802
824	20 04	BRA L3	; go to common code
826	1D 08 02 40	L2: BCLR \$0802,#64	; clear bit 6 in byte \$802
82A	1E 08 02 40 02	L3: BRSET \$0802,#64,L4	; common code: test bit 6
82F	20 FE	BRA *	; if bit 6 is 0, wait here
831	20 FE	L4: BRA *	; if bit 6 is 1, wait here

Figure 2.18. Program Segment for Setting, Clearing, and Testing of a Bit

Location	Contents	Mnemonic	Comment
810	CC 00 05	LDD #5	; put number 5 in D
813	04 34 FD	L: DBNE D,L	; decrement D until 0
816	00	BGND	; halt

Figure 2.19. Program Segment for a Wait Loop

We now consider the control instructions other than conditional branches. See Table 2.9. Some instructions combine a logical or arithmetic test with a conditional branch and do not modify condition codes. `BRCLR` branches if all "1" bits in the mask are "0" in the word read from memory. Similarly, `BRSET` branches if all masks "1" bits are "1" in the word read from memory. These bits can be set and cleared using `BSET` and `BCLR` listed in Table 2.6. Figure 2.18 illustrates such setting, clearing, and testing of individual bits in memory. If this program segment is entered from location L1 ($820), then bit 6 of location $802 is set there, and the instruction at L3 ($82A) branches to location L4 ($831). However, if this program segment is entered from location L2 ($826), then bit 6 of location $802 is cleared there, and the instruction at L3 ($82A) doesn't branch but falls through to location $82F.

`DBEQ`, `DBNE`, `IBEQ`, and `IBNE`, which have a postbyte and a relative offset, decrement or increment a counter, which may be A, B, D, X, Y, or SP, and branch if the result in the counter is zero or nonzero, as indicated by the mnemonic and coded in the post byte. `TBEQ` and `TBNE` similarly test a register without incrementing or decrementing it and branch if the result is zero or nonzero. The low-order post byte bits indicate which register is used as a counter or test register (0, A; 1, B; 4, D; 5, X; 6, Y; and 7, SP) and the high-order three bits indicate the operation (000, `DBEQ`; 001, `DBNE`; 010, `TBEQ`; 011, `TBNE`; 100, `IBEQ`; 101, `IBNE`). Post-byte bit 4 is appended as high bit to the instruction's third byte to give a 9-bit offset.

The program segment in Figure 2.19 wastes time while an I/O operation takes place. (Calculation of the last byte, the offset $FD, will be discussed in §3.3.) The `DBNE` instruction takes three clock cycles, where each clock cycle is 125 nanoseconds. This instruction loops to itself five times in a *delay loop*, which wastes 1.875 μsec.

The simple jump instruction is the simplest control instruction; the effective address is put into the program counter. `JMP $899` puts $899 into the program counter, and the next opcode byte is fetched from $899. It simply "jumps to location $899."

You commonly encounter in programs a repeated program segment. Such a segment can be made into a *subroutine* so it can be stored just once but executed many times. Special instructions are used to branch to and return from such a subroutine. For example, if the subroutine begins at location $812, the instruction `JSR $812` (for *jump to subroutine)* causes the PC to be loaded with $812 and the address immediately after the `JSR` instruction (say it is $807) to be pushed onto the hardware stack, low byte first. Figure 2.20a shows this *return address,* saved on the stack. `BSR` (for *branch to subroutine)* similarly pushes the program counter but locates the subroutine using relative addressing (§3.3). At the end of the subroutine, the l-byte instruction `RTS` (for *return from subroutine)* pulls the top two bytes of the hardware stack into the PC, high byte first. `JSR`, `SUB`, and `RTS`, efficiently call, and return from, the subroutine.

a. Subroutine Addresses b. Subroutine stack c. Handler stack

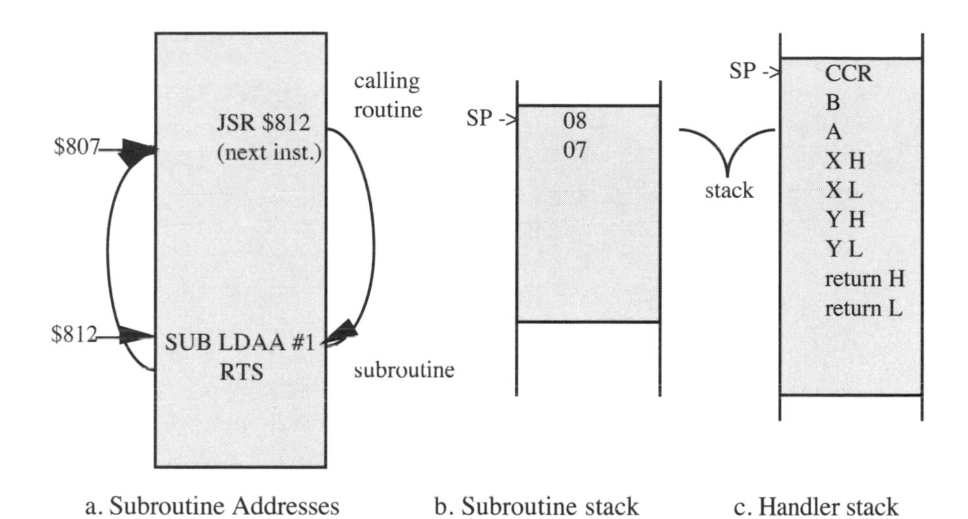

Figure 2.20. Subroutine and Handler Addresses

Figure 2.20 illustrates the use of the stack for holding temporary results as discussed in 2.1, with subroutine return addresses as illustrated in §2.5. We suggest that you step through this program using the simulator or debugger and watch the stack expand and compress. A constant in X, 1, is pushed on the stack before the subroutine and restored by pulling X after the subroutine is executed and has returned. This is commonly done when the calling routine needs the saved value later. The subroutine return address is saved on the stack by the `BSR` instruction and restored by the `RTS` instruction at the end of the subroutine. Inside the subroutine, the X and Y registers are saved and restored in order to exchange them. Pushing and pulling is often done to hold intermediary results.

Location	Contents	Mnemonic	Comment
800	CF 0A 00	LDS #$A00	; initialize to top of SRAM
803	CE 00 01	LDX #1	; put some constant in X
806	34	PSHX	; save it on the stack
807	07 03	BSR SUB	; call the subroutine
809	30	PULX	; restore the saved value of X
80A	20 FE	BRA *	; wait until user stops
80C	34	SUB: PSHX	; push constant on stack
80D	35	PSHY	; push another value on stack
80E	30	PULX	; pull other value in X
80F	31	PULY	; pull constant into Y
810	3D	RTS	; return to calling routine

Figure 2.21. Program Segment for Swap Subroutine

A special CALL instruction permits saving and then loading a page register when saving and then loading the program counter. That extends the addressing capability to over 16 bits in some 6812s, such as the 'A4. The corresponding RTC instruction returns from a subroutine called by a CALL instruction.

As noted earlier, the stack pointer is to be initialized at the beginning of a program, with an instruction like LDS #$C00. It must be initialized before any instruction, such as JSR or CALL, uses the stack pointer. If it is not, the RTS or RTC does not work because the return address is "saved" in a location that is not RAM, so it is lost.

The *(hardware or I/O) interrupt* is very important to I/O interfacing. Basically, it is evoked when an I/O device needs service, either to move some more data into or out of the device or to detect an error condition. *Handling* an interrupt stops the program that is running causes another program to be executed to service the interrupt and then resumes the main program exactly where it left off. The program that services the interrupt (called an *interrupt handler or device handler*) is very much like a subroutine, and an interrupt can be thought of as an I/O device tricking the computer into executing a subroutine. An ordinary subroutine called from an interrupt handler is called an *interrupt service routine*. However, a handler or an interrupt service routine should not disturb the current program in any way. The interrupted program should get the same result no matter if or when the interrupt occurs.

I/O devices may request an interrupt in any memory cycle. However, the data operator usually has bits and pieces of information scattered around in hidden registers. It is not prepared to stop the current instruction because it doesn't know the values of these registers. Therefore, interrupts are always recognized at the end of the current instruction, when all the data are organized into accumulators and other registers that can be safely saved and restored. The time from when an I/O device requests an interrupt until data that it wants moved is moved or the error condition is reported or fixed is called the *latency time*. Fast I/O devices require low latency interrupt service. The lowest latency that can be guaranteed must exceed the duration of the longest instruction because the I/O device could request an interrupt at the beginning of such an instruction's execution.

The SWI instruction is essentially like an interrupt. It saves all the registers as shown in Figure 2.20c and puts the contents of $fff6, $fff7 into the program counter, to begin an SWI handler at that address. All TRAP instructions (there are over 200 of them) save all the registers as the SWI instruction does and put the contents of $fff8, $fff9 into the program counter to begin a trap handler at that address. RTI pulls the contents of the registers saved on the stack and fetches the next opcode at the address that is the returned program counter. WAI stacks all the registers and waits for an interrupt to occur. STOP stacks the registers and stops all the 6812 clocks to conserve power. A system reset or an interrupt will cause the computer to resume after these instructions. Two *interrupt inhibit* bits (also called an *interrupt mask* bit) I and X are kept in the condition code; when they are set, interrupts are not permitted. A *stop disable* bit S is used to prevent execution of the STOP instruction. BGND places the MPU in a background mode to permit the background debug module to examine memory and registers and possibly modify some of them. If background debugging is not enabled, BGND can be made to act exactly like an SWI instruction.

The condition code register, accumulators, program counter, and other registers in the controller and data operator are collectively called the *machine state* and are saved whenever an interrupt occurs as shown below, resulting in the stack in Figure 2.20c.

SP - 2 -> SP; PC -> (SP):(SP+1);
SP - 2 -> SP; X -> (SP):(SP+1);
SP - 1 -> SP; CCR -> (SP);
SP - 2 -> SP; Y -> (SP):(SP+1);
SP - 2 -> SP; B -> (SP); A -> (SP+1);

After completion of a handler entered by a hardware interrupt or similar instruction, the last instruction executed is *return from interrupt* (`RTI`). All handlers end in an `RTI` instruction. `RTI` pulls the top nine words from the stack, replacing them in the registers the interrupt took them from. The `RTI` instruction executes the operations:

(SP) -> CCR; SP + 1 -> SP
(SP):(SP+1) -> X; SP + 2 -> SP
(SP):(SP+1) -> PC; SP + 2 -> SP
(SP) -> B; (SP+1) -> A; SP + 2 -> SP
(SP):(SP+1) -> Y; SP + 2 -> SP

You can modify the program in Figure 2.21 to see how the trap instruction saves and restores the machine state. Replace the BSR instruction at location \$807 with an SWI instruction whose opcode is \$3F (and a NOP, \$A7) and the RTS instruction at location \$810 with RTI whose opcode is \$0B; put the adddress \$80C into locations \$FFF6 and \$FFF7; and rerun this program. You should see that changing the registers inside the trap handler has no effect on the returned values of the registers, because they are saved on the stack and restored by the RTI instruction.

We have covered both the conditional and unconditional branch instructions. We have also covered the jump and related instructions together with subroutine branch and jump instructions. Control instructions provide the means to alter the pattern of fetching instructions and are the second most common type of instruction. If you use them wisely, they will considerably enhance the static and dynamic efficiency.

2.6 Input-Output Instructions

The last class of instructions for the 6812, the input-output or I/O class, is easy to describe because there aren't any! With the 6812 a byte is transferred between an accumulator and a register in an I/O device through a memory location chosen by hardware. The `LDAA` instruction with that location then inputs a byte from the register of the I/O device to accumulator A, while the `STAA` instruction with that location does the corresponding output of a byte. Other instructions, such as `MOVB`, `MOVM`, `ROL`, `ROR`, `DEC`, `INC`, and `CLR`, may be used as I/O instructions, depending on the particular device. We look more closely at all of these issues in Chapter 11.

2.7 Special Instructions

Table 2.10 lists 6812's special instructions, which are arithmetic instructions of primary interest in fuzzy logic. They use index addressing, which is discussed in the next chapter. Fuzzy logic uses minimum and maximum functions to logically AND and OR fuzzy variables. The 6812 has instructions `MAXA`, `MAXM`, `MINA`, `MINM`, `EMAXD`, `EMAXM`, `EMIND`, and `EMINM` to determine the maximum or minimum of an 8-bit or a 16-bit pair of unsigned numbers, one of which is in a register (A or D) and the other of which is one or two bytes at the effective address, and to put the maximum or minimum

Table 2.10. Special Instructions

EMAXD opr	MAX(D, (E):(E+1)) -> D; MAX Unsigned 16-Bit
EMAXM opr	MAX(D, (E):(E+1)) -> (E):(E+1); MAX Unsigned 16-Bit
EMIND opr	MIN(D, (E):(E+1)) -> D; MIN Unsigned 16-Bits
EMINM opr	MIN(D, (E):(E+1)) -> (E):(E+1); MIN Unsigned 16-Bits
MAXA	MAX(A, (E)) -> A
MAXM	MAX(A, (E)) -> (E)
MINA	MIN(A, (E)) -> A; MIN Unsigned 8-Bit
MINM	MIN(A, (E)) -> (E); MIN Unsigned 8-Bit
ETBL opr	16-Bit Table Lookup and Interpolate
TBL opr	8-Bit Table Lookup and Interpolate
EMACS opr	(X):(X+1) x (Y):(Y+1) + (E):(E+1):(E+2):(E+3) -> (E):(E+1):(E+2):(E+3); 16x16 Bit -> 32 Bit; Multiply and Accumulate (signed).
MEM	u (grade) -> (Y); X + 4 -> X; Y + 1 -> Y; A unchanged
REV, REVW	MIN-MAX rule evaluation
WAV, wavr	Weights for Weighted Average Calculation

in either the register or the memory at the effective address. `EMACS` is a multiply-and-accumulate instruction similar to such instructions used in digital signal processors (DSPs).

The following examples use pointer addressing, in which the effective address is the contents of an index register, without adding any other value to it.

We are about to output the contents of accumulator D to an output device, but the output must be limited to be at least Vmin, and at most Vmax. Suppose location $803 has the address of Vmax and location $805 has the address of Vmin. Pointer addressing (§3.2) is one of the modes usable with the `EMAXD` and `EMIND` instructions. Figure 2.22 limits the value of accumulator D to be Vmin ≤ D ≤Vmax.

Figure 2.23 illustrates the use of index registers in the multiply-and-accumulate instruction to evaluate the expression: A = A + (B * C), where A is a signed 32-bit number at $910, B is a signed 16-bit number at $914, and C is a signed 16-bit number at $916. Pointer addressing is the only mode that can be used with the `EMACS` instruction. From Table 2.10, the `EMACS` instruction executes the expression (X):(X+1) x (Y):(Y+1) + (E):(E+1):(E+2):(E+3) -> (E):(E+1):(E+2):(E+3); from the CPU12RG/D manual, it executes this operation in thirteen clock cycles, a little under 2 μsec.

Location	Contents	Mnemonic	Comment
82A	CE 08 03	LDX #$803	; get address of Vmax
82D	18 1B 00	EMIND 0,X	; obtain minimum
830	CD 08 05	LDY #$805	; get address of Vmin
833	18 1A 40	EMAXD 0,Y	; obtain maximum

Figure 2.22. Program Segment for Ensuring a Value Is between Limits

Location	Contents	Mnemonic	Comment
830	CE 09 14	LDX #$914	; get address of 1st
833	CD 09 16	LDY #$916	; get address of 2nd
836	18 12 09 10	EMACS $910	; multiply and accumulate

Figure 2.23. Program Segment for a Multiply and Add Operation

The TBL and ETBL instructions perform 8-bit and 16-bit table lookup and interpolation. TBL puts into accumulator A the value (E) + (B * ((E+1) - E)) where E is the effective address, which can be the pointer address as in EMAXD, and B is accumulator B, considered as an unsigned fraction. CMPB can search a list of values, which are in increasing order, for the nearest value just below, and B and TBL can interpolate between that value and the next higher value. ETBL is similar but is for 16-bit unsigned number interpolation: It puts into accumulator D the value (E):(E+1) + (B * ((E+2):(E+2) - E:(E+1))). The CPD instruction can search a list of values to use ETBL.

The instructions MEM, REVW, REV, and WAV are used for fuzzy logic rule evaluations, which are developed in §7.6. These highly specific and efficient operations make the 6812 singularly well-suited to fuzzy logic control applications.

The fuzzy logic membership instruction MEM uses accumulator A as the current input, and X points to a 4-byte data structure that describes a trapezoidal membership function (P1, P2, S1, S2). The instruction puts into the byte pointed to by Y, the function value, and then adds 4 to X and 1 to Y to access the next trapezoid and output value. If A < P1 or A > P2, then the output function value is 0, or else the output function value is MIN(A - P1) * S1, (P2 - A) * S2, $FF).

REV and REVW perform min-max rule evaluation for 8-bit and 16-bit unsigned numbers. For REV, each rule input is an 8-bit offset from the base address in Y. Each rule output is an 8-bit offset from the base address in Y. $FE separates rule inputs from rule outputs, and $FF terminates the rule list. REV may be interrupted. For REVW, each rule input is the 16-bit address of a fuzzy input. Each rule output is the 16-bit address of a fuzzy output. The value $FFFE separates rule inputs from rule outputs, and $FFFF terminates the rule list. REV and REVW use this MIN-MAX rule: Find the smallest rule input (MIN), and store to rule outputs unless fuzzy output is already larger (MAX).

WAV calculates the sum-of-products and sum-of-weights for a list of 8-bit unsigned elements. Accumulator B is the number of elements in both lists, X points to the first list, and Y points to the second list. The sum-of-products is put in registers Y (high-order 16 bits) and D (low-order 16 bits), and the sum of weights, pointed to by Y, is put into register X. The instruction wavr resumes the execution of the WAV instruction if it is interrupted in the middle of its execution.

2.8 Remarks

One might wonder why some move instructions, such as LDAA, TSTA, and STAA, always put V = 0 rather than leaving V unchanged as they do C. The reason is that doing so allows all of the signed branches to work after these instructions as well as after the arithmetic type of instruction. For example, suppose that one wants to look at the contents of some memory location, say \$811, and branch to location L if the contents of location 458, treated as a signed number, are greater than 0. The sequence

```
TST   $811
BGT   L
```

does exactly this. If the `TST` instruction had left V unaffected, we would have had to use the longer sequence:

```
LDAA   $811
CMPA   #0
BGT    L
```

A little more experience will show that the designer's choice here is quite reasonable, because we will find a more frequent use of signed branches for load instructions than for checking for signed overflow, as we will do in the next chapter.

Do You Know These Terms?

See the end of Chapter 1 for instructions.

stack
push
pull
hardware stack
buffer
post byte
hidden register
prefix byte
subroutine
jump to subroutine
return address
branch to subroutine
return from subroutine
hardware interrupt
I/O interrupt
interrupt handling
interrupt handler
device handler
interrupt service routine
service routine
latency time
interrupt inhibit
interrupt mask
stop disable
machine state
return from interrupt

PROBLEMS

When a program (ending in BGND or SWI) or program segment is asked for in the problems below, use the format that is used for the examples in the text.

1. Assume the following instruction is missing from the 6812 instruction set. Show the shortest program segment that will accomplish *exactly* the same effect as this missing instruction. That is, this program segment must give the same results in all the registers, including the condition code register.

(a) XGDX (or EXG X,D) (b) TFR X,Y (c) PSHD

2. Assume MOVB is missing from the 6812 instruction set. Show the shortest program segment that accomplishes *exactly* the same effect as MOVB $803,$822.

3. Show the shortest program segment that will push the following 32-bit constant on the stack. The most significant byte must be at the lowest address.

(a) 0 (b) 1 (c) –1

4. Write a shortest program to evaluate a quadratic polynomial. Let a be at $810, b be at $812, c be at $814, x be at $816; the program is to put a x^2 + b x + c into $818. All numbers, including the result, are 16-bit two's-complement signed numbers.

5. Write a shortest program to execute an inner product. Let x[0] be at $810, x[1] be at $812, y[0] be at $814, y[1] be at $816; the program is to put x[0] y[0] + x[1] y[1] into $818. All numbers are 16 bit unsigned numbers.

6. Write a shortest program to compute the resistance of a pair of resistors connected in parallel. Let r1 be at $810 and r2 be at $812; the program is to put r1 || r2 into $814. All values are 16-bit unsigned numbers.

7. If a count C is obtained, the frequency is 8,000,000 / C. Write a shortest program to compute the 16-bit frequency corresponding to the 16-bit count C in location $81a, putting the result into $81c. Show this program, beginning at $81e, in hexadecimal.

8. Why doesn't DAA work after an INC, DEC, or ASR instruction?

9. Assume the following instruction is missing from the 6812 instruction set. Show the shortest program segment that will accomplish *exactly* the same effect as the missing instruction. For part (c) assume that locations $813 and $814 are able to be used as scratch bytes. (Scratch means the location is available for storing temporary results.)

(a) BSET $810,#$aa (b) BCLR $811,#$f (c) EORA $812

10. ASCII characters are defined by Table 4.1. Write a single instruction for the following:

(a) Accumulator A is an upper- or lower-case character. Convert it to lower case.
(b) Accumulator A is an upper- or lower-case character. Convert it to upper case.
(c) Accumulator A is a single BCD number. Convert it to an ASCII character.
(d) Accumulator A is an ASCII character. Convert it to a BCD number.

11. Write a fastest program segment to put the following property, of a number in Accumulator A, into accumulator B (do not use branch instructions). For part (c), assume that location $822 is a scratch byte.

(a) the count of the number of 1's (b) the parity (c) the number of leading zeros

12. A 32-bit number is in accumulator D (low-order) and index register Y (high-order). Write a shortest program segment to execute the following on these same registers.

(a) logical shift left 1 bit (b) logical shift right 1 bit (c) arithmetic shift right 1 bit

13. Illustrate the differences between `BLT` and `BMI` with an example that branches to location L if accumulator A is less than the contents of the byte at $869.

14. Will a signed branch work after a `DEC` or `INC` instruction? Explain. What about unsigned branches?

15. Assume the following instruction is missing from the 6812 instruction set. Show the shortest program segment that will accomplish the same effect as the missing instruction, except that the condition codes will be changed.

(a) `BRCLR $811,#$f,L` (b) `BRSET $810,#$aa,L` (c) `DBNE A,L`

16. The 6812 doesn't have an `LBSR` instruction. Compare the static efficiency of `JSR`, using program counter relative addressing, to an `LBSR`, which will be coded like `LBRA`.

17. Figure 2.16 shows a delay loop for up to about 25 msec. Write a shortest delay loop for up to 27 minutes. You do not need to compute two constants used in this loop.

18. Assume the following instruction is missing from the 6812 instruction set. Show the shortest program segment that will accomplish the same effect as the missing instruction, except that the condition codes will be changed differently. For part (c), assume that index registers X and Y can be modified.

(a) `MINM 0,X` (b) `EMAXD 0,Y` (c) `EMACS $810`

19. Write a shortest program to execute an inner product using `EMACS` and `EMULS`. Let x[0] be at $810, x[1] be at $812, y[0] be at $814, y[1] be at $816; the program is to put x[0] y[0] + x[1] y[1] into $81a, and $818 and $819 are scratch bytes. All numbers are 16-bit signed numbers.

20. Write a shortest program segment to extract the three bits that were inserted by Figure 2.20, leaving the extracted bits right-justified in accumulator B.

21. Write a shortest program segment to convert temperature from Fahrenheit (±300°), in Accumulator D, to Celsius. The output value is left in Accumulator D. You may preload constants into registers to shorten your machine code, but show their values.

22. The 32-bit binary number at location $822 is the number of ticks, where a tick time is 1/60th second, and zero represents Saturday midnight. Write a shortest program to put the day-of-week in location $826, (military time) hour in $827, minute in $828, seconds in $829, and tick-within-a-second in $82a.

23. Write a shortest program to write problem 22's 32-bit binary number tick count into location $822, for input values written in its day-of-week, hour, minute, seconds, and tick-within-a-second memory words (locations $826 to $82a). $82b to $832 is scratch.

The MC68HC812A4 die.

3

Addressing Modes

In the past two chapters, we have introduced the instruction cycle and the instruction set. We have used a few addressing modes in order to understand those ideas. However, we did not attempt to convey the principles of addressing modes. We now complete our understanding of the instruction by describing the addressing modes used in the 6812.

Recall from Chapter 1 that an instruction generally consists of an operation with one address in memory for an operand and/or result. How that address is determined is called *addressing,* and the different ways that the address is determined are called addressing *modes*. The data are *accessed* in a program, that is, read or written, by an instruction with the addressing modes available in the computer. These modes correspond to the data structures that can be easily managed in that computer. If you want to handle a particular structure, such as a string of characters, an addressing mode such as postincrement is very useful, as we discuss in more detail in Chapter 9. This chapter introduces the 6812's addressing modes, which provide the tools that make handling the most useful data structures so easy on this machine. Learning the rich set of addressing modes here will also make it easier later to learn about the common data structures.

This chapter introduces the following general aspects of addressing. We first discuss addressing modes that are determined by bits in the operation code byte, which is generally the first byte of the instruction. Indexing modes use a post byte and are discussed next. Relative modes are then discussed to show the important concept of position independence. We give examples that rework the addition program of Chapter 1 to illustrate data structure ideas and position independence using these addressing modes. Finally, we consider some architectural thoughts about addressing such as multiple address instructions and the effective address computation in the fetch execute cycle. We also discuss the level of addressing that indicates how many times an address must be read from memory to get the actual or effective address of the operand or result used with the instruction.

Upon completion of this chapter, you should be able to use the addressing modes described here with any instruction that has been introduced in Chapter 2. You should be able to determine what has been done to compute the effective address, what that effective address will be, and what side effects are generated where some modes are used. This will prepare you to use good data structures in your programs and thus to write shorter, faster, and clearer programs as you progress through this material.

3.1 OP Code Byte Addressing Modes

This section discusses addressing that is selected in the opcode byte, which is generally the first byte of the instruction. We have already introduced this idea in an ad hoc manner in Chapter 1 when we discussed implied, immediate, and direct addressing. Now we add page zero addressing and explain when each address mode should be used.

Some instructions do not involve any address from memory for an operand or a result. One way to avoid going to memory is to use only registers for all the operands. The `DEC` instruction decrements (subtracts one from) the value in an accumulator so that `DECA` and `DECB` are really the same operation, with the registers A and B serving as the addresses for the operand and result. Motorola considers `DECA` and `DECB` to be different instructions, whereas other manufacturers would call them the same instruction with a register address that indicates which register is used. Either case has some merits, but we will use Motorola's convention.

There is also an instruction

```
DEC 100
```

that recalls the word at location 100, decrements that word, and writes the result in location 100. That instruction uses direct addressing (as discussed in Chapter 1), whereas `DECA` does not use direct addressing. Because the instruction mnemonic for instructions such as `DECA` makes it clear which registers are being used, at least for simple instructions, Motorola calls this type of addressing *inherent* or *implied*. It is a zero-level mode. For instance, `CLRA` clears accumulator A (puts its contents to zero) and uses inherent addressing, whereas

```
CLR 1000
```

clears the word at location 1000 and uses direct addressing. Several other instructions, such as `SWI` and `BGND`, which we are using as a halt instruction, have been included in the inherent category because the operation code byte of the instruction contains all of the addressing information necessary for the execution of the instruction.

We have used the immediate addressing mode in Chapter 1, where the value of the operand is part of the instruction, as in

```
LDAA #67
```

which puts the number 67 into accumulator A. We use the adjective "immediate" because when the instruction is being fetched from memory the program counter contains the address of the operand, and no further memory reads beyond those required for the instruction bytes are necessary to get its value.

You should use inherent addressing wherever it will shorten the program storage or speed up its execution, for example, by keeping the most frequently used data in registers as long as possible. Their use will involve only inherent addressing. Immediate addressing should be used to initialize registers with constants or provide constants for other instructions, such as `ADDA`.

Other modes allow the data to be variable, as opposed to the fixed data. The 6812 has two such modes, direct and page zero, to allow for accessing any word in memory, but they allow accessing the most common words more efficiently.

We introduced the direct mode in Chapter 1, and we merely review it here. It is really the only mode required for any program that we would write if we were not concerned about efficiency and if we permitted the program to modify one of its own instructions. Indeed, that was the way the first computer was programmed. However, if one examines a program that changes its instructions, it is very unclear. An example of this type of program, using what is called *self-modifying code,* is given in the problems at the end of this chapter. To avoid self-modifying code and to improve efficiency, other addressing modes will be introduced. In the direct mode, the address of the operand or result is supplied with the instruction. For example, as discussed before,

```
LDAA $803
```

puts the contents of location $803 into accumulator A. The opcode for this instruction is $B6, and the instruction is stored in memory as the sequence:

$B6
$08
$03

In the direct mode, a full 16 bits are always used to describe the address even though the first byte may consist of all zeros. Unfortunately, when they first developed the MC6800, a predecessor of the 6812, Motorola called page zero addressing "direct addressing," most likely because they envisioned only a 256-byte RAM. Motorola called direct addressing "extended addressing." This nonstandard use of the term has continued through Motorola's 8-bit and 16-bit microcontrollers and confuses everyone who uses or studies a variety of machines, including other Motorola microprocessors. Because we intend to teach general principles, using the 6812 as an example rather than teaching the 6812 only, we will stick to the traditional term. But when you read Motorola's literature, remember to translate their "extended addressing" into "direct addressing." Do not confuse Motorola's "direct addressing" with our use of the term "direct addressing."

Experience has shown that most of the accesses to data are to a rather small number of frequently used data words. To improve both static and dynamic efficiency, the 6812 has a compact and fast version of one-level addressing to be used with the most commonly accessed data words. The *page zero mode* is an addressing mode that forms the effective address by forcing the high-order byte to be 0 while the lower-order byte is supplied by the eight bits in the instruction. For example,

```
LDAA $67
```

will put the contents of location $0067 into accumulator A. This instruction is stored in memory as:

$96
$67

1	reg.	address mode	opcode

Figure 3.1. Op Code Byte Coding

Clearly, page zero addressing uses fewer instruction bits, and thus the instruction can be fetched faster than with direct addressing. In the 6812, the I/O registers occupy page zero (unless the address map is changed). Therefore I/O instructions use page zero addressing, and your programs can't use page zero addressing for anything else.

The symbol "<" can be used in several addressing modes. It will generally mean that a short 8- or 9-bit number in the instruction is used in the calculation of the effective address. It will be used here for page zero addressing. The symbol ">" can be used for direct addressing. It will generally denote that a 16-bit number in the instruction is used in the calculation of the effective address. In Chapter 4 we will find that these symbols can usually be dropped when the instruction mnemonics are automatically translated into machine code, because the computer that does the translation can figure out whether an 8-bit or a 16-bit value must be put in the instruction. Until then, to enhance your understanding of how the machine works and to simplify hand translation of mnemonics into machine code, we will use the symbol "<" to designate forced 8-bit direct address values and the symbol ">" to designate forced 16-bit direct address values.

Coding of the opcode byte for almost half of the 6812 instructions follows a simple pattern (see Figure 3.1). In instructions in which the most significant bit is 1, the opcode is generally `SUB`, `SBC`, `AND`, `BIT`, `LDAA`, `EOR`, `ADC`, `OR`, `ADD`, or a compare opcode. For these instructions, the next most significant bit generally indicates the accumulator register used. For instance (using immediate addressing), `SUBA` is $80, while `SUBB` is $C0. The next two bits indicate the addressing mode: 00 is immediate, 01 is page zero, 11 is direct, and 10 is index (using a post byte to distinguish among different index modes as discussed in the next section). For instance, `SUBA #` is $80, `SUBA` <0 is $90, `SUBA` >0 is $B0, and its index addressing modes use the opcode $A0. The four least significant bits are the opcode. About half of the 6812 instructions follow this pattern. Many of the other 6812 instructions similarly encode their opcode bytes to systematically derive the opcode and addressing mode from bits in the opcode byte. However, do not attempt to memorize these decoding rules. The best way to encode an instruction is to look up its coding in the CPU12RG/D manual.

This section introduced some of the simpler addressing modes: inherent, direct, and page zero. We saw that inherent addressing should be used when data is kept in registers, typically for the most frequently used data. Page zero addressing, where data is kept on page zero, should be used for the rest of the frequently used data and is used for I/O registers in the 6812. We now turn to the next group, which is based on decoding the post byte and the use of other registers in the addressing mode.

3.2 Post-Byte Index Addressing Modes

This section introduces a collection of addressing modes that are encoded in a post byte and that use index registers in the address calculation. To improve efficiency, the controller is often provided with a few registers that could be used to obtain the effective address. These registers are called *pointer* or *index registers*. Obtaining an address from such an index register would be faster because the number of bits needed to specify one of a few registers is much less than the number of bits needed to specify any word in memory that holds the address. Moreover, index addressing is the most efficient mode to handle many data structures, such as character strings, vectors, and many others, as we discuss later. With this potential, index registers have been used in a number of similar modes, called collectively *index addressing* modes, which are introduced below.

Before we get into the modes of index addressing we have to discuss the idea of a *post byte*. As noted earlier, the 6812 is a successor of the 6811. The latter had only the modes inherent, immediate, page zero, direct, and one form of index addressing discussed below. To keep the same customers that they had for the 6811 happy with the newer machine, Motorola opted to make the 6812 as similar as possible to its predecessors. But to introduce more addressing modes, they needed more room in the instruction. The 6812 is as similar to its predecessors as possible, using the same opcodes in many cases. The extra addressing modes were provided by including an extra byte, right after the opcode byte, for addressing information only and then only for variations of index addressing that are used on the 6812. This byte is the post byte.

The 6812 uses index addressing with two index registers, X and Y, the stack pointer SP and program counter PC. Although these have equivalent addressing capabilities, the SP register and program counter have special uses that are discussed in later sections of this chapter. Generally, all the addressing modes described for X below also apply to the other registers. First, there are load instructions that can load these registers. For example, the instruction

```
LDX #$843
```

will load the 16-bit X register with $843. It is machine coded very much like the `LDAA` immediate instruction. (See the CPU12RG/D manual.) In the following examples, assume X is $843.

$CE
$08
$43

The other registers can be loaded using similar instructions, and other addressing modes can be used to get the two bytes to be put in the index register. In all cases, the effective address determined by the instruction is used to get the first byte to be put into the high byte of the index register. The effective address plus one is used to get the second byte to be put into the low byte of the index register.

Coding of the post byte is shown in Figure 3.2. You can read the tree shown therein from left to right to decode a post byte, or from right to left to encode an index mode into a post byte. To decode a post byte, start at the tree root, and proceed right if bit 5 is zero, otherwise go down. Then check the bit(s) indicated at the next branching point, to determine where to go next, and so on. To encode an index mode into a post byte, locate the index mode on the right, then follow the tree from there to the root, noting the settings of the bits along the way that constitute the post byte code. This information is also shown in the CPU12RG/D manual, in Table 1 and Table 2, using other formats.

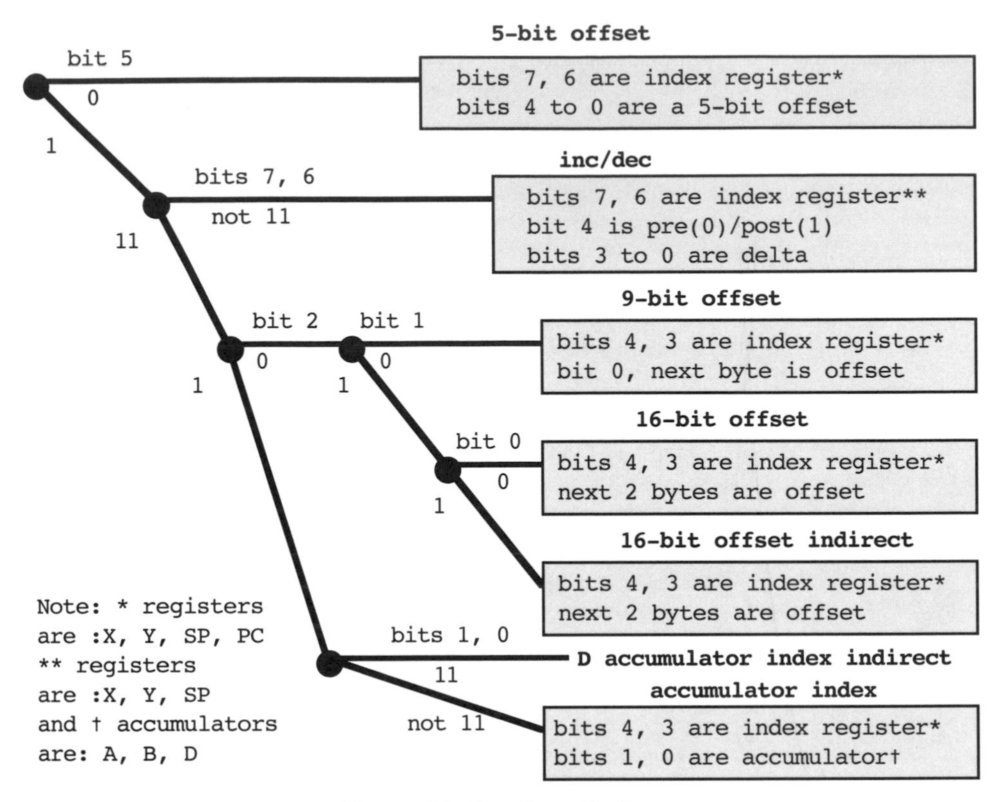

Figure 3.2. Post Byte Coding

Index addressing uses a signed *offset* in the post byte or the post byte and one or two bytes following it. When executed, the offset is added to the index register, say X, to get the effective address of the operand or result in memory. See Figure 3.4.

Effective addresses are frequently within ±16 locations of the index register address, and many others are within ±256 locations. Thus, for greater efficiency, a shortest 5-bit or a 9-bit option is used for some cases, but a full 16-bit index option is also available for cases that do not fall in the range of ±256 locations. The 5-bit offset is entirely contained in the post byte. The 9-bit offset's sign bit is in the post byte, and the remaining eight bits are in the following byte. The 16-bit offset is in the two bytes following the post byte.

The shortest mode with a 5-bit offset will always be used when the offset is between –16 and +15. The instruction `LDAA 1,X` loads the number contained in location 1 + $843 into accumulator A. The post byte for this 5-bit offset mode (see Figure 3.2) has a zero in bit 5, the index register in bits 7 and 6 (00 is X, 01 is Y, 10 is SP, and 11 is PC), and the offset in bits 4 to 0. `LDAA 1,X`'s machine code is:

$A6
$01

Location	Contents	Mnemonic	Comment
820	CE 08 43	LDX #$843	; get address
823	A6 01	LDAA 1,X	; get 1st byte
825	AB 02	ADDA 2,X	; add 2nd byte
827	6A 03	STAA 3,X	; store in 3rd byte

Figure 3.3. Program Segment to Add Two Bytes Using Vector Indexing

$A6 is the opcode byte for any `LDAA` index mode, and $01 is the post byte. The saved offset is sign extended and added to the index register (see Figure 3.4).

The program segment in Figure 3.3 adds the word at $844 to the word at $845, putting the sum in $846. No instruction's execution changes the contents of X.

The 9-bit option will be used when the offset is between –256 and +255 or when the offset is between –16 and +15 and a "<" symbol, as it is used in the page zero mode, is written preceding the offset. The instruction

```
LDAA <$11,X
```

loads the number contained in location $11 + $843 = $854 into accumulator A. The post byte for this 9-bit offset mode (see Figure 3.2) has ones in bits 7 to 5, the index register in bits 4 and 3 (00 is X, 01 is Y, 10 is SP, and 11 is PC), a zero in bits 1 and 2, and the sign bit of the offset in bit 0. Like the 5-bit offset case, the saved offset is sign extended and added to the index register to get the effective address, as illustrated by Figure 3.4. The machine code is

$A6
$E0
$11

where $A6 is the opcode byte for any index option with `LDAA`, `$E0` is the post byte, and bit 0 of the post byte and the next byte $11 are the offset.

When a larger offset is needed, the full 16-bit offset option can be used. The 16-bit option will be used when the offset is outside the range –256 and +255 or when the offset is in this range and a ">" symbol, as it is used in the direct mode, precedes the offset. The instruction

```
LDAA >$3012,X
```

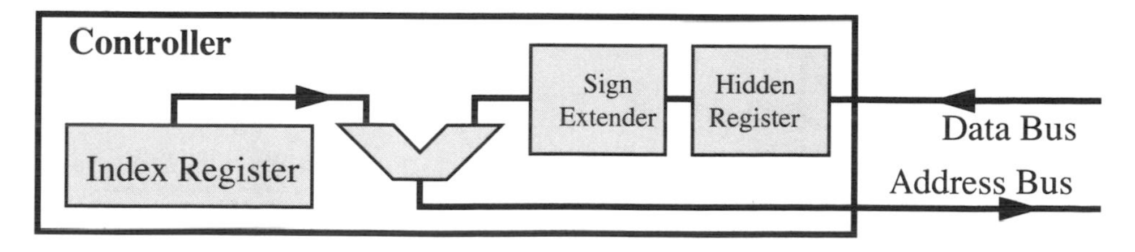

Figure 3.4. Offset Calculation

loads the number contained in location \$3012 + \$843 = \$3855 into accumulator A. The post byte for this 16-bit offset mode (see Figure 3.2) has ones in bits 7 to 5, the index register in bits 4 and 3 (00 is X, 01 is Y, 10 is SP, and 11 is PC), and 010 in bits 2 to 0. The machine code is given by

\$A6
\$E2
\$30
\$12

where \$A6 is the opcode byte for any index option with `LDAA`, `$E2` is the the post byte, and \$3012 is the 16-bit two's-complement offset. The saved offset is added to the index register to get the effective address, as illustrated by Figure 3.4.

In short, addresses requiring several accesses are kept in index registers, if possible, and utilize the more efficient index addressing. Shorter offsets produce more efficient programs and can be used if the index register value is close to the effective addresses that will be used. But while negative offsets are mechanically as usable as positive offsets to more often use shorter offsets, positive offsets are often preferred for clarity.

The 5-, 9-, and 16-bit offset index addressing modes are useful for reading data out of a vector. Suppose a 10-element vector of 8-bit items has element 0 at \$843, element 1 at \$844, element 2 at \$845, and so on. Then if X has \$843, then

```
LDAA 2,X
```

puts element 2 into accumulator A. Suppose now that a 10-element vector of 8-bit items has element 0 at \$872, element 1 at \$873, element 2 at \$874, and so on. Then if X has \$872, this instruction still gets element 2 out of the vector. This instruction uses the efficient 5-bit offset mode. The following instruction gets element i from the vector beginning at \$843 into accumulator A, where the vector index i is in index register X:

```
LDAA $843,X
```

This instruction uses the less efficient 16-bit offset mode, but it lets the variable index be in the X index register.

Index registers can be either *autoincremented* or *autodecremented* before or after being used in effective address calculations. It is denoted by a delta value between 1 and

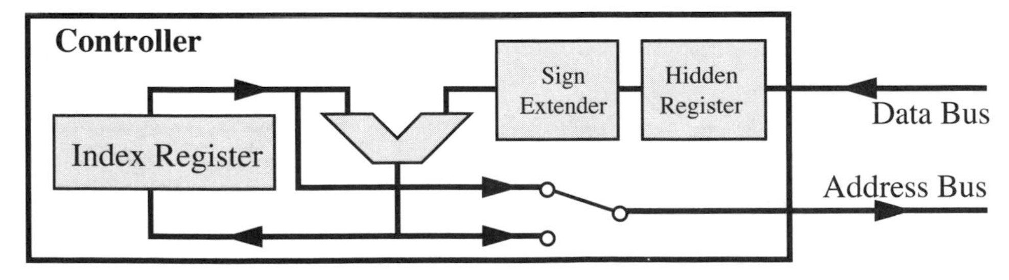

Figure 3.5. Autoincrement Address Calculation

8, a comma, and the register name with a "+" or "–" symbol. If "+" appears, the index register is incremented by the delta value, and if "–" appears, the index register is decremented by the delta value; if this symbol appears before the register name, incrementing or decrementing is done before effective address calculation, and if the symbol appears after the register, incrementing or decrementing is done after the calculation. Consider an example of postincrementing by 1; if X had $843,

```
LDAA 1,X+
```

loads the contents from location $843 into A and then increments the contents of X by 1 to make it $844. For an example of preincrementing by 1, if X had the value $843,

```
LDAA 1,+X
```

increments the contents of X by 1, and then loads the contents from location $844 into A. An example of postincrementing by 2, if X had the value $843,

```
LDD 2,X+
```

loads the contents from locations $843 and $844 into D and then increments the contents of X by 2 to make it $845. For an example of predecrementing, if X had the value $843,

```
LDAA 1,-X
```

decrements the contents of X by 1 to make it $842 and then loads the contents from location $842 into A. Delta can be as high as 8.

These addressing modes are encoded in the post byte as follows (see Figure 3.2): Bits 7 and 6 identify the register (00 is X, 01 is Y, and 10 is SP, but 11 is not used for this mode), bit 5 is 1, bit 4 is 0 if the index value changes before address calculation and 1 if after, bit 3 is 1 if decrementing and 0 if incrementing. For incrementing, bits 2 to 0 are the value of delta minus 1 (or equivalently, delta is the low-order three bits plus 1). For decrementing, bits 2 to 0 are the value of delta, as a negative two's-complement number, to be added to the index register. For example, for `LDAA 1,X+` the post byte is $30, for `LDAA 1,+X` the post byte is $20, for `LDD 2,X+` the post byte is $31, for `LDAA 1,-X` it is $2F, and for `LDAA 2,-X` it is $2E, and so on.

Figure 3.5 illustrates how the delta value fetched from memory can be added to the index register. The index value before or after modification can be used as the effective address by appropriately setting the switch that determines the effective address.

Consider addition again. If you want to add the word at location $843 with the word at location $844, putting the result at location $845, execute the code in Figure 3.6.

Location	Contents	Mnemonic	Comment
820	CE 08 43	LDX #$843	; get address
823	A6 30	LDAA 1,X+	; get 1st byte
825	AB 30	ADDA 1,X+	; add 2nd byte
827	6A 30	STAA 1,X+	; store in 3rd byte

Figure 3.6. Program Segment to Add Two Bytes Using Autoincrementing

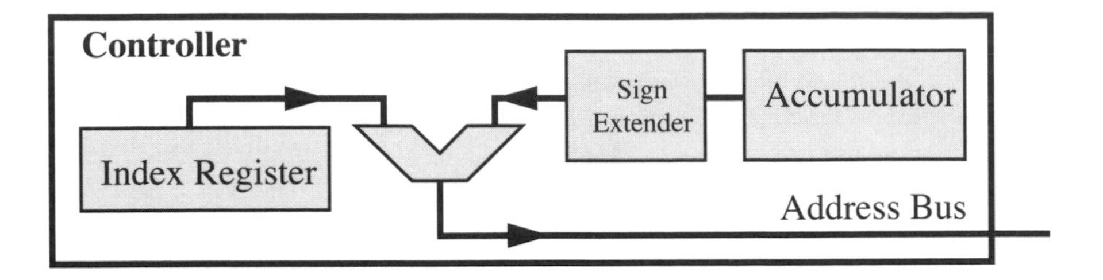

Figure 3.7. Accumulator Index Address Calculation

Note that these increment and decrement modes produce a side effect. They not only compute the effective address, they also change the value in the index register used to get the address. No other addressing mode has such a side effect. We will see how useful these options are when we look at some examples later in this chapter.

Sometimes, the effective address is the sum of two variable numbers. The index register can be used to hold one of the variables, and an accumulator, A, B, or D, can hold the other number, as in `LDAB A,X.` The sign-extended accumulator (A) is added to the index register (X) to provide the effective address. The effective address can be obtained as in Figure 3.7. This is called accumulator indexed addressing.

The contents of the registers A and B are treated as 8-bit two's-complement numbers in these instructions while the contents of D may be treated as a 16-bit two's-complement number or an unsigned 16-bit number, because the sum of the contents of D and the contents of any 16-bit index register, truncated to 16 bits, is the same unsigned 16-bit number in either case. The post byte for accumulator index addressing is as follows: Bits 7 to 5 and bit 2 are 1, the index register is encoded in bits 4 and 3 (00 is X, 01 is Y, 10 is SP, and 11 is PC), and the accumulator is encoded in bits 1 and 0 (00, A; 01, B; and 11, D). The instruction `LDAB A,X` is encoded as follows:

$E6
$E4

Accumulator index addressing modes are useful for reading data out of a vector where the location of the vector in memory, as well as the vector index, are determined at run time. Suppose a 10-element vector of 8-bit items has element 0 at $843, element 1 at $844, element 2 at $845, and so on. Then if X has $843, and accumulator A is 2, then

```
LDAA A,X
```

puts element 2 into accumulator A. Suppose now that a 10-element vector of 8-bit items has element 0 at $872, element 1 at $873, element 2 at $874, and so on. Then if X has $872, and accumulator A is 2, then this instruction still gets vector element 2.

Finally, *indirect addressing* can be combined with accumulator D and 16-bit offset forms of index addressing discussed above. Indirect addressing goes to memory to get the address of the operand, as we describe with examples below. In the 6812, indirect addressing may only be used with these two forms of index addressing. The instruction

```
LDAA [D,X]
```

will use the sum of accumulator D and the index register X as an effective address to read two bytes and then use these two bytes as another effective address, to load accumulator A with the word at the latter address. For instance, if D is clear, X contains the value $843, location $843 contains $08, and location $844 contains $67, `LDAA [D,X]` will load the word at $867 into accumulator A. The post byte for indirect D accumulator index addressing has ones in bits 7 to 5 and 2 to 0, and the index register is specified in bits 4 and 3 (00 is X, 01 is Y, 10 is SP, and 11 is PC). The post byte for the instruction `LDAA [D,X]` is $E7. The instruction

```
LDAA [$12,X]
```

will use the sum of the 16-bit offset $0012 and the index register X as an address to read two bytes, use these two bytes as another address, and load accumulator A with the word at the latter address. Note that even though the offset of this instruction is an 8-bit number, only 16-bit index addressing is permitted when indirect addressing uses an offset. For instance, if X contains the value $843, location $855 contains $08, and location $856 contains $23, `LDAA [$12,X]` will load the word at $823 into accumulator A. The post byte for indirect 16-bit offset index addressing has ones in bits 7 to 5 and 1 and 0, a zero in bit 2, and the index register is specified in bits 4 and 3 (00 is X, 01 is Y, 10 is SP, and 11 is PC). The post byte for the instruction `LDAA [$12,X]` is $E3.

The `LEAX`, `LEAY`, and `LEAS` instructions can use only index addressing modes, but not index indirect modes. They can be used like a transfer instruction; `LEAX 0,Y` will transfer Y to X. More generally, they can be used to add a signed number constant or variable to an index register and possibly put the result in a different register. The instruction `LEAX -3,X` subtracts 3 from index register X, while `LEAY A,X` adds accumulator A to the value of X and puts the result in Y. These instructions are alternatives to arithmetic instructions such as `ADDD` or `SUBD` and are especially useful when the result will eventually be put in an index register.

The idea of using a register to form the effective address is very powerful. Several addressing modes were introduced that use this idea. The index mode doesn't modify the contents of the register, but can add a 5-, 9-, or 16-bit offset to get the effective address. The most common change to an address is to increment or decrement it. The instruction can automatically increment the value in the index register before or after it is used, by one to eight. This will be quite common in some data structures that we meet later. A mode that adds the values of an accumulator to the value of an index register permits one to compute addresses that are derived from two variable values, rather than from a variable and a fixed value. Finally, these modes may be combined with indirect addressing for some special applications. With these modes of addressing, the 6812 is a very powerful microprocessor. With this power, we can show you how to use data structures intelligently to make your programs shorter, faster, and clearer.

3.3 Relative Addressing and Position Independence

The microcomputer is very much like any other computer; however, the use of ROMs in microcomputers raises an interesting problem that is met by the last mode of addressing that we discuss. The problem is that a program may be put in a ROM such that the

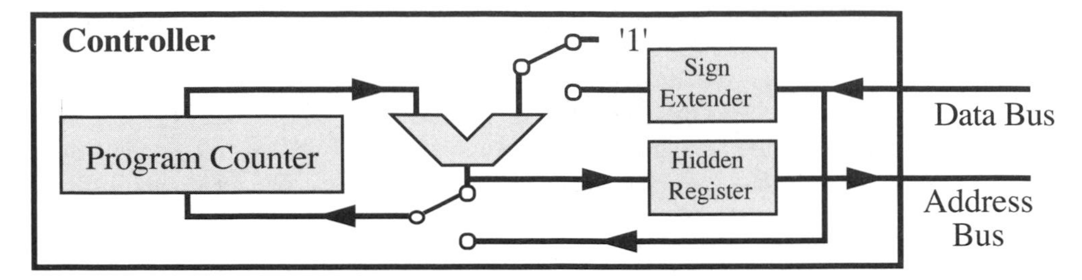

Figure 3.8. Simplified Control Hardware for Relative Addressing

program starts at location \$1000 and ends at \$2000. Suppose that someone buys this ROM, but his/her microcomputer has another program in a ROM that starts at location \$1000 and ends at \$2000. We would like to be able to use this new ROM so that the new program would start at location \$4000 and end at location \$5000, for instance, or wherever there is room in the address space of the microcomputer. However, because the program is in a ROM, it cannot be changed by the buyer. Similarly, a compiler or assembler that downloads a particular program into different parts of memory won't have to change addresses if the program is position independent. A program that works the same way, wherever it is loaded in memory, is said to be *position independent*. Position independent programs can be written by an assembler or compiler to run anywhere in memory without modification. Programs we have seen so far are position independent when the location of the data is fixed, and, in fact, most program segments that do not use `JMP` or `JSR` instructions using direct addressing are position independent.

Program counter relative addressing, or simply *relative* addressing, adds a two's-complement number, called a *relative offset,* to the value of the program counter to get the effective address of the operand. Figure 3.8 illustrates a simplified implementation of a controller. The top switch can add "1," or the sign-extended data bus, to the program counter. The former is used to increment the program counter each time an instruction byte is fetched, and the latter is used for relative branches. The bottom switch permits the adder's output or the data bus value to be put into the program counter. The bottom switch selects the latter when a `JMP, JSR, RTS,` or `RTI` instruction or interrupt loads the program counter. The adder's output can also be used as an effective address.

The relative addressing mode is used to implement position independence. If the program segment at \$1000 to \$2000 was in a ROM and that ROM was installed so that the instruction following the `BNE` was at \$4000, the `BNE` instruction would still have the relative offset \$20. If Z is 0 when the instruction is executed, the program counter would be changed to \$4020. That would be the address of the instruction that had the label L. The program would execute the same way whether it was stored at location \$1000 or \$4000. This makes the program position independent.

Branching instructions all use relative addressing. For example, the instruction `BRA L` for "branch always" to location L will cause the program counter to be loaded with the address L. An example of a branch is illustrated in Figure 3.9. Observe that label L is two bytes below the end of the `BRA L` instruction. The program counter PC has the address \$834 of the next instruction, `LDAA #4`, when it is executing the `BRA L` instruction. The second byte of the `BRA L` instruction, the offset, 2, is added to the program counter, to make it \$836, and then the next byte is fetched.

Location	Contents	Mnemonic	Comment
830	86 03	LDAA #3	; put number 3 in A
832	20 02	BRA L	; skip to store
834	86 04	LDAA #4	; put number 4 in A
836	7A 08 43	L: STAA $843	; store number in A

Figure 3.9. Program Segment Using BRA, Illustrating Position Independence

The example in Figure 3.10 constantly flips bits in location 1. It might be used in Chapter 12; location 1 is an output port, and this program segment outputs a square wave on all the bits. The two's-complement offset is negative because the branch is backwards. Observe that, after `BRA L` is fetched, the program counter is on location $816; the offset $FB is –5, so the program counter becomes $811 after it is executed.

Many programmers have difficulty with relative branch instructions that branch backwards. We recommend using *sixteen's complement* arithmetic to determine the negative branch instruction displacement. The sixteen's complement is to hexadecimal numbers as the two's complement is to binary numbers. To illustrate this technique, the displacement used in the branch instruction, the last instruction in the program in Figure 3.10, can be determined as follows. When the branch is executed, the program counter has the value $816, and we want to jump back to location $811. The difference, $816 – $811, is $05, so the displacement should be –$05. A safe way to calculate the displacement is to convert to binary, negate, then convert to hexadecimal. Because $5 is 00000101, the two's complement negative is 11111011. In hexadecimal, this is $FB. That is not hard to see, but binary arithmetic gets rather tedious. A faster way takes the sixteen's complement of the hexadecimal number. Just subtract each digit from $F (15), digit by digit, then add 1 to the whole thing. Then –$05 is ($F – 0),($F – 5) + 1 or $FA + 1, which is $FB. That's pretty easy, isn't it!

If the relative offset is outside the 8-bit range, one uses the long branch equivalent, `LBRA L`, which uses a 16-bit two's-complement relative offset.

Program counter relative addressing can be used to read (constant) data that should be stored with the program. Relative addressing can be implemented using a 5-bit, 9-bit, or 16-bit signed relative offset. Nine bit offset relative addressing is denoted by the "<" before and "`,PCR`" after the offset and 16-bit offset by ">" symbol before and "`,PCR`" after the offset. (This mode's machine code uses a post byte as it is an index option.)

Location	Contents	Mnemonic	Comment
810	87	CLRA	; clear A
811	41	L: COMA	; invert bits
812	5A 01	STAA $1	; output to a port
814	20 FB	BRA L	; repeat forever

Figure 3.10. Program Segment to Put a Square Wave on an Output Port

For example,

```
LDAA <L,PCR
```

can load any word into A that can be reached by adding an 8-bit signed number to the program counter. (Recall that the PC is pointing to the next instruction just below the LDAA instruction when the effective address L is calculated.) The instruction

```
LDAA >L,PCR
```

can be used to access words that are farther away than –128 to + 127 locations from the address of the next instruction; it adds a 16-bit offset to the current value of the program counter to get the effective address. Although the machine coding of relative addressed instructions is the same as that of index addressed instructions, do not dwell too much on that similarity because the offset put in the machine code is determined differently.

Program counter relative indirect addressing can be used to access locations such as I/O ports as in

```
LDAA [L,PCR]
```

Assuming that L is 18 bytes below this instruction, the machine code is given by

$A6
$FB
$00
$12

where $A6 is the opcode byte for any LDAA index mode; the post byte $FB indicates indirect index addressing with 16-bit offset, but using the program counter as the "index register", and the last two bytes are added to the program counter. The indirect address ($12 in the example above) is in a location relative to the program. If the program is loaded into a different location, the offset $12 is still used to get the indirect address. Such use of relative and indirect relative addressing lets the program have one location and only one location where a value is stored, so that a downloaded file can insert the value in one place to run the program anywhere it is stored.

Branch and long branch instructions do not need the ",PCR" symbol in the instruction because they only use relative addressing with 16-bit relative offsets. However, the BSR L, having an 8-bit offset, doesn't have a corresponding long branch to subroutine. But JSR L,PCR is a 16-bit position independent subroutine call that has the same effect as the missing LBSR L.

A 16-bit position independent indirect subroutine call, JSR [L,PCR], can jump to a subroutine whose address is in a "jump table," as discussed in a problem at the end of this chapter. Such jump tables make it possible to write parts of a long program in pieces called sections and compile and write each section in EEPROM at different times. Jumps to subroutine in a different section can be made to go through a jump table rather than going directly to the subroutine. Then when a section is rewritten and its subroutines appear in different places, only that section's jump table needs to be rewritten, not all the code that jumps to subroutines in that section. The jump table can be in EEPROM at the beginning of the section, or in RAM, to be loaded at run time.

A program is not position independent if any instruction in it causes it to do something different when the program is moved, intact, to a different location. The only real test for a program's position independence is to show that it can be moved without changing its operation. One necessary condition, however, is that all changes to the program counter be position independent, and using branch instructions in place of jump instructions, or JMP and JSR instructions with program counter relative addressing, will generally make that possible. The relative addressing mode is generally used with data that move with the program, such as constants that are on the same ROM as the program, and with instructions that compute the address to jump to in a manner to be introduced later. Listed with other instructions, then, the relative mode allows programs to be position independent, and that may be very important in a microcomputer that uses a lot of ROMs.

3.4 Stack Index Addressing, Reentrancy, and Recursion

The stack pointer may be used with all the index addressing modes, but its uses have special meaning. These uses correspond to pushing and pulling, and they support reentrancy and recursion. Also, the index registers X and Y may be used as auxiliary stack pointers. This section shows these variations of index addressing.

The instruction `LDAA 1,SP+` is essentially the same as `PULA` because both pull a byte from the (hardware) stack into accumulator A. Similarly, the instruction `LDD 2,SP+` is essentially the same as `PULD;` the instruction `STAA 1,-SP` is essentially the same as `PSHA;` and the instruction `STD 2,-SP` is essentially the same as `PSHD.` The differences between these pairs of instructions is in their length and execution time and in the effect they have on condition codes. The instructions `PULA, PSHA, PULD,` and `PSHD` are usually preferred because they are faster and shorter, but the instructions `LDAA 1,SP+, STD 2,-SP,` and so on may be used if pulling or pushing needs to set the condition codes for a future conditional branch instruction.

Moreover, autoincrement addressing can be used with other instructions to pull a byte or 16-bit word and simultaneously use the pulled data in an instruction. The sequence:

```
PSHB
ADDA 1,SP+
```

is often used in code generated by C and C++ compilers to add accumulator B to accumulator A (equivalent to the simpler instruction ABA). However, this push-and-pull-into-an-instruction technique can be used with other instructions like ANDA and ADDD. The sequence:

```
PSHX
ADDD 2,SP+
```

is often used in code generated by C and C++ compilers to add index register X to accumulator D.

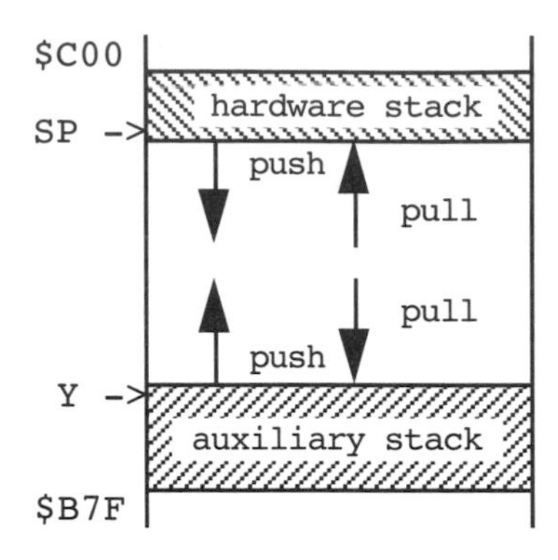

Figure 3.11. A Stack Buffer for Two Stacks

The hardware stack, pointed to by SP, is useful for holding a subroutine's arguments and local variables. This will be discussed at the end of this section. However, because return addresses are saved on and restored from the hardware stack, we sometimes need a stack that is not the same as that hardware stack. For instance, we may push data in a subroutine, then return to the calling routine to pull the data. If we use the hardware stack, data pushed in the subroutine need to be pulled before the subroutine is exited, or that data will be pulled by the `RTS` instruction, rather than the subroutine's return address. A second stack can be implemented using an index register such as Y. If index register Y is also needed for other purposes in the program, this second stack pointer can be saved and restored to make it available only when the second stack is being accessed.

Figure 3.11 illustrates that a second auxiliary stack may use the same buffer as the hardware stack. The hardware stack pointer is initially loaded with the address of (one past) the high address of the buffer, while the second auxiliary stack pointer (Y) is loaded with (one below) the low end of the same stack buffer. The second stack pointer is initialized as: `LDY #$B7F.` Accumulator A can be pushed using `STAA 1,+Y.` A byte can be pulled into accumulator A using `LDAA 1,Y-.` A 16-bit word can be pushed and pulled in an obvious way. Observe that autoincrementing and autodecrementing are reversed compared to pushing and pulling on the hardware stack, because, as seen in Figure 3.11, their directions are reversed.

The advantage of having the second stack in the same buffer area as the hardware stack is that when one stack utilizes little of the buffer area, the other stack can use more of the buffer, and vice versa. You only have to allocate enough buffer storage for the worst-case sum of the stack sizes, whereas if each stack had a separate buffer, each buffer would have to be larger than the worst case size of its own stack.

A *recursive subroutine* is one that calls itself. The procedure to calculate n factorial, denoted n!, is recursive; for any positive nonzero integer n, if n is one, n! is 1, otherwise n! is (n–1)! * n. The subroutine in Figure 3.12 calculates n!; upon entry, n is in accumulator D, and upon exit, n! is in accumulator D.

Location	Contents	Mnemonic	Comment
803	04 44 09	FACT: TBEQ D,FAC1	; if input zero, output 1
806	3B	PSHD	; save parameter on stack
807	83 00 01	SUBD #1	; reduce by one
80A	07 F7	BSR FACT	; call self to compute (n-1)!
80C	31	PULY	; restore parameter on stack
80D	13	EMUL	; multiply n * (n-1)!
80E	3D	RTS	; return to caller
80F	CC 00 01	FAC1: LDD #1	; generate 1, which is 0!
812	3D	RTS	; return to caller

Figure 3.12. Subroutine to Compute n! Recursively

Although recursive subroutines implement a scientist's induction mechanism, they are not always useful. Consider the alternative in Figure 3.13 that uses a loop. The alternative is significantly more efficient. The recursive solution uses the hardware stack as a counter, pushing a 2-byte return address and 2-byte saved parameter value each time it calls itself to reduce its parameter by 1. If n is 5, this subroutine uses up 20 bytes of stack storage. This is not efficient. But there are efficient recursive subroutines, especially for following linked list data structures, as we will see in Chapter 9.

A subroutine is *reentrant* if it can be stopped, for instance because of an interrupt, and then resumed, and it will get the same result as if it were not stopped, even though the interrupt handler might call this subroutine during its execution. Also, a time-sharing computer uses interrupts to switch between tasks or threads that share a computer. Reentrant subroutines can be used by each task or thread, without concern that, when a thread or task is stopped during execution of the subroutine, another thread will execute the subroutine. The subroutine in Figure 3.14 is nonreentrant, and following it is a reentrant subroutine; both clear five bytes beginning at location $824.

Location	Contents	Mnemonic	Comment
803	B7 45	FACT: TFR D,X	; put number in D and X
805	20 03	BRA FAC2	; go to end of loop to test
807	B7 56	FAC1: TFR X,Y	; copy interation number
809	13	EMUL	; multiply
80A	04 35 FA	FAC2: DBNE X,FAC1	; repeat until count is zero
80D	3D	RTS	; return to caller

Figure 3.13. Subroutine to Compute n! in a Loop

Location	Contents	Mnemonic	Comment
803	86 05	CLEAR: LDAA #5	; put number 5 in A
805	7A 08 22	STAA $822	; store in fixed location
808	CE 08 24	LDX #$824	; set pointer to begin
80B	69 30	CLR1: CLR 1,X+	; clear byte pointed to
80D	73 08 22	DEC $822	; decrement location
810	26 F9	BNE CLR1	; until it becomes zero
812	3D	RTS	; return to caller

Figure 3.14. Nonreentrant Subroutine to Clear Memory

This program fails to work correctly if it is interrupted after the STAA instruction is executed, and before the RTS instruction is executed and the interrupt handler calls this subroutine. The second call to this subroutine will wipe out the counter at location $822 because the second call will also use this same location and will leave it cleared when the first call to this subroutine resumes execution. The first call to this subroutine will not work the same way if the interrupt occurs as it does if the interrupt doesn't occur. However, the subroutine in Figure 3.15 will work correctly if it is similarly interrupted.

The key idea behind both recursion and reentrancy is to keep data on the stack. The stack provides new storage locations for each instantiation of the subroutine to keep its variables separate from the variables of other instantiations, so they aren't destroyed.

Note that the decrement instruction accesses the counter on the stack without pulling the byte. If three 1-byte items were pushed on the stack, the instruction LDAA 2,SP will read into accumulator A the first byte pushed without removing it from the stack. In general, items can be pushed on the stack at the beginning of a procedure and pulled off at the end of the procedure. Within the procedure the items can be read and written using offsets from the stack pointer.

Location	Contents	Mnemonic	Comment
803	86 05	CLEAR: LDAA #5	; put number 5 in A
805	36	PSHA	; save on the hardware stack
806	CE 08 24	LDX #$824	; set pointer to begin
809	69 30	CLR1: CLR 1,X+	; clear byte pointed to
80A	63 80	DEC 0,SP	; decrement byte on the stack
80C	26 FA	BNE CLR1	; until it becomes zero
80E	1B 81	LEAS 1,SP	; remove item on stack
810	3D	RTS	; return to caller

Figure 3.15. Reentrant Subroutine to Clear Memory

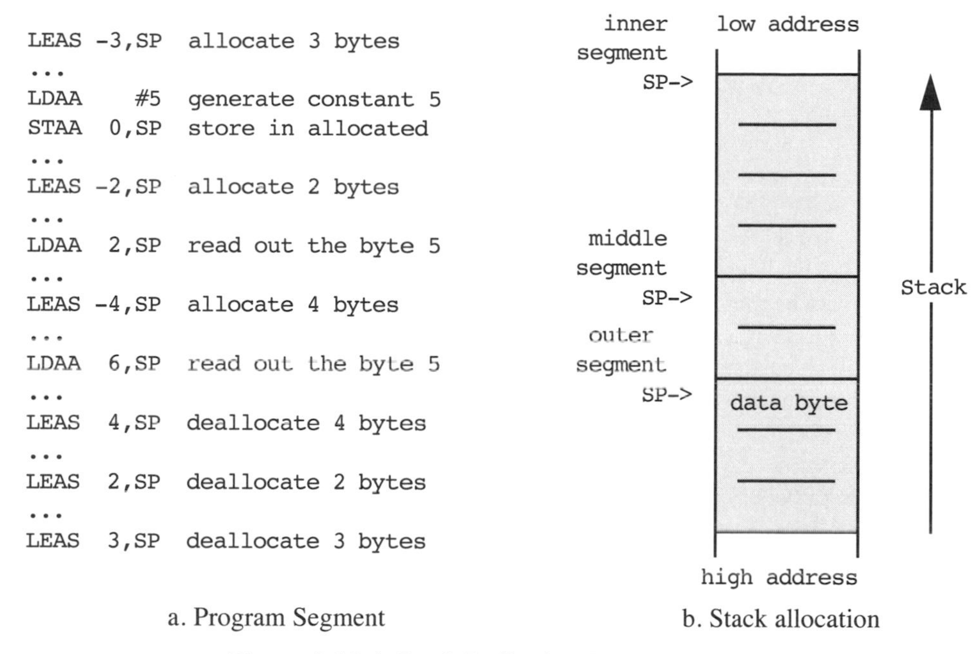
LEAS -3,SP allocate 3 bytes
...
LDAA #5 generate constant 5
STAA 0,SP store in allocated
...
LEAS -2,SP allocate 2 bytes
...
LDAA 2,SP read out the byte 5
...
LEAS -4,SP allocate 4 bytes
...
LDAA 6,SP read out the byte 5
...
LEAS 4,SP deallocate 4 bytes
...
LEAS 2,SP deallocate 2 bytes
...
LEAS 3,SP deallocate 3 bytes
inner segment SP->
low address
middle segment SP->
outer segment SP->
data byte
Stack
high address

a. Program Segment b. Stack allocation

Figure 3.16. A Stack Buffer for Nested Segments

The concept of storing data on the stack leads to *nested allocation*, access, and deallocation of local variables. Nested segments are commonly used in C and C++ programs to call procedures; the outer segment holds parameters passed to the procedure, and the inner segment stores some of the local variables of a procedure. Further, C and C++ programs nest segments within a procedure to hold temporary variables needed to evaluate a C statement. This concept is fully explained in terms of a program trace, which is introduced first. Then we consider a simple trace, and then a nested trace.

One can record a program's instructions in the exact order they are executed to obtain a trace. Simple program segments without branching statements are the same as their traces. If a program has a loop that is executed five times, the trace has five copies of the instruction sequence in the loop.

In a program trace, one can allocate local variables by pushing items on the stack, and one can deallocate them by pulling them from the stack, as already illustrated in this section. Once allocated, the data on the stack can be accessed by reading or writing the contents as discussed above. Moreover, one can allocate several bytes in one instruction, using the `LEAS` instruction. For instance, to allocate five bytes, execute `LEAS -5,SP`. By moving the stack pointer SP five locations toward lower memory, five bytes of data can be stored in these bytes that were skipped over. The `LEAS` instruction can deallocate several words at a time. To deallocate five bytes, execute the instruction `LEAS 5,SP`. A stack is said to be balanced in a simple trace, which has no alternative branches in it; so it is linear, if the number of allocated bytes equals the number of deallocated bytes, and at no step in the trace, between allocation and deallocation, are more bytes deallocated than were allocated. If, due to conditional branches, there are several possible simple traces to take from when space has been allocated to a given point in the program, the stack is balanced at that point if it is balanced in every possible simple trace to that point. To balance the stack means to deallocate all the bytes that have been allocated, so that it becomes balanced. We usually allocate space for variables at the beginning of a subroutine and deallocate the space just before we exit the subroutine to balance the stack, but we can have program segments that are not subroutines in which space is allocated at the beginning of the program segment and deallocated at the end of that program segment.

It is possible for a program to have a segment that has space allocated at its beginning and deallocated at its end and to have within it another segment that has space allocated at its beginning and deallocated at its end. This is called a nested allocated segment. Figure 3.16a illustrates a program that has a nested segment where the outer segment allocates three bytes, an inner segment allocates two bytes, and an inner segment of it allocates another four bytes. The outer segment writes the number 5 in its lowest addressed byte, the next inner segment reads this byte into accumulator A, and the innermost segment reads this byte into accumulator B. Note that different offsets are used with the stack pointer to access the same byte of data, due to intervening allocations, but the outer data is available to each inner program segment, even though they are nested.

3.5 Examples

We now tie together some of the ideas that were introduced above using some examples. These examples give you some experience with addressing modes and loops.

One of the most common operations is to clear a block of memory. The program segment in Figure 3.17 clears 39 bytes starting at location $910.

This example can be sped up by using `STD`, where accumulator D is zero. A similarly common operation is to move a block of data from one area to another. The following program segment moves 15 bytes from a block starting at location $930 to a block starting at location $921. The `MOVW` instruction can move data twice as fast.

We now extend an example started in Chapter 1. Suppose that we want to add N 1-byte numbers that are stored consecutively beginning in location $843. The value of N is stored in location $841, and the result is to be placed in location $842. The program segment in Figure 3.19 does this for either unsigned or signed (two's-complement) numbers. If the numbers are unsigned, the result will be correct as long as there is no *unsigned overflow,* that is, the sum can be expressed with an 8-bit unsigned number. If the numbers are signed, the result will likewise be correct as long as there is no *signed*

Location	Contents	Mnemonic	Comment
820	CC 00 27	LDD #39	; put number count in D
823	CE 09 10	LDX #$910	; put start address in X
826	69 30	L1: CLR 1,X+	; clear byte, autoincrement
828	04 34 FB	DBNE D,L1	; count down and loop

Figure 3.17. Program Segment to Clear a Block of Memory

Location	Contents	Mnemonic	Comment
840	CC 00 0F	LDD #15	; put count in D
843	CE 09 30	LDX #$930	; put source address in X
846	CD 09 21	LDY #$921	put dest. address in Y
849	18 0A 30 70	L: MOVB 1,X+,1,Y+	; move byte, autoincrement
84D	04 34 F9	DBNE D,L	; count down and loop

Figure 3.18. Program Segment to Move a Block of Memory

overflow; that is, the result can be expressed with an 8-bit two's-complement number. Note that accumulator B is initially loaded with the number of times that the loop is to be executed. This loop counter (accumulator B in this case) is decremented by `DBNE`, which branches back to the location L if the (B) is greater than zero after it is decremented. The loop from location L to the `DBNE` instruction is repeated N times, where N is the initial value in accumulator B.

In the program in Figure 3.19, register A is used to accumulate the result, register B holds the number of remaining bytes to be added, and the index register X contains the address that points to the next byte to be added.

We extend the multiply-and-accumulate example of §2.7 to evaluate the expression: $\Sigma^4_{i=0} B_i * C_i$, where a vector of five signed 16-bit numbers B_i are at $914, and a vector of five signed 16-bit numbers C_i is at $91E. This expression is called the *inner product* of the vectors B and C. The two vectors B and C can have as many elements as you want, but the two vectors have the same number of elements. It is widely used in signal processing and data compression. See Figure 3.20. Note that EMACS only uses pointer addressing, so the index registers X and Y must be moved using LEA instructions in order to pick up the elements B_i and C_i. This procedure is very similar to the WAV instruction but is for 16-bit elements while WAV is for 8-bit elements.

This section illustrates several examples of the great value of the synergetic combination of autoincrement index addressing and counting using the `DBNE` instruction. The combination of accumulator index addressing and counting using the `DBNE` instruction, whose counter register is the accumulator used with the index, is also widely used. We seem to run into such a combination in every other program that we write.

Location	Contents	Mnemonic	Comment
820	CE 08 43	LDX #$843	; point to first number
823	F6 08 41	LDAB $841	; get count
825	87	CLRA	; initialize sum
826	AB 30	L: ADDA 1,X+	; add 2nd byte
829	04 31 FB	DBNE B,L	; count down and loop
82B	7A 08 42	STAA $842	; store result

Figure 3.19. Program Segment to Add Vector Elements

Location	Contents	Mnemonic	Comment
820	CC 00 05	LDD #5	; get number of elements
823	CE 09 14	LDX #$914	; get address of 1st
826	CD 09 1E	LDY #$91e	; get address of 2nd
829	18 12 09 10	L: EMACS $910	; multiply and accumulate
82D	1A 02	LEAX 2,X	; move pointer
82F	19 42	LEAY 2,Y	; move pointer
831	04 34 F5	DBNE D,L	; count down, loop until zero

Figure 3.20. Program Segment to Compute an Inner Product

3.6 Architectural Notions of Addressing

The particular computer that we are studying, the 6812, is a one-address computer. Have you thought, perhaps, that a computer that has instructions with two addresses may be better than a one-address computer? In some cases, it would be, and a three-address computer would be even better; but in other cases, it would not. We will compare the static efficiency of one-address and three-address computers to help you look beyond the particular machine that we are studying, to understand the general principle of addressing, and at the same time to reassure you that the 6812 is a good machine for most applications. Next, we will review the detailed fetch/execute cycle to expose some possible ambiguities in the addressing operation of the 6812. This may help you to understand some very useful addressing techniques. Although this discussion does not show you how to apply specific addressing modes as the previous section did, it will further your general understanding of addressing and programming.

We might want to add the contents of location 511 to the contents of 512 and put the result into 513. In the 6812, we would execute the program segment

```
LDAA  511
ADDA  512
STAA  513
```

The same effect could be obtained in a different computer that had a *three-address instruction*. The instruction

```
ADD 511,512,513
```

would add the contents of location 511 to that of 512, putting the result into 513. The 6812 program segment used nine bytes, while this three-address machine might use only seven bytes. The three-address machine is more efficient for this example. However, if we want to add the numbers in locations 511 through 515 and put the result in 516, the three-address machine must use something like:

```
ADD 511,512,516
ADD 513,516,516
ADD 514,516,516
ADD 515,516,516
```

while the one-address 6812 uses

```
LDAA 511
ADDA 512
ADDA 513
ADDA 514
ADDA 515
STAA 516
```

A comparison now shows that the three-address machine takes 28 bytes while the one-address 6812 takes 18. Of course, this computation is very inefficient for the three-address machine, but it may represent a more typical computation than the one that the particular instruction directly handles.

In §12.5, we will see a three-address architecture in the fast and powerful 500 series of Motorola microcomputers. The three-address architecture is actually the method of choice for these powerful microcomputers because this architecture actually permits several instructions to be executed in parallel if the instructions' registers are mutually distinct. Nevertheless, there are applications in which the three-address architecture is justifiable based on static or dynamic efficiency.

You may have already run into confusing addressing modes. If you haven't yet, we would like to offer the following discussion to help you when you do. Consider the instruction

```
LDX 0,X
```

that loads a register using an address that is calculated using the same register. Is this like a definition of a term that uses the term to define itself? No. It is quite legal and very useful in handling data structures such as linked lists, which you will study in Chapter 10. Let us review the fetch/execute cycle again, with this particular instruction as an example. First, the opcode and then the post byte are fetched. The opcode is decoded, and then the address is calculated. Predecrementing, if needed, is done at this point. Finally, the operation is carried out. Note that the address is calculated using the *old* value in the index register X. Then the two words recalled from that address are put into the index register to become the new value of the register. For example, if X contained 100, location 100 contained 0, and location 101 contained 45, then, after the instruction is executed, the X register contains 45.

There are some further ambiguities with the last load instruction and the corresponding store instruction when postincrementing is used. For example, with the instruction

```
LDX  2,X+
```

it is not clear whether the load is executed before the + or after the +. Note that if the latter is true, the + would have no effect on the instruction. Indeed, in the 6812, the + is carried out before the operation; in this case a load, so that

```
LDX   2,X+
```

is the same as

```
LDX   2,X-
```

For any load instruction involving the same index register for the location of the operand and the location of the result, the general rule is that postincrementing has no effect on the instruction. However, the fact that the postincrementing is carried out before the operation produces an unexpected result in the store counterpart of the load instruction just discussed. For example, with

```
STX   2,X+
```

suppose that X initially contains 373. After the instruction is executed, one will find that X becomes 375, and 375 has been stored in locations 373 and 374. We conclude this discussion by noting that predecrementing has none of these ambiguities. For example, if X initially contains 373 before the instruction

```
STX 2,-X
```

is executed, then 371 will be stored in locations 371 and 372.

There is often considerable confusion about `LDX` (direct), `LDX #`, and `LEAX`. Consider the following examples, assuming location $820 stores $1234.

```
LDX   $820
```

will load $1234 into `X`. Direct addressing loads the data located at the instruction's address. However, immediate addressing loads part of the instruction into the register, as

```
LDX   #$820
```

will load $820 into `X`. Sometimes, immediate addressing is used to load an address into memory so that pointer addressing (index addressing with zero offset) can access the data:

```
LDX   #$820
LDX   0,X
```

will eventually load $1234 into X. Also, the `LEAX` instruction loads the effective address into an index register. When it is used with program counter relative addressing, it has the same effect as `LDX #` but is position independent.

```
LEAX   $820,PCR
LDX    0,X
```

will eventually load $1234 into X. But LEAX can be used with other addressing modes for other effects; for instance `LEAX 5,X` adds 5 to X, and `LEAX D,X` adds D to X.

3.7 Summary

In this chapter we looked at the addressing modes in the 6812. We saw four general themes: the use of page zero, the use of index registers, the use of relative addressing for position independence, and the use of stack addressing for reentrancy and recursion.

With the first theme, we saw inherent and page zero addressing are useful for improving static and dynamic efficiency over direct addressing. Put the most commonly accessed variables in registers, using inherent addressing to access them, and put the next most common variables in page zero, using page zero addressing to access them.

For the second theme, we saw that index registers may be used efficiently to handle addresses that require several accesses and that index registers may be useful for data structure accesses. Index addressing is the fastest and shortest index addressing option and index addressing using 5-bit offsets is available for locations close to that pointed to by the register, while 16-bit offsets are available for all accesses. We also saw that the accumulators may be used, in lieu of an offset, to combine a variable in an index register with a variable in an accumulator to get the effective address. Index registers and their addressing modes provide a lot of power, which we explore further throughout this book.

With the third theme, the program counter is used as a kind of index register and the same steps used to carry out index addressing are used to carry out relative addressing using the program counter in place of an index register. Although the mechanics are the same, the effect is quite different, and the representation of the address is different. In particular, the address in the instruction using relative addressing is the effective address, not an offset, while the machine code for the instruction uses a relative offset, which is the amount that must be added to the program counter to get the effective address. This mode is useful in making programs position independent, so that they may be mass produced in ROMs and many different systems can use the same ROM.

The last theme showed how the stack pointer can be used with an offset to access local variables and parameters passed on the stack. The reentrancy and recursion techniques are shown to be easily implemented using stack pointer addressing.

This chapter covered the rich collection of addressing modes in the 6812. These correspond to the modes in most of the other microcomputers and to most of the useful modes in any computer. Now that you know them, you should be prepared to use them with any instruction in the 6812 (where they are permitted) as we discuss these instructions in the next chapter. You should know which mode to use, based on our study of the themes above so that you can produce shorter, faster, and clearer programs.

Do You Know These Terms?

See the end of Chapter 1 for instructions.

addressing
addressing modes
accessed
inherent
implied
self-modifying code
page zero mode
pointer register
index register
index addressing
post byte
offset
autoincrement
autodecrement
indirect addressing
position independent
program counter relative
relative
relative offset
sixteen's-complement
recursive subroutine
reentrant subroutine
nested allocation
unsigned overflow
signed overflow
inner product
three-address instruction
jump vector

PROBLEMS

1. Identify all instructions that have a direct mode of addressing but do not have a page zero mode of addressing.

2. Identify all instructions that have both direct and page zero addressing, in which the direct addressing opcode byte is the page zero addressing opcode byte plus $20. Which instructions have both direct and page zero addressing, in which the direct addressing opcode byte is not the page zero addressing opcode byte plus $20?

3. We often write a constant to an output port, which is a byte on page zero. Compare the static and dynamic efficiency, and clarity, of the `MOVB #$12,$0034` instruction to the instruction sequence `LDAA #$12 STAA $34`. When should you use the MOVB instruction, and when should you use the `LDAA` – `STAA` sequence?

4. Suppose that we have a vector of 1-byte signed numbers whose first byte is at location $840 and whose length is at location $83f and is less than 32 bytes. Write a shortest program to search through the vector, using autoincrement addressing, putting all those numbers that are negative and even into a vector beginning at location $860, keeping the order of the numbers in the second vector the same as the original vector, and putting the length of the new vector in location $85f.

5. Suppose that we have N 16-bit two's-complement numbers stored beginning at location $850. The two bytes of each number are stored consecutively, high byte first. Write a shortest program, using autoincrement addressing, that puts the maximum number in locations $84e and $84f, high byte first. Do not use "special" instructions. The variable N is stored in location $84d. How would your program change if the numbers were unsigned?

6. Write a shortest program that adds a 3-byte number at locations $832 through $834 to a 3-byte number at locations $835 through $837, putting the sum in locations $838 through $83a. Each number is stored high byte first and other bytes at higher addresses. When the program finishes, condition code bits Z, N, V, and C should be set correctly. *Hint:* Use just one index register to read in a byte from each number and also write out a byte, and obtain the final condition code Z by ANDing Z bits obtained after each add.

7. A ten-element 16-bit per element vector at location $844 is initially clear. Write a shortest program segment that increments the vector element whose index is in accumulator B and that is a positive integer less than 10. After the program segment is executed several times, each vector element has a "frequency-of-occurrence" of the index. This vector is called a histogram.

8. Write a shortest program segment that sets a bit in a bit vector having 256 bits. Location $856 and the following 7 bytes contain $80, $40, $20, $10, 8, 4, 2, and 1. Index register X points to the byte that contains the leftmost (lowest-numbered) bit of the bit vector. Bits are numbered consecutively from 0, the sign bit of the byte pointed to by X, toward less significant bits in that byte, and then toward bytes at consecutively higher addresses. If accumulator A is a bit number n, this program segment sets bit n in the bit vector pointed to by X.

9. Write a shortest program segment, which is to be executed only once, that adds a 24-bit number at locations $811 to $813, to a 24-bit number at $814 to $816 to get a 24-bit result at $817 to $819 but that does not use any index registers; it uses only self-modifying code. Each address's low byte is decremented after each time it is used.

10. Write a shortest program segment that adds a 24-bit number to a 24-bit number to get a 24-bit result but that does not use any index register, only indirect addressing. Use locations $811 and $812 to hold the pointer to the first 3-byte number (which is the address of its least significant byte), locations $813 and $814 to hold the pointer to the second number (which is the address of its least significant byte), locations $815 and $816 to hold the pointer to the result (which is the address of its least significant byte). Assume that no byte of any of the 24-bit numbers spans a page discontinuity, where the low byte of the address is zero. This program segment need be executed only once.

11. Suppose that Y = 613 and X = 918 before each of the following instructions is executed. Give the contents of the registers X and Y after each is executed, in decimal. Then explain what is stored where if `STY 2,-Y` is executed with Y = 613.

(a) `LEAX 2,-Y` (b) `LEAX 2,-X` (c) `LEAX 2,Y+`

12. Give the shortest 6812 instruction sequences that carry out the same operation as the following nonexistent 6812 instructions. Condition codes need not be correctly set.

(a) `AAX` (b) `ADX` (c) `LSLX`

13. A section is a collection of n subroutines that are assembled together and written together in a PROM, EPROM, or EEPROM. The first 2 n bytes of storage for each section contain the direct address of each subroutine in the section, in a *jump vector*. The first two bytes are the address of the first subroutine, and so on. Suppose section 1 begins at $F000, so subroutine 3's address would be in $f006. In another section, a call to subroutine m in section 1, puts the 16-bit number from location 2 m + $f000 into the program counter. Show the machine code for parts (a) and (b).

(a) Write a single instruction, at location $d402, to call subroutine 3

(b) How do we fill the "jump table" at location $f000 with addresses of subroutines at run time (assuming the jump table is in RAM). In particular, if the subroutine at location f is a label at the beginning of the third subroutine whose address is at location $f006, write a program sequence to generate and write this address in the vector.

(c) How does this capability simplify the debugging of large programs?

14. The jump vector of problem 13 is to be made position independent. Each element is a relative offset to the subroutine. Repeat part (a), (b), and (c) of problem 15 for this jump vector. Write parts (a) and (b) as a program segment, where X points to the jump table's beginning.

15. Write a shortest program segment beginning at $866 to call subroutine PRINT, at $852, with the address of the character string to be printed in index register X, first for a string stored at location $876 and then for one at $893. However, the calling program segment, the subroutines, and the strings may be in a single ROM that can be installed anywhere in memory. They are in locations fixed relatively with respect to each other (position independence). Show your machine code.

16. Write a shortest program segment to put square waves on each output port bit, at location 0 so bit i's square wave's period is 2^i times the period of bit 0's square wave.

17. Write a shortest program segment to add index register X to accumulator D, transferring the data on the auxiliary stack pointed to by Y, as shown in Figure 3.11.

18. Write a shortest program segment to exclusive-OR accumulator A into accumulator B, transferring the data on the auxiliary stack pointed to by Y, as shown in Figure 3.11.

19. The Fibbonacci number of 0, $\mathcal{F}(0)$ is 1, and $\mathcal{F}(1)$ is 1, otherwise $\mathcal{F}(i)$ is $\mathcal{F}(i-1) + \mathcal{F}(i-2)$ for any positive integer i. Write a subroutine `FIB` that computes the Fibbonacci number; the input i is in index register X and the result $\mathcal{F}(i)$ is left in accumulator D.

(a) Write a recursive subroutine. (b) Write a nonrecursive (loop) subroutine.

20. Write a subroutine `POWER`, with input signed number n in accumulator D and unsigned number m in index register X that computes n^m leaving the result in accumulator D.

(a) Write a recursive subroutine. (b) Write a nonrecursive (loop) subroutine.

21. In Figure 3.16a, the instruction `MOVW #$18bc,1,SP` writes to a local variable on the stack in the outer loop. Write an instruction to load this value into index register X, which is just inside the next inner loop, where the instruction `LDAA 2,SP` is. Write an instruction to load this value into index register X, which is just inside the innermost loop, where the instruction `LDAA 6,SP` is.

22. Assume that an overflow error can occur in an `ADD` instruction in the innermost loop in Figure 3.16a, just after the instruction `LDAA 6,SP`. The following instruction `BVS L`, after the `ADD` instruction, will branch to location L. Write an instruction at this location L to deallocate stacked local variables such that the stack pointer will be exactly where it was before the first instruction of this figure, `LEAS -3,SP`, was executed.

23. Write a shortest subroutine that compares two n-character null (0) terminated ASCII character strings, s1 and s2, which returns a one in accumulator B if the strings are the same and zero otherwise. Initially, X points to the first member of s1 (having the lowest address), Y points to the first member of s2, and n is in accumulator A.

24. Figure 3.21 shows a table where the first column is a 32-bit Social Security number; other columns contain such information as age; and each row, representing a person, is 8-bytes wide. Data for a row are stored in consecutive bytes. Write a shortest program segment to search this table for a particular social security number whose high 16 bits are in index register Y, whose low 16 bits are in accumulator D, and for which X contains the address of the first row minus 8. Assume that a matching Social Security number will be found. Return with X pointing to the beginning of its row.

SS Number	Age	Sex	Phone Number
653931754	19	M	555-1000
546539317	18	F	555-8720
...	...	...	...

<-   32 bits   ->

<-- 8 bytes -->

Figure 3.21. A Table

Technological Arts' **Adapt812** is a modular implementation of the 68HC812A4, in single-chip mode, which includes all essential support circuitry for the microcontroller. A well designed connector scheme groups the dedicated I/O lines on one standard 50-pin connector, while routing the dual-purpose I/O lines to a second 50-pin connector, to form the address and data bus for use in expanded memory modes.

4

Assembly Language Programming

In the examples presented so far, you have probably noticed some real programming inconveniences, such as finding the operation code bytes, computing the addresses (particularly relative addresses), and associating the variable names with their memory locations. Furthermore, if you change the program, much of this routine work may have to be done again. What we have been doing is sometimes called *hand assembly,* in that we generate all of the machine code ourselves. Certainly, hand assembly is appropriate to the understanding of computer fundamentals. Beyond this we need to know hand assembly to remove the errors without reassembly. In this chapter we study the assembler and the skill of assembling programs using the computer.

Before the success of C and C++ compilers, when most programs were written in assembly language, more knowledge of assembly language was needed than is needed now because most programs are written in C and C++. The programmer needs to know how to read an assembly-language listing, which assembles code written by a compiler, and how to insert critical assembly-language statements in a C or C++ program. This chapter discusses critical assembler concepts that a programmer writing in C and C++ must know. The next chapter will delve deeper into assembly-language concepts, to enable the programmer to write large assembly-language programs.

Table 4.1. ASCII Codes

	00	10	20	30	40	50	60	70
0	'0'		' '	0	@	P	`	p
1			!	1	A	Q	a	q
2			"	2	D	∏	b	r
3			#	3	C	S	c	s
4			$	4	D	T	d	t
5			%	5	E	U	e	u
6			&	6	F	V	f	v
7			'	7	G	W	g	w
8			(	8	H	X	h	x
9			)	9	I	Y	i	y
A	'\n'		*	:	J	Z	j	z
B			+	;	K	[	k	{
C	'\f'		,	<	L	\	l	|
D	'\r'		-	=	M	]	m	}
E			.	>	N	^	n	~
F			/	?	O	_	o	

An *assembler* is a program someone else has written that will help us write our own programs. We describe this program by how it handles a line of input data. The assembler is given a sequence of *ASCII characters*. (Table 4.1 is the table of ASCII characters.) The sequence of characters, from one carriage return to the next, is a *line of assembly-language code* or an *assembly-language statement*. For example,

(space) `LDAA` (space) `#$10` (carriage return) (1)

would be stored as source code in memory for the assembler as:

$20
$4C
$44
$41
$41
$20
$23
$24
$31
$30
$0D

The assembler outputs the machine code for each line of assembly-language code. For example, for line (1), the assembler would output the bytes $86 and $10, the opcode byte and immediate operand of (1), and their locations. The machine code output by the assembler for an assembly-language program is frequently called the *object code*. The assembler also outputs a *listing* of the program, which prints each assembly-language statement and the hexadecimal machine code that it generates. The assembler listing also indicates any errors that it can detect *(assembly errors)*. This listing of errors is a great benefit, because the assembler program tells you exactly what is wrong, and you do not have to run the program to detect these errors one at a time as you do with more subtle bugs. If you input an assembly-language program to an assembler, the assembler will output the hexadecimal machine code, or object code, that you would have generated by hand. An assembler is a great tool to help you write your programs, and you will use it most of the time from now on.

In this chapter you will look at an example to see how an assembly-language program and assembler listing are organized. Then you will look at assembler directives, which provide the assembler with information about the data structure and the location of the instruction sequence but do not generate instructions for the computer in machine code. You will see some examples that show the power of these directives. The main discussion will focus on the standard Motorola assembler in their MCUez freeware.

At the end of this chapter, you should be prepared to write programs on the order of 100 assembly-language lines. You should be able to use an assembler to translate any program into machine code, and you should understand how the assembler works. Although you may not be able to understand how to write an assembler, you will be prepared from now on to use an assembler as a tool to help you write your programs.

4.1 Introductory Example and Assembler Printout

We now consider a simple example to introduce you to assembly-language programs. Consider a program that obtains the maximum of a sequence of numbers. We will assume that this sequence consists of 16-bit unsigned numbers stored consecutively in memory, high byte first for each number. This data structure is called a *vector* or (one-dimensional) array. The name of the vector will be the location of the first byte of the vector, so that the high byte of the ith number in the vector (i = 0, 1, 2, . . .) can be found by adding 2*i to the vector name. Suppose then that Z is a vector of four 16-bit two's-complement numbers beginning in location $86a with N stored in location $868. The ith number will be denoted Z(i) for i = 0 through N – 1. We want a program that finds the maximum of these numbers, putting it in locations $868 and $869.

One possible program for this, following the style of previous examples, is shown in Figure 4.1. We have arbitrarily started the program at address $89C.

Looking at the preceding program, we certainly would like to use just the mnemonics column with the variable addresses and the labels for the branches and let the assembler generate the other two columns, that is, do what we have been doing by hand. We would also like to be able to use *labels,* also called *symbolic addresses* (or just *symbols*) for the memory locations that hold the values of variables. The meaning of symbolic addresses is explored in greater detail in the next chapter. We use them in this section to get the main idea (they are used before dissecting them carefully). The use of symbolic addresses allows program segment (2) to be replaced by program segment (3).

```
LDX #$86A                    (2)
STD $868
```

```
LDX #Z                       (3)
STD RESULT
```

Location	Contents	Mnemonic	Comment
89C	CE 08 6A	LDX #$86A	; point X to the vector Z
89F	CD 00 03	LDY #3	; set count
8A2	EC 31	LDD 2,X+	; get first element
8A4	18 1A 31	L: EMAXD 2,X+	; get max with next element
8A7	04 36 FA	DBNE Y,L	; decrement, loop
8AA	7C 08 68	STD $868	; store answer in result
8AD	00	BGND	; halt

Figure 4.1. Program MAX

```
         ORG    $868
N:       EQU    3
RESULT:  DS.B   2
Z:       DS.B   50
*
         LDX    #Z       ; Point X to the vector Z
         LDY    #N       ; get count
         LDD    2,X+     ; Z(0) into D
LOOP:    EMAXD  2,X+     ; D- Z(i)
         DBNE   Y,LOOP   ; Another number?
         STD    RESULT   ; Store result
         BGND            ; Halt
```

Figure 4.2. Assembler Source Code for the Program MAX

Program segment (3) is clearer than program segment (2). An assembly-language source code for the program in Figure 4.1 is shown in Figure 4.2.

Putting this assembly-language program into the assembler yields the output listing shown in Figure 4.3. Although some new mnemonics have crept in, we can nevertheless see that we do not have to refer to actual addresses, only labels. We can see that we have not had to calculate relative offsets for the branching instructions, and we have not had to find memory locations or machine code. We now look into some of the details.

An assembly-language source statement takes the following form, where the fields, which are groups of consecutive characters, are separated by one or more one spaces:

Label Field Operation Field Operand Field Comment

The *label field* and the comment field may be empty and, depending on the operation, the operand field may be empty.

Label Field

A label (or symbolic address), if present, must have a letter as the first character and continue with letters, numbers, periods, or underscores. If a line's first character is an asterisk (*) or semicolon (;), the whole line is treated as a comment. Finally, labels that are identical to register names are not allowed (e.g., A, B, CC, X, Y, S, SP, PC, D, and PCR). The label ends in a colon (:). In some assemblers the colon is mandatory after a label; in some it cannot be used; and in others it is optional.

Operation Field

Except for comment lines, the *operation field* must consist of an instruction mnemonic or assembler directive (more about these later). The mnemonic must be written with no spaces: `CLRA, TSTB, ADDD,` and so on.

Operand Field

The *operand field* contains the addressing information for the instruction. Although numbers can be used to specify addresses, you will find that symbolic addresses are generally much easier to use in the operand field. For example, using the symbolic

Table 4.2. Addressing Modes

Mode	Example	Notes
Inherent	No Operands	
Immediate	#<expression>	
Page 0	<expression>	1
Direct	<expression>	2
Relative	<label>	
Indexed	<expression>,X	3
Preincrement	<expression>,+X	4
Predecrement	<expression>,-X	4
Postincrement	<expression>,X+	4
Postdecrement	<expression>,X-	4
Double indexed	A,X	3,5
Indirect indexed	[<expression>,X]	3
Indirect double indexed	[D,X]	3

Notes: 1: Prefix "<" or postfix .B forces Page 0. **2:** Prefix ">" or postfix .W forces Direct. **3:** Can substitute Y, SP, or PC for X. **4:** Can substitute Y or SP for X. **5:** Can substitute B or D for A

address or label ALPHA, the addressing modes in Table 4.2 can now all use symbolic addresses in place of numbers in the previous examples.

The assembler understands the use of addition, multiplication, and the like, using symbolic addresses in *expressions*. If ALPHA is location 100 and the operand field contains ALPHA+1, the assembler will put in the value 101. In simplest terms, an expression is just the usual algebraic combination of labels, numbers, and C language operations +, -, *, /, %, <<, >>, &, | , ~, !, <, >, <=, >=. !=, ==. Pascal operators = and <> are also recognized. Parenthesis are allowed, and precedence and evaluation are exactly as they are in C. Some examples of expressions are:

```
JUMP          JUMP*(8 + TAB)          ((RATE-2)*17)-TEMP
```

Comment Field

In the *comment field*, the programmer can insert short comments stating the purpose of each instruction. The comment must begin with a semicolon (;). In other assemblers, the comments begin one or more blanks after the operand field and are printed in the assembler listing but are otherwise ignored by the assembler.

In summary, writing an assembly-language program is a lot easier than writing machine code by hand. You can use symbolic addresses, letting the assembler determine where to put them and letting the assembler make sure that the instructions have the right operand values. You do have to conform to the rules of the language, however, and you have to spell the mnemonics exactly the way the assembler wants to recognize them. Although it would be nice to be able to just talk to the computer and tell it what you want it to do using conversational English, an assembler can barely understand the mnemonics for the instructions if you write them correctly and carefully. Nevertheless, writing assembly-language programs is easier than writing hexadecimal machine code.

```
1   1 0000                    ORG    $868
2   2 0868 0000 0003 N:       EQU    3
3   3 0868           RESULT:  DS.B 2
4   4 086A           Z:       DS.B 50
5   5 089C CE086A             LDX    #Z         ; Point X to Z
6   6 089F CD0003             LDY    #N         ; get count
7   7 08A2 EC31               LDD    2,X+       ; Z(0) into D
8   8 08A4 181A31    LOOP:    EMAXD 2,X+        ; D- Z(i)
9   9 08A7 0436FA             DBNE   Y,LOOP     ; Another number?
10 10 08AA 7C0868             STD    RESULT     ; Store result
11 11 08AD 00                 BGND              ; Halt
```

Figure 4.3. Assembler Listing for the Program MAX

The listing, shown in Figure 4.3, generally mirrors the source code but includes machine code and storage information. The listing line begins with a pair of line numbers. The first number is an absolute line number used for error messages, and the second is a relative line number used for *include files* and macro expansions discussed in the next chapter. The hexadecimal location of the instruction is given next; then the hexadecimal machine code is displayed. Finally, the source code line is shown.

4.2 Assembler Directives

Before looking more closely at how the assembler works, we describe the simplest *assembler directives*. These are instructions to the assembler that do not result in any actual executable machine coded instructions but are, nevertheless, essential to providing information to the assembler. A number of these will be introduced in this section and are listed in Table 4.3 for your convenience.

If we go back to the example at the beginning of the chapter, we recall that what we wanted was to just write down the mnemonics column and let the assembler generate the memory locations and their contents. There must be some additional information given to the assembler; in particular, you have to tell the assembler where to start putting the program or store the variables. This is the purpose of the `ORG` (for ORiGin) directive. The mnemonic ORG appears in the operation column, and a number (or expression) appears in the operand column. The number in the operand column tells the assembler where to start putting the instruction bytes, or reserved bytes for variables, that follow. For example, if the assembler puts the three bytes for `LDX #123` in locations 100, 101, and 102, the bytes for the instructions that follow are put consecutively in locations 103, 104, ... The operand can be described in decimal, hexadecimal, or binary, following Motorola's usual conventions. Thus we could replace the `ORG` directive above by

```
ORG 256
```

If there is no `ORG` directive at the beginning of your program, the assembler will start at memory location 0. There can be more than one `ORG` directive in a program. ABSENTRY sets the entry point, the initial value of the PC, in the HIWAVE debugger, when a program is loaded, so you don't have to enter the PC each time you load it.

Table 4.3. Assembler Directives

Mnemonic	Example	Explanation
ORG	ORG $100	Puts the next byte in location $100
ABSENTRY	ABSENTRY ALPHA	Initializes PC to ALPHA in HIWAVE
EQU	ALPHA: EQU $10	Makes the symbol ALPHA have value $10
DS	ALPHA: DS 10	**Define Space** Increments location counter by 10
DCB	ALPHA: DCB 3,55	**Define Constant Block** Fills 3 bytes with constant 55
DC		**Define Constant**
	ALPHA: DC.B $20,$34	Initializes the current location to $20 and the next location to $34
	ALPHA: DC.B 'ABC'	Initializes the word at this location to the ASCII letter A the next location to the ASCII letter B, and the next C
	ALPHA: DC.W $1234	Initializes the current location to $12 and the next location to $34

In all of our examples, we have set aside memory locations for variables. In the last example, we set aside bytes for `N`, `RESULT`, and `Z`. The way we tell the assembler to do this is with the `DS` (define space) directive. An optional postfix `.B` indicates bytes are allocated. Here `DS` appears in the operation field and the number n in the operand field tells the assembler that n bytes are being allocated. If no postfix is used, `.B` is assumed by default. Alternatively, a postfix of `.W` indicates that words are allocated so the number of bytes is $2n$, and a postfix of `.L` indicates that long words are allocated so the number of bytes is $4n$. The label in the `DS` directive is the variable name that the allocated space is given. The label is given the value of the address of its first, and perhaps only, byte. In the program of Figure 4.3, `RESULT` is given the value $868, and `Z` is given the value $86A.

The symbolic address `N`, which was introduced in §4.1, appears to have a split personality, especially for data. The symbol `N` is being used in two different ways here, as an address and as the value of a variable. The way to understand this is by analogy to a glass of water. When you say "drink this glass," the glass is the container, but you mean to drink the contents of the container. You do not expect the listener to be confused because he or she would never think of drinking the container. So too, the symbolic address `N` stands for the container, variable `N`'s location, whereas the contents of the container, the word at the address, is variable `N`'s value. If you think hard enough, it is generally clear which is meant. In the instructions `LDX #L` or `LEAX L,PCR`, the symbolic address `L` is the address of the variable, which is the container. In the instruction `LDAA L`, the symbolic address represents the contents, in that it is the contents of location `L` that goes into A. But if you are confused about what is meant, look to see if the symbolic address is being used to represent the container or its contents.

The `DS` assembler directive assigns a number to the symbolic address or container. In the preceding example, `N`'s container has the value $868 because $868 is the address of

N. However, the contents of N are not assigned a value, in contrast to a directive DC discussed later. The contents of N are *undefined;* they are the data that happen to be at location $868 at the time the program is started. The DS directive does not do anything to the value of a variable. We say the memory is *allocated* but is not *initialized*.

We have covered everything in the program of Figure 4.2 except the label LOOP, which appears in the label field for a machine instruction, not assembler directives. When a label is used with a machine instruction, it is given the value of the address of the first byte of that instruction. Notice that the value of LOOP in Figure 4.3 is $8a4. This value is also the address of the opcode byte of the EMAXD instruction. Thus the container LOOP is the address $8A4, while the contents of LOOP are the bits of the opcode byte for the EMAXD instruction.

Looking at other common assembler directives, the EQU directive (for EQUate) assigns a specific value to a label. In Figure 4.2, the label N is given the value 3 by the EQU directive. Generally, equates can be used to assign values to containers. Used this way, they are like DS directives, where the programmer assigns an address to the container rather than letting the assembler choose the value automatically. The EQU directive enables you to control where variables are stored, as in hand coding, but allows symbolic addresses to be used, as in assembly-language coding to improve readability. We will find EQU directives useful in fixing addresses in monitor programs and in fixing the addresses of I/O devices. These directives are often used to replace constants, to improve readability, and to simplify the modification of programs. For example, the instruction LDY #3 has been replaced, in Figure 4.2, by the lines

```
N: EQU  3
   • • •
   LDY  #N
```

where the EQU directive is put near the top of the program. Using EQU directives makes the program more readable and self-documenting, to an extent. It also makes it easier to modify the program if a different count N is used. The value of the count is in the EQU directive near the beginning of the program. If all changeable parts are kept in this area, it is fairly easy to modify the program for different applications by rewriting the EQU statements in this area. With an EQU directive, the label field cannot be empty, and the operand field can be an expression as well as a number. As we shall see later, there is a small restriction on the labels used in an expression of an EQU directive.

The DC (define constant) directive puts the values in the operand field into successive memory locations starting with the next available memory location. DC.B (define constant byte) allocates and initializes an 8-bit word for each item in the list in its operand field. The suffix .B is the default; DC is the same as DC.B. A label, if used, is assigned the address of the first value in the operand field. As an example

```
TABLE:    DC.B  14,17,19,30                (4)
```

appearing in a program generates four consecutive bytes whose values are 14, 17, 19, and 30 and whose locations are at TABLE, TABLE+1, TABLE+2, and TABLE+3, as shown.

```
TABLE      ->   $0E
TABLE+1    ->   $11
TABLE+2    ->   $13
TABLE+3    ->   $1E
```

The actual value of the container TABLE will depend on where it is placed in the program. Note that, in contrast to the DS directive, this directive initializes or assigns values to the container (the address) as well as allocating room for its contents (the word at that address). Beware, however, that the contents are given this value only when the program is loaded into memory. If the program is rerun without being loaded again, the value of the contents is what was left there as a result of running the program the last time. When rerunning the program, you should check these values and possibly rewrite them before you assume they are the contents specified by the program.

The DC.W (define constant word) directive does exactly the same thing as DC.B in (4) except that now two bytes are used for each value in the operand field, where, as usual, the high byte is first. For example, the directive

```
TABLE     DC.W 14,17,19,30
```
(5)

puts the values in memory as shown.

\$00
\$0E
\$00
\$11
\$00
\$13
\$00
\$1E

The DC.L directive allocates and initializes a 32-bit memory block for each item in the directive's operand list. Its mechanism is essentially like that for DC.W in (5).

The DC.B directive can have a sequence of ASCII characters in the operand field. (See Table 4.1 for a table of ASCII characters and their representations.) The ASCII codes for the characters are now put in the successive memory locations. The most convenient form is

```
LIST DC.B    "ABC"
```

where quotes enclose all the ASCII characters to be stored, namely, A, B, and C. Single quotes can be used instead of these quotes, especially where a character is a quote.

The define constant block DCB.B directive has a number n and a value v in the operand field; n copies of v are now put in the successive memory locations. Suffixes .B, .W, and .L can be used in an obvious way, and .B is the default.

To see how these directives might be used, suppose that we wanted to store a table of squares for the numbers between 0 and 15. The program, whose assembler listing is shown in Figure 4.4, uses this table to square the number N, returning it as NSQ. With the given data structure, the location of N^2 equals the location TABLE plus N. Thus if X contains the location TABLE and B contains the value of N, the effective address in the instruction LDAA B,X is the location of N^2.

```
 1  1 0000                         ORG    $868
 2  2 0868                 * this program squares the number N between 0 and 15
 3  3 0868 0001       N:          EQU  1
 4  4 0868            NSQ:         DS.B 1
 5  5 0869 00010409  TABLE:        DC.B    0,1,4,9,16,25,36,49,64,81
        086D 10192431
        0871 40516479
        0875 90A9C4E1
 6  6 0879 CE0869                  LDX     #TABLE      ; POINT X TO TABLE
 7  7 087C C601                    LDAB    #N          ; PUT N INTO B
 8  8 087E A6E5                    LDAA     B,X        ; PUT N**2 INTO A
 9  9 0880 7A0868                  STAA    NSQ         ; STORE RESULT
12 12 088F 00                      BGND
```

Figure 4.4. Assembler Listing for the Program Square

4.3 Mechanics of a Two-Pass Assembler

Some questions will soon arise about how symbolic addresses can be used without error. These questions have to be answered in terms of forward references, and their answers have to be understood in terms of how an assembler generates its output in two passes. Although we do not study how to write an assembler program (except in problems at the end of the chapter), we do want you to get a feeling for how it works so that you can understand how forward references are limited by what a two-pass assembler can do.

How does an assembler work? We begin by reading down through the instructions, called a *pass*. The first pass builds a *symbol table,* a list of all the symbolic addresses for labels and their values. The second pass will generate both the listing that shows the effects of each assembler line and the object code that is used to run the program.

We have earlier used the symbol "*" for the *location counter*. The location counter keeps track of the address where the assembler is when it is reading the current assembly-language statement, somewhat like the program counter does when the program runs. The location counter symbol "*" is always the address of the first byte of the instruction. In both passes, the location counter advances as code is generated.

The assembly-language program of Figure 4.5 finds all the odd, negative, 1-byte integers in the array `COLUMN` and puts them into the array `ODD`. On the first pass, the `ORG` statement sets the location counter to $800. Thus the label N has the value $800, the label `M` has the value $801, the label `COLUMN` has the value $802, and the label `ODD` has the value $834. The instruction `CLR M` will take three bytes (and we know what they are), the instruction `LDAB N` will take three bytes (and we know what they are), and so forth. Similarly, we see that the first byte of instruction

```
LOOP: LDAA 1,X+
```

will be at location $872. Thus the symbolic address (the container) `LOOP` has the value $872. Continuing in this way, we come to

```
BPL   JUMP
```

```
*         This program searches the array COLUMN looking for odd, negative,
*         one-byte numbers which then are stored in array ODD. The length of
*         COLUMN is N and  the length of ODD is M, which the program calculates.
*
          ORG   $800
N:        DS    1
M:        DS    1
COLUMN:   DS    50
ODD:      DS    50
*
          CLR   M                  ; initialize M
          LDAB  N                  ; Put N into B
          LDX   #COLUMN            ; Point X to COLUMN
          LDY   #ODD               ; Point Y to ODD
LOOP:     LDAA  1,X+               ; Next number of COLUMN into A
          BPL   JUMP               ; Go to next number if positive
          BITA  #1                 ; Z = 1 if, and only if, A is even
          BEQ   JUMP               ; Go to next number if even
          STAA  1,Y+               ; Store odd, negative number
          INC   M                  ; Increment length of ODD
JUMP:     DBNE  B,LOOP             ; Decrement counter; loop if not done
          BGND                     ; Halt
```

Figure 4.5. Program to Select Negative Odd Numbers

which takes two bytes in the program. We do not know the second byte of this instruction because we do not know the value of the address `JUMP` yet. (This is called a *forward reference,* using a label whose value is not yet known.) However, we can leave this second byte undetermined and proceed until we see that the machine code for `DBNE` is put into location \$87f, thus giving `JUMP` the value \$87f. As we continue our first pass downward, we allocate three bytes for `DBNE B,LOOP`. We do not find this instruction's offset yet, even though we already know the value of `LOOP`.

Scanning through the program again, which is the second pass, we can fill in all the bytes, including those not determined the first time through, for the instructions `BPL JUMP`, `BEQ JUMP`, and `DBNE B,LOOP`. At this time, all object code can be generated, and the listing can be printed, to show what was generated.

What we have described is a *two-pass assembler*. On the first pass it generates the symbol table for the program, and on the second pass it generates the machine code and listing for the program.

We have been using the prefix "<" in instructions like `LDAA <N` or a postfix "`.B`" such as in `LDAA N.B` to indicate an 8- or 9-bit addressing mode. If the prefix "<" or postfix "`.B`" is omitted, the assembler will still try to use 8-bit or 9-bit addressing when possible. Specifically, on the first pass, if the assembler knows the value of `N` when the instruction `LDAA N` is encountered, it will automatically use page zero addressing if `N` is on page zero. If it does not know the value of `N` yet, or if `N` is known but is not on page zero, it will then use direct addressing.

We have also been using the inequality symbols with index addressing to indicate whether the constant offset is to be described with eight or sixteen bits. The 6812 actually has another choice for the offset that we have not discussed before now because there is no special symbol for it. This is the 5-bit offset option. In this case, one can actually squeeze the offset into the post byte as described in the instruction set summary of the CPU12RG/D manual. The assembler chooses between the three offset options in exactly the same way that it chooses between page zero and direct addressing. On the first pass, if it knows the values in all the labels used in an expression for the offset, it will choose the shortest possible offset option or, if the expression is zero, it will take the zero offset option, which is pointer addressing. If it does not know some of the labels used in the expression for the offset, the assembler will default to the 16-bit offset option, determining these bytes on the second pass. From now on, we will drop the use of inequality signs in all addressing modes, except where it is needed in the relative mode to designate a short forward reference. Generally, it is best to let the assembler choose the appropriate option.

We sometimes observe an error message "phasing error," or "labels changed values on second pass." Such an error occurs when an instruction's length is computed on the first pass but is computed to have a different value on the second pass. Following such an instruction, successive labels will have a different value on the second pass than they had on the first pass. To fix such an error, read backward from the first instruction with the line that has such an error until you see an instruction whose length changes in the second pass, due to its using a different addressing mode. Put a prefix or suffix on its operand to force the instruction's first pass length to its second pass length.

```
        ORG     $800
K:      DS      M
M:      EQU     2
*
        LDD     K
        ADDD    #3
        STD     K
        SWI
```

Figure 4.6. Program with Illegal Forward Reference

As we have discussed earlier, an assembler does several things for us. It allows us to use instruction mnemonics, labels for variable locations, and labels for instruction locations while still providing the machine code for our program. As Figure 4.6 shows, however, we must be careful with forward references when using assembler directives.

The assembler reads the assembly-language program in Figure 4.6 twice. In pass one, the symbol table is generated and, in pass two, the symbol table, the instruction set, and assembler directive tables are used to produce the machine code and assembly listing. On each pass, each line of assembly language is processed before going to the next line so that some undetermined labels may be determined on the second pass. For example, in the program in Figure 4.6 the assembler will not determine the length of `M` on the first pass because the `DS` directive makes a forward reference to `M`, that is, uses a symbol in the expression for `K` that has not been determined yet. Suppose now that we change the program a little bit. See Figure 4.7.

```
        ORG     $800
M:      EQU     2
K:      DS      M
*
        LDD     K
        ADDD    #3
        STD     K
        SWI
```

Figure 4.7. Program without Forward Reference

When the line `K: DS M` is assembled, the value of `M` is known. Usually, it is easy to see which programs with forward referencing are assembled correctly just by examining how the assembler works with that particular program. An "undefined symbol" error occurs when `K: DS M` is assembled and `M` is not yet defined.

By now it should be obvious that for correct assembly a label can appear only once in the label field. Multiple occurrences are given an error message. (However, in the next chapter we will see a `SET` directive in which labels can be redefined.)

Looking at the instructions

```
      BNE JUMP
      ...
JUMP: ADDA M
```

in a particular program, one might wonder what happens if the location `JUMP` is more than 127 bytes below the `BNE` instruction. Does the assembler still proceed, not knowing location `JUMP`, and then give an error message when it finds that `JUMP` is beyond the 127-byte range on the second pass? Or does it immediately put in the long branch equivalent

```
LBNE   JUMP
```

and determine the right 2-byte relative address on the second pass? It might seem reasonable to expect the latter, but the first possibility has been chosen because the latter choice would force all forward branches to be long branches. In other words, the assembler leaves the burden of picking the shortest branching instruction to the programmer. For exactly the same reason, the programmer will want to use the inequality sign "<" with forward references for relative addressing used with other instructions. As an example, you should use `LDAA <L,PCR` instead of `LDAA L,PCR` when the effective address `L` is a forward reference which is within 127 bytes of the next byte after the `LDAA` instruction. Otherwise, the assembler will choose the 2-byte relative offset option.

4.4 Character String Operations

Before we look into an assembler, we will study some operations that copy, search, and manipulate character strings. These operations make it easier to understand how an assembler works, which we cover in the next section. They also provide an opportunity to show how assembly-language source code is written, in order to simplify your

```
         ORG  $800
K:       Dc.b "ALPHA",0  ; a NULL-terminated character string for part (b).
OUTPUT:  Ds.b 10         ; storage buffer for output characters for part (c).
OUTPTR:  Dc.w OUTPUT     ; pointer to the above buffer
```

a. Data

```
PRINT:  LDX #K           ; get address of string
NEXT:   LDAA 1,X+        ; get a character of string, move pointer
        BEQ  END         ; if it is NULL, exit
        BSR  PUT         ; otherwise print the character in A
        BRA  NEXT        ; repeat the loop
END:    SWI              ; return to the debugger
```

b. Calling PUT

```
PUT:    PSHX             ; save
        LDX OUTPTR       ; get pointer to output string
        STAA 1,X+        ; save character, move pointer
        STX OUTPTR       ; save pointer
        RTS              ; return
        PULX             ; restore
```

Figure 4.8. Print Program

programming effort. From now on, we will not write machine code, but we will write (ASCII) source code and use the assembler to generate the machine code.

The first three examples illustrate character string processing. The first example prints out a character string. The second transfers a character string from one location to another. The third compares two strings, returning 1 if they match. These examples are similar to `PUT`, `STRCPY`, and `STRCMP` subroutines used in C.

Figure 4.8b's program prints a string of characters using a subroutine `PUT`, like problem 3.15. Such strings often end in a NULL (0) character. The program reads characters from the string using `LDAA 1,X+`, and calls `PUT` to print the character in A. This also sets the condition code Z bit if the byte that was loaded was NULL, which terminates execution of the loop. An analogous program inputs data from a keyboard using the subroutine `GET` and fills a vector with the received characters until a carriage return is received. These programs can be generalized. Any subroutine that uses characters from a null-terminated character string can be used in place of `PUT`, and any subroutine that puts characters into a string can be used instead of `GET`.

`PUT` and `GET` are actually I/O procedures we show in §11.8, which require considerable understanding of I/O hardware. We don't want to pursue the actual `PUT` and `GET` subroutines quite yet. Instead, we replace the actual `PUT` and `GET` subroutines with a *stub* subroutine (Figure 4.8c). After stopping the computer, examine the string `OUTPUT` to see what would be output. Similarly, a stub subroutine can be used instead of `GET`, to "input" characters. The sequence of input characters is preloaded into a string.

Our second example (Figure 4.9) copies a null-terminated character string from one location to another. The original string is generated by the assembler and downloaded into memory, using `Src Dc.b`. The program copies it to another part of memory at `Dst Ds.b`. Note that the NULL is also copied to the destination string.

```
Src:   Dc.b "ALPHA",0 ; initialization of the source string (downloaded)
Dst:   Ds.b 16        ; allocation of the destination string

COPY:  LDX  #Src      ; get address of source string
       LDY  #Dst      ; get address of destination string
NEXT:  LDAA 1,X+      ; get a character of string, move pointer
       STAA 1,Y+      ; store it in the destination
       BNE  NEXT      ; if it is not NULL, reexecute the loop
```

Figure 4.9. Character Move Program Segment and Data

Our third example (Figure 4.10) compares one null-terminated character string with another. Both strings are downloaded into memory starting at label Src and Cmprd. We examine several cases of execution right after this program listing.

We examine several cases with this comparison program. Consider the case where Src is "BETA." The first time after label NEXT, the CMPA instruction clears the Z condition code bit, and the program goes to BAD to clear A and exit. Consider the case where Src is "ALPH." The fifth time at label NEXT, the LDAA instruction sets the Z condition code bit, and the program goes to EXIT where it tests the byte pointed to by Y, which is the ASCII letter A, so it goes to BAD to again clear A and exit. Consider the case where Src is "ALPHAS." The sixth time after label NEXT, the CMPA instruction clears the Z condition code bit, and the program goes to BAD to clear A and exit. Finally, consider the case where Src is "ALPHA." The fifth time at label NEXT, the LDAA instruction sets the Z condition code bit, and the program goes to EXIT where it tests the byte pointed to by Y, and because it is zero, the program sets A to 1 and exits. If the two strings are identical, the program ends with 1 in A, otherwise it ends with 0 in A.

The next example illustrates a comparison of a letter provided in accumulator A, to find a match among a collection of four letters L, A, S, and D, assuming it may not be any other letter. The number left in B is 0 if the letter is L, 1 if it is A, 2 if S and 3 if D. Two ways to do this are (1) execute compares with immediate operands, and (2) store the letters in a string and compare the unknown letter to each letter in the string. See Figure 4.11. Except for small collections of letters, the string method is best.

```
Src:   Dc.b "ALPHA",0 ; source string (downloaded)
Cmprd: Dc.b "ALPHA",0 ; comparand string (downloaded)

CMPR:  LDX  #Src      ; get address of source string
       LDY  #Cmprd    ; get address of comparand string
NEXT:  LDAA 1,X+      ; get a character of source string, move pointer
       BEQ  EXIT      ; if it is NULL, exit the loop
       CMPA 1,Y+      ; compare it to comparand character, move pointer
       BEQ  NEXT      ; if it is the same, reexecute the loop
BAD:   CLRA           ; otherwise exit; A is cleared to indicate mismatch
       SWI            ; return to the debugger
EXIT:  TST  0,Y       ; see if compare character is also NULL
       BNE  BAD       ; if it is not NULL, terminate indicating failure
       LDAA #1        ; it must be identical - end with A set to 1
```

Figure 4.10. Character String Compare Program Segment and Data

```
SRCH:  CLRB          ; initialize result in case first branch is taken
       CMPA #'L'     ; compare to  a character of string, move pointer
       BEQ  EXIT     ; if it is 'L', exit the loop
       INCB          ; increase result to 1 in case next branch is taken
       CMPA #'A'     ; compare to  a character of string, move pointer
       BEQ  EXIT     ; if it is 'A', exit the loop
       INCB          ; increase result to 2 in case next branch is taken
       CMPA #'S'     ; compare to  a character of string, move pointer
       BEQ  EXIT     ; if it is 'S', exit the loop. Otherwise it is S
       INCB          ; increase result to 3 since all other cases tested
EXIT:  SWI           ; return to the debugger
```

a) Using immediate operands

```
       ORG  $800
Cmprd: Dc.b "DSAL"   ; comparand string (downloaded)

CMPR:  LDX  #Cmprd   ; get address of comparand string
       LDAB #3       ; loop counter, and also the return value
NEXT:  CMPA 1,X+     ; compare to a character of string, move pointer
       BEQ  EXIT     ; if it is a matching character, exit the loop
       DBNE B,NEXT   ; decrement B, count out four loop executions
EXIT:  SWI           ; return to the debugger
```

b) Using a character string of comparison values

Figure 4.11. Character Search in a String

```
       ORG  $800
Dict:  Ds.b 32       ; storage for the dictionary
Ptr:   Dc.w Dict     ; address beyond the end of the dictionary
Cntr:  Dc.b 0        ; size of the dictionary

INSRT: LDX  Ptr      ; get the pointer
       STD  2,X+     ; store letter that is in A and value that is in B
       STX  Ptr      ; save the pointer
       INC  Cntr     ; increment the count of the dictionary size
```

a) Build

```
CMPR:  LDX  #Dict    ; get address of beginning of the dictionary
       LDAB Cntr     ; loop counter, and also the return value
NEXT:  CMPA 2,X+     ; compare to a character of string, move pointer by 2
       BEQ  EXIT     ; if it is NULL, exit the loop
       DBNE B,NEXT   ; decrement B, count out four loop executions
EXIT:  LDAB -1,X     ; get the numerical value of the letter into B
```

b) Search

Figure 4.12. Dictionary Program Segments

Location	Contents	Mnemonics
0	03	L 3
1	44	A 4
2	85	S 5

a. Machine code

```
  L A
  A B
  S C
A D 12
B D 34
C D 00
```

b. Source code

Figure 4.13. Machine and Source Code

Finally, we show a pair of programs that will build and search a dictionary (of letters). Figure 4.12a's program segment is executed each time a letter in accumulator A is inserted into the dictionary. The letter has a numerical "value" associated with the letter in B. The program segment in Figure 4.12b searches the dictionary. On entry, a letter is put in A. The "value" of the letter is left in B when the segment is completed.

These simple search programs can be fairly easily expanded for searching for longer strings of characters, or counting characters, rather than testing for a NULL, to determine when to terminate the search. Variations can handle the case where the comparand is not found. However, these are all linear searches in which the time to search for an item in the dictionary is linearly related to the number of elements in the dictionary. For large dictionaries, linear searches are entirely too slow; linked lists (§10.4) are much faster. However, linear searches are adequate for the simple assembler in §4.5.

4.5 A Simplified Two-Pass Assembler

An assembler is really a simple program. To illustrate how it works and to gain valuable experience in assembly-language techniques, we write parts of a "Simple Assembler" SA1 for a simple computer, with the overall specifications shown in Figure 4.14. Figure 4.13a shows SA1's machine code for a program to add two numbers (like Figure 1.5), and Figure 4.13b similarly shows how its source code might appear.

1. The target computer has only one 8-bit accumulator and 64 bytes of memory.
2. The target computer's opcodes will be L (binary 00), A (01), and S (10), coded as bits 7 and 6 of the opcode byte, which have a 6-bit direct address coded in the low-order 6 bits of the opcode byte. The assembler has a directive D, for "define constant byte," that has one two-digit hexadecimal number operand.
3. A source code line can have (1) a label and one space or else (2) two spaces. Then it has an opcode or assembler directive. Then it has a space and an operand, ending in a carriage return.
4. The assembler is to be run on the 6812 host. Assume the source code does not have errors. The source code will be stored in a constant ASCII string SOURCE, which is null terminated; and the object code, stored in 8-byte vector OBJECT, is indexed using an 8-bit variable LCNTR. No listing is produced.
5. All labels will be exactly one character long. The symbol table, stored in the 8-byte vector LABELS, consists of four 2-byte rows for each symbol, each row comprising a character followed by a one-byte address.

Figure 4.14. Simple Computer and Assembler SA1 Specifications

```
ORG $800
JSR PASS1
JSR PASS2
SWI
```

Figure 4.15. Assembler Main Program

The first instruction, which will be stored in location 0, loads the contents of location 3. The left two bits, the opcode, are 00, and the address of location 3 is 000011, so the machine code is 03 in hexadecimal. The next instruction's opcode is 01 for add; its effective address is 000100. The last instruction's opcode is 10 for store; its effective address is 000101. The source code shown in Figure 4.13b includes directives to initialize location 3 to $12, location 4 to $34, and location 5 to 0.

The assembler is written as two subroutines called `PASS1` and `PASS2`. This program segment illustrates the usefulness of subroutines for breaking up a large program into smaller subroutines that are easier to understand and easier to debug.

The data are defined by assembler directives, generally written at the beginning of the program. See Figure 4.16. They can be written just after the program segment shown in Figure 4.15. The first directive allocates a byte to hold the object pointer (which is the location counter). The second directive allocates and initializes the ASCII source code to be assembled. The next two lines allocate two eight-element vectors, which will store the machine code and symbol table.

```
LCNTR:    Ds.b 1   ; index used to store object code, which is the location counter
SOURCE:  Dc.b " L A",$d," A B",$d," S C",$d,"A D 12",$d,"B D 34",$d,"C D 00",$d,0;
OBJECT:   Ds.b 8   ; machine code
LABELS:   Ds.b 8   ; symbol table
```

Figure 4.16. Assembler Directives

```
PASS1:  CLR   LCNTR ; clear index to object code vector
        LDX   #SOURCE ; begin source scan: x-> first letter in source string
        LDY   #LABELS ; y-> first symbol

P11:    LDAB  1,x+ ; get the line's first character to B and move x to next character
        BEQ   P14  ; exit when a null character is encountered
        CMPB  #' ' ; if B is a space
        BEQ   P13  ; get opcode by going to P13
        STAB  1,y+ ; move character to symbol table
        MOVB  LCNTR,1,y+  ; put label value into symbol table
P13:    LDAB  1,x+ ; load B with character, move pointer
        CMPB  #$d  ; compare to carriage return which ends a line
        BNE   P13  ; until one is found. Note that x-> next character after this.
        INC   LCNTR ; increment location counter (we are processing the next line)
        BRA   P11  ; go to P11 to process the next line
P14:    RTS
```

Figure 4.17. Assembler Pass 1

```
PASS2:  CLR   LCNTR ; clear location counter, which is object code index
        LDX   #SOURCE ; begin source scan: x-> first letter in source string
P21:    LDAB  2,+x ; move past label and space, to get opcode character
        LEAX  2,X   ; skip mnenmonic and space after it
        CMPB  #'D' ; if mnemonic is a directive D,
        BEQ   P22   ; go to get the hex value
        JSR   GETOPCD  ; otherwise get the opcode, returns opcode in A
        LDAB  1,x+ ; get symbolic name which is instruction effective address
        JSR   FINDLBL  ; search labels, OR label value into opcode in A
        BRA   P23
P22:    BSR   GETHEX  ; get hexadecimal value into A
P23:    LDAB  LCNTR     ; get location counter which is 8-bits, into B
        EXG   b,y       ; expand B by filling with zero bits, to get 16-bit Y
        STAA  OBJECT,y  ; store the opcode-address or the hex value
        INC   LCNTR       ; increment location counter
        LDAB  #$d   ; skip to end of the current source code line
P24:    CMPB  1,x+
        BNE   P24
        TST   0,x   ; get first character of next line; if null, exit
        BNE   P21   ; otherwise loop again
        RTS
```

Figure 4.18. Assembler Pass 2

PASS1 (Figure 4.17) simply reads the characters from the source listing and inserts labels and their values into the symbol table. As is typical of many programs, an initial program segment initializes the variables needed in the rest of the subroutine, which is a loop. This loop processes one line of assembly-language source code. If the line begins with a label, it inserts the label and the current location counter into the symbol table. If the line begins with a space, it skips the character. It then scans the characters until it runs into the end of the line, indicated by a carriage return. Then it repeats the loop. When a NULL character is encountered where a line should begin, it exits.

PASS2 (Figure 4.18) simply reads the characters from the source listing and generates machine code, which is put into the object code vector. As in PASS1, an initial program segment initializes the variables needed in the rest of the subroutine, which is a loop. This loop processes one line of assembly-language source code. The program skips the label and space characters. If the mnemonic is a D for define constant, it calls a subroutine GETHEX to get the hexadecimal value; otherwise, it passes the opcode mnemonic to a subroutine GETOPCD that searches the list of mnemonic codes, returning the opcode. In the latter case, the subroutine FINDLBL finds the symbolic label, ORing its value into the opcode. The machine code byte is then put into the object code OBJECT.

GETOPCD (Figure 4.19) searches until it finds a match for the mnemonic. As it searches for a match in B, it generates the machine code in A. Because there are no errors in our source code, this extremely simple search procedure will always succeed in returning the value for the matching mnemonic. Because the directive D has been previously tested, if the opcode mnemonic is not an "L" or an "A" it must be "S."

```
*************************
*          Get an Opcode
*          entry: B is mnemonic opcode character
*          exit:   A is opcode
*                  X is unchanged
*
GETOPCD:  CLRA           ; if exit next, return zero as machine opcode
          CMPB  #'L'  ; Load
          BEQ   GO1
          LDAA  #$40  ; if exit next, return $40 as machine opcode
          CMPB  #'A'  ; Add
          BEQ   GO1   ; if it isn't D, L, or A, it must be S, for Store
          LDAA  #$80  ; return $80 as machine opcode
GO1:      RTS
```

Figure 4.19. Subroutine to Get the Opcode

`FINDLBL` (Figure 4.20) begins by setting up X for a loop; the loop searches each row of the symbol table, which was created in `PASS1`, until it finds a match. It searches each row of the symbol table. Because we assume there are no errors in our source code, it will always succeed in returning the value for the matching label.

`GETHEX` (Figure 4.21) calls an internal subroutine `G1` to translate an ASCII character to a hexadecimal value. `G1` uses the fact that ASCII letters "0" to "9" are translated into hexadecimal numbers by subtracting the character value of "0" from them, and the remaining ASCII characters "A" to "F" are translated into hexadecimal by further subtracting 7 from the result (because there are seven letters between "9" and "A" in the ASCII code). The first value obtained from the first letter is shifted to the high nibble and pushed on the stack. When the second value is obtained from the second letter, it is combined with the value pulled from the stack.

```
*************************
* Find Label
* entry: label character in B, OP code byte (from GETOPCD) in A
* exit: ORs symbol's value into A
*                  x is unchanged
*
FINDLBL:  LDY   #LABELS ; y-> first symbol table row
F1:       CMPB  2,y+ ; compare first character in B and move to next symbol table entry
          BNE   F1   ; if mismatch, try next by going to F1
          ORAA  -1,y ; OR previous row's value into the OP code in A
          RTS        ; return to caller
```

Figure 4.20. Subroutine to Insert a Label as an Operand

```
************************
*         Get hexadecimal value
*         entry:  X->first character of hex number
*         exit:   A:value, X->next character after hex number
*         saved:  B, Y
*
GETHEX:   BSR   GH1   ; convert ascii character to a nibble
          LSLA        ; move to high nibble
          LSLA
          LSLA
          LSLA
          PSHA        ; save on stack
          BSR   GH1   ; convert ascii character to a nibble
          ORAA  1,sp+ ; pop and combine
          RTS
*
GH1:      LDAA  1,x+ ; get next symbol
          CMPA  #'9'
          BLS   GH2
          SUBA  #7
GH2:      SUBA  #'0' ; subtract ascii 0
          RTS
```

Figure 4.21. Convert ASCII Hex String to a Binary Number

The reader should observe that this subroutine, `PASS2`, is broken into subroutines `GETOPCD`, `GETHEX`, and `FINLBL`. Each of these subroutines is more easily understood and debugged than a long program `PASS2` that doesn't use subroutines. Each subroutine corresponds to an easily understood operation, which is described in the subroutine's header. This renders the subroutine `PASS2` much easier to comprehend.

The contents of the vector `OBJECT` will be downloaded into the target machine and executed there. The assembler permits the programmer the ability to think and code at a higher level, not worrying about the low-level encoding of the machine code.

The reader should observe the following points from the above example. First, the two-pass assembler will determine where the labels are in the first pass. Thus, labels that are lower in the source code than the instructions that use these labels will be known in the second pass when the instruction machine code is generated. Second, these subroutines further provide many examples of techniques used to convert ASCII to hexadecimal, used to search for matching characters, and used to insert data into a vector.

4.6 Summary

In this chapter, we learned that an assembler can help you write much larger programs than you would be able to write by hand coding in machine code. Not only are the mnemonics for the instructions converted into instruction opcode bytes, but also symbolic addresses are converted into memory addresses. However, every new powerful tool also has some negative aspects. To use an assembler, you have to spell the mnemonics correctly and use the symbolic addresses exactly the same way throughout the program. You have to be concerned about the rules of writing a line of assembly-language code and the rules about forward references. But once these are mastered, you can use this powerful tool to help you write larger assembly-language programs.

The middle of this chapter explored some techniques for handling character strings to prepare you for the simple assembler. These techniques are pervasively used in microcontrollers. The `PUT` subroutine and the corresponding `GET` subroutine are used whenever we need to output or input characters. While they are actually discussed in §11.8, a stub can be used in the meantime to simulate output as shown in §4.4, and a similar stub can be used to simulate input. The string copy subroutine can be modified to make a string concatinate subroutine to append strings, and variations of the search and dictionary subroutines can recognize strings of characters, to respond to them.

At the end of this chapter, we presented a simple assembler. This program is larger than those that we found in Chapters 1 to 3. Scanning over this program, you should become aware of the need for a tool like the assembler to write longer programs. Consider the effort of writing such a long program manually, as we did in Chapters 1 to 3. We will also note, in Chapter 6, the need for subroutines. Our assembler used subroutines to break up a long program into shorter subroutines, which were easier to understand and to debug. This program and preliminary material in §4.4 also introduced techniques in handling ASCII character strings. You will use these techniques in most of the programs that you write from now on.

The assembler is just one such tool for converting your ideas into machine instructions. High-level languages can be used too, using compilers and interpreters to convert your language into the machine's language. High-level languages let you write even larger programs with a similar degree of effort, but they move you away from the machine, and it is difficult to extract the full power of the computer when you are no longer in full control. While high-level languages are used extensively to program most computers, especially larger computers, you will find many instances where you will have to program small computers in assembly language in your engineering designs.

This section has introduced the essential ideas of the assembler. The next chapter further expands the capabilities of the conditional and macro assembler and the linker. However, this chapter contains all the reader needs to know to read the assembly-language source code that is generated by a C compiler.

Do You Know These Terms?

See the end of chapter 1 for instructions.

hand assembly
assembler
ASCII character
line of assembly-language code
assembly-language statement
object code
listing
assembly errors
vector
labels
symbolic address
symbol
label field
operation field
operand field
expressions
comment field
include file
assembler directive
undefined data
allocate
initialize
pass
symbol table
location counter
forward reference
two-pass assembler

PROBLEMS

1. Suppose that the ORG statement is removed from the assembler. How can such statements be handled by other directives, and what assumptions have to be made to make it possible to completely replace the ORG statement?

2. Although the assembler has an ALIGN, an EVEN, and a LONGEVEN assembler directive, the ORG directive can align the location counter to an integer multiple of a constant. Write such an ORG directive that aligns the next location to the next multiple of four. Do not move the location counter if it is already aligned (low two bits are zero).

3. An I/O port at location $cb is to be loaded with a constant. If bit 2 is asserted, input hardware is activated; if bit 3 is asserted, output hardware is activated; if bit 5 is asserted, input hardware can cause an interrupt; and if bit 7 is asserted, output hardware can cause an interrupt. Write EQU directives that define four constants that can be ORed together to generate a constant to be stored into the memory location or can be used with BSET, BCLR, BRSET or BRCLR instructions. The constant ION turns on the input, OON truns on the output, IINT enables input interrupts, and OINT enables output interrupts. Write a MOVB instruction that turns on the input and output hardware and enables input interrupts. Write an instruction that subsequently enables output interrupts and another that disables output interrupts, neither of which change any other bits in location $cb except bit 7. Comment on the use of symbols to improve clarity.

4. A two dimensional array of 8-bit elements is to have R rows and C columns. Write EQU statements to define R to 3 and C to 4. Write a DC.B statement to allocate enough storage for the array for any R and C. Write a shortest program segment that reads the byte at row i column j into accumulator B, assuming rows and columns are numbered starting with zero, and elements of a row are stored in consecutive memory locations.

5. A vector of four 16-bit constants is to be initialized after location RATES, each of which are calculated as 8,000,000 divided by 16 times the desired rate. For example, to get a rate of 9600, put 52 into the element. The first element is to have a rate of 9600; the next, of 1200; the next, of 300; and the last of 110. Write this DC directive.

6. A vector of eight 8-bit constants is to be initialized after location PTRNA, each of which is a bit pattern displayed on consecutive lines on a screen to draw a letter A. For instance, the top row of eight bits will be $10, the next row will be $28, and so on. Write this DC directive.

7. Write a directive to clear all bytes from the current location counter to the location whose four low address bits are zero.

8. Write a directive to fill all bytes from the current location counter to location $FFF6 with value $FF.

```
        ABSENTRY   ENT
        ORG        $800
ENT:    LDD        K,PCR
        ADDD       #3
        STD        K,PCR
        SWI
K:      DS         1
```

a) Illegal Forward Reference

```
        ABSENTRY   ENT
        ORG        $800
ENT:    LDD        <K,PCR
        ADDD       #3
        STD        <K,PCR
        SWI
K:      DS         1
```

b) Legal Forward Reference

Figure 4.22. Program with Relative Address Reference

9. Figure 4.22 shows two programs that attempt to add 3 to variable K. However, Figure 4.22a may not assemble because of an illegal forward reference; or, if it does assemble, it produces less efficient code than you can produce by hand. Explain why this problem might occur. Figure 4.22b illustrates a solution to the problem, by using the "<" operator. Explain why this will assemble to produce efficient code.

10. Figure 4.23 shows three programs that attempt to add 3 to the variable K. However, Figure 4.23a may not assemble because of an illegal forward reference; or, if it does assemble, it produces less efficient code than you can produce by hand. Explain why this problem might occur. Figure 4.23b illustrates a solution to the problem by changing the forward to a legal backward reference. Explain why this assembles correctly. Figure 4.23c illustrates another solution to the problem, by using the "<" operator. Explain why this will assemble directly. Finally, explain why programmers generally put their data in front of (at lower addresses than) the program(s) that use the data.

11. Write a shortest assembly-language (source code) program that calls subroutine GET, which inputs a character, returning it in accumulator A, and stores these characters in the vector *K* as in Figure 4.8. Consecutively input characters are put in consecutive bytes until a carriage return is input; then a null (0) byte is written in *K*.

12. Write a shortest assembly-language (source code) subroutine that moves characters as in Figure 4.9 but converts lower case letters to upper case letters as it moves them.

```
     ORG    $800
ENT: LDD    K
     ADDD   #3
     STD    K
     SWI
     ORG    0
K:   DS     1
```

a) Illegal Forward Reference

```
      ORG    0
K:    DS     1
*
ENT:  LDD    K
      ADDD   #3
      STD    K
      SWI
```

b) Legal Backward Reference

```
      ORG   $800
ENT:  LDD   <K
      ADDD  #3
      STD   <K
      SWI
      ORG   0
K:    DS    1
```

c) Legal Forward Reference

Figure 4.23. Program with Direct or Page Zero Reference

13. Write a shortest assembly-language (source code) subroutine that concatenates one null-terminated string onto the end of another null-terminated string, storing a null at the end of the expanded string. Assume that on entry, X points to the first string, Y points to the second string, and there is enough space after the second string to fit the first string into this space. This program is essentially the C procedure *strcat()*.

14. Write a shortest assembly-language (source code) subroutine to compare at most n characters of one null-terminated string to those of another null-terminated string, similar to Figure 4.10. Assume X points to the first string, and Y points to the second string, and A contains the number n. Return carry set if and only if the strings match.

15. Write a shortest assembly-language (source code) subroutine that builds a symbol table as in Figure 4.12a but stores six-letter symbolic names and a two-byte value in each symbol table row. Upon entry to the subroutine, X points to the first of the six letters (the other letters follow in consecutive locations), and accumulator D contains the two-byte value associated with this symbol. The symbol table is stored starting at label `LABELS`, and the number of symbols (rows) is stored in one-byte variable `SIZE`.

16. Write a shortest assembly-language (source code) subroutine that searches a symbol table as in Figure 4.12b but searches six-letter symbolic names having a two-byte value in each symbol table row. Upon entry to the subroutine, X points to the first of the six letters (the other letters follow in consecutive locations). The symbol table is stored starting at label `LABELS`, and the number of symbols (rows) is stored in one-byte variable `SIZE`. The subroutine returns with carry bit set if and only if a matching symbol is found; then Y points to the beginning of the row where the symbol is found.

17. Write a shortest assembly-language (source code) program that finds the maximum MAX of `N` 4-byte signed numbers contained in array `Z` where `N` < 100. Your program should have in it the assembler directives

```
N       DS      1
MAX     DS      4
Z       DS.L    100
```

and be position independent. How would your program change if the numbers were unsigned?

18. Write an assembly-language program that finds the sum `SUM` of two 4-byte signed magnitude numbers `NUM1` and `NUM2`. The result should also be in signed-magnitude form. Your program should include the assembler directives

```
        ORG     $800
N       DS      1
NUM1    DS      4
NUM2    DS      4
SUM     DS      4
```

19. Write an assembly-language program that adds N 2-byte signed numbers stored in the vector Z, N < 10. Your program should sign-extend each number as it is being added into the current sum so that the result SUM is a 3-byte signed number. Your program should have the assembler directives

```
        ORG     $800
N       DS.B    1
Z       DS.W    10
SUM     DS      3
```

Can overflow occur in your program?

20. Write an assembly-language program to find the smallest nonzero positive number `NUM` in the array `Z` of `N` 2-byte signed numbers. If there are no nonzero positive numbers, the result should be put equal to zero. Your program should have the following assembler directives:

```
        ORG
N       DS      1
Z       DS.W    100
NUM     DS      2
```

21. Write an assembly-language program that finds the sum `SUM` of two N-byte numbers `NUM1` and `NUM2` and, when the `SWI` is encountered, has the condition code bits Z, N, V and C set properly. Your program should have the directives

```
N       DS      1
NUM1    DS      20
NUM2    DS      20
SUM     DS      20
```

```
        ORG     $800            Begin program
        LEAX    N,VECTOR        Point X to VECTOR
        LDAB    N               Length of VECTOR into B
LDDP    LD      A,X+            A vector element into A
        BLO     L1              if negative.
        COM     A               replace with two's-complement
L1      ANDA    #~1             Make contents of A even
        STAA    -1,X            Put number back
        DECB                    Counter is in accumulator B
*       BCS     \0
        SWI                     End of program

VECTOR  DS      N
N       EQU     5               Number of elements in VECTOR
```

Figure 4.24. Program with Errors

```
        ORG     0
COUNT   DS      0                   Number of characters changed
STRING  DC.B    "A b c . 1          String to be converted
BEGIN   DC.B    STRING              Starting address of string
S       DC.B    STRING-BEGIN        The length of the string
*
        ORG     #$800
        LD      A,S                 String size into A
        LEAY    #COUNT              Counter address into Y
        LDX     BEGIN               X points to STRING
        CLR     COUNT               initialize counter
LOOP    LDAB    0,X                 Get next character
        CMPB    #'a                 Compare character with "a"
        BLOW    L                   if lower, go to L
        CMPB    #97                 Compare character with "z"
        BLS     L                   if higher, go to L
        ANDB    #~$20               Change by clearing bit 5
        INC     0,Y                 increment counter
L       STAB    ,X+                 Put back letter
        DEC     A                   Decrement number left
        BNE     LOOP                Check next character
        SWI
```

Figure 4.25. Another Program with Errors

22. Correct the assembly-language program in Figure 4.24 so that each line, and the whole program, would be accepted by the Motorola assembler without an assembler error. Do not change any lines that are already correct. The program replaces each of the N 8-bit two's-complement numbers with the absolute value of the number, rounded down to the next lower even number. For example, + 4 is replaced with + 4, – 4 with + 4, – 5 with + 4, + 5 with + 4, and so on.

23. Correct the assembly-language program in Figure 4.25 so that each line would be accepted by the Motorola assembler without an assembler error. The program takes a sequence `STRING` of ASCII characters and converts all of the lowercase letters in `STRING` to uppercase letters while finding the number `COUNT` of letters that were converted. Do not change any lines that are already correct.

Location	Contents	Mnemonics
0	0003	LD 3
1	0104	AD 4
2	0205	ST 5

a. Machine code

```
    LD ALF
    AD BET
    SS GAM
ALF DC 0012
BET DC 0034
GAM DC 0000
```

b. Source code

Figure 4.26. Machine and Source Code for SA2

For problems 4.24 to 4.30, we will write parts of a "Sample Assembler" SA2 with the following overall specifications:

a) The target computer has only one 16-bit accumulator (A), one condition code (N), and 256 16-bit memory words. Each memory address is for a 16-bit word.

b) The target computer's opcodes will be LD (0), AD (1), ST (2), and BM (3) (similar to 6812 opcodes `LDAA`, `ADDA`, `STAA`, and `BMI`). The assembler has a directive DC, for "define constant byte," that has one four-digit hexadecimal number operand, and EN, which is the end of the source code.

c) A source code line can have a 3-character label or not (DC must have a label), one or more spaces, an opcode or assembler directive, one or more spaces, an operand, and optionally, one or more spaces and comments, ending in a carriage return. Permissible addressing modes are 8-bit direct for LD, AD, and ST, and signed 8-bit relative addressing for BM. So each instruction is 2 bytes.

d) The assembler is to be run on the 6812 host. The source code can have errors, and error numbers are printed. The source code will be stored in a constant ASCII string `SOURCE`, which is null terminated, the object code, stored in 16-byte vector `OBJECT`, is indexed by an 8-bit variable `LCNTR`; and the listing, stored in 80-byte vector `LISTING`, is pointed to by 16-bit variable `LPTR`. Finally, 8-bit variable `LINE` is the source code line number being read.

e) All labels will be exactly three characters long. The symbol table, stored in 16-byte vector `LABELS` consists of four 4-byte rows for each symbol, each row comprising three characters followed by a one-byte address. The number of symbols currently in the table is in 8-bit variable `SIZE`.

24. Write DS assembler directives needed to declare all storage locations for "SA2."

25. Correct the assembly-language program in Figure 4.27. Do not change any lines that are already correct. The calling routine, part of pass 1 of an assembler, checks the symbol table for a matching label and enters the label if no match is found. The symbol table contains this label, which is three ASCII characters, and the associated value, which is 1 byte. This program consists of a calling routine and a subroutine `FINDLBL` for "SA2," used in Problems 28 and 29, to compare a string of assembler source code characters at `SPTR` against known labels in the symbol table; if a match is found, the symbol table row number in which it is found is returned in accumulator A with carry clear; if not found, it returns with carry set. The program must assemble in Hiware and run correctly in the 6812 'B32 chip.

26. Write a shortest subroutine `GETOPCD` for "SA2," used in Problem 29, to compare a string of assembler source code characters against the permissible opcodes, returning the opcode or value of the defined constant in accumulator D, with carry set if found, and returning with carry clear if no match is found.

```
          origin 800      ; put at the beginning of SRAM
          entry SOURCE    ; provide entry address to Hiwave
*
LCNTR:    ds.w  1         ; 8-bit location counter
SIZE:     ds.b            ; 8-bit number of symbols
SOURCE:   dc.b  "ALF DC 7A",$d,"LB1 LD LB2",$d,"LB2 AD LB1,$d
LABELS    dc.b  32        ; allocate 32-byte symbol table
*
          ldx   SOURCE    ; get source code string address
          bsr   FINDLABEL ; look for label
          bvs   FOUND     ; if none found, skip. Note: y -> next symbol entry
          movb  1,x+,y+1  ; if found, copy first char from source to symbol table
          movd  2,x+,2,y+ ; then copy 2nd and 3rd characters to symbol table
          movb  #LCNTR,1,y+ ; and copy location counter into symbol table
          add   SIZE,1    ; increase label count
FOUND:    bra   *         ; wait for debugger to stop program
*
FINDLBL:  pshx            ; save caller's pointer on the stack
          leax  #LABELS   ; x-> first symbol table row
          ldab  1,X+      ; get a character from source file
          ldy   +1,X      ; get next two characters from source file
          lda   SIZE      ; A = number of symbols
          branch F2       ; go to end at F2 (in case there are zero items)
*
F1:       cpb   4,x+      ; compare first character in B and move to next row
          bne   F2        ; if mismatch, try next by going to F2
          cpy   3,x       ; compare second, third character in Y
          clra            ; clear A and carry indicating success
          ldx   -4,x      ; make x-> beginning of row found
          bra   F4        ; exit after balancing stack, putting pointer in Y
F2:       dec   a         ; count down accumulator A
          bpl   F1        ; go to F1 to do search if more to come
          setc            ; set carry to indicate failure
F3:       leay  SOURCE    ; make y-> beginning of row found
          pula            ; restore caller's pointer from the stack
          rts             ; return to caller
```

Figure 4.27. FINDLBL Subroutine with Errors

27. Correct the assembly-language program in Figure 4.28. Do not change any lines that are already correct. This shortest subroutine `LIST` for "SA2" prints a line of the listing at the end of `PASS2`, after two bytes have been put into the object code. It prints the decimal line number (< 10), the hexadecimal object code, and the source code. `OUTHEX` prints a binary number *n* in accumulator A as two digits (the hexadecimal representation of *n*). `OUTCH` places the ASCII character in accumulator A into a line of the listing, which is pointed to by the X register.

```
         org     800      ; put at the beginning of SRAM
OPTR:    ds.w    1        ; pointer to object code that was generated
LINE:    ds.b    1        ; line number ( line < 10 )
LINEBGN: ds.w    1        ; address of first character in line of source code
*
* listing subroutine : X points to the listing line
LIST:    ldab    #0       ; generate ASCII character "zero"
         addb    #LINE    ; add binary line number ( line < 10 )
         bsr     OUTCH    ; output line number
         ldb     #$20     ; output space
         bsr     OUTCH
         ldy     Optr     ; get pointer to object code
         lda     1,y+     ; y points to where next byte will be put,
         bsr     OutHex   ; output two previously stored bytes of object code
         ldaa    1,y+
         bsr     OutHex
         styr    OPTR     ; save pointer to object code
         ldab    #$20     ; output space
         bsr     OUTCH    ;
         ldy     LINEBEGIN ; get beginning of source line
Loop:    ldab    2,y+     ; get character, advancer to next character
         bsr     OUTCH    ; output character
         cpb     #d       ; up to and including c.r.
         bne     LOOP
         return           ; return
*
OUTHEX:  tab              ; duplicate byte to be printed
         lsrb             ; shift right four bits
         bsr     OUTHEX1  ; output hex number that was in high nibble of A
         tab   ; fall through, output low nibble of A
OUTHEX1: andb    #$f0     ; clear away all but low nibble
         cmpb    #9       ; if low nibble is a letter (A - F)
         bls     OUTHEX2  ; then adjust the output
         addb    #7       ; by adding 7 to the input
OUTHEX2: addb    #0       ; convert to ASCII character
OUTCH:   stab    1,x      ; store the character in the listing, move pointer
         rts              ; return to caller
```

Figure 4.28. Line Print Subroutine with Errors

28. Write a shortest subroutine PASS1 for "SA2" that will fill the symbol table, if a symbol appears more than once; it prints the line number, the word "error," and the error number 1; and it terminates with carry clear; otherwise, it terminates with carry set. The line number is to be printed in hexadecimal using OUTHEX (Problem 27). Assume the answers to Problems 24, 25, and 27 are "included" (don't write them).

29. Write a shortest subroutine **PASS2** for "SA2" that writes in the object and listing, reporting error 2 if an illegal opcode appears, and error 3 if a symbol is not found. "Include" the answers to Problems 24 to 27 (don't write them). A listing line consists of a hexadecimal line number, the two-byte hexadecimal code, and the line of source code. If an error occurs, the next line shows "error" hexadecimal error number.

30. Write a shortest subroutine **SYMLTBL** for "SA2" that prints its symbol table. Assume the answers to Problems 24 and 27 are "included," and use **PUT** to print the ASCII character in accumulator A (don't write them, and assume they do not change the registers used in the subroutine **SYMLTBL**). Each line lists a symbolic name and its hexadecimal value that was stored in a row of the symbol table.

The Motorola M68HC12B32EVB board can implement all the experiments and examples in this book, except those of Chapter 10. When used without another 6812 board, the debugger called DBUG_12 will use half of SRAM, permitting the other half to be used for an experiment.

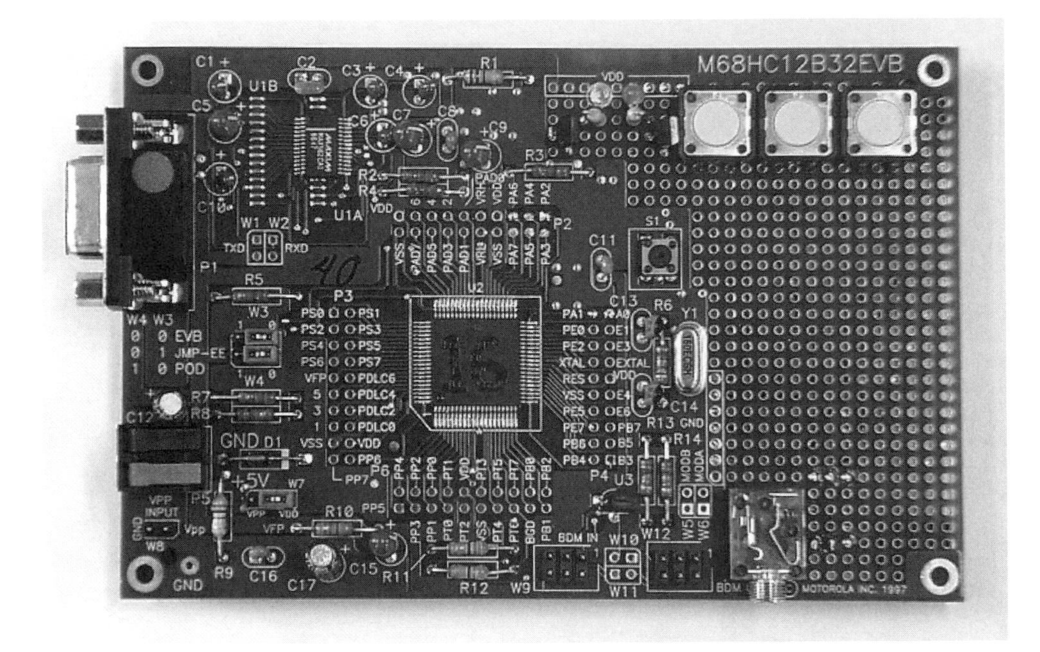

5

Advanced Assemblers, Linkers, and Loaders

This chapter discusses the advanced assembler and the linker, which are tools needed to assemble large programs; it is written for the reader who intends to write a lot of assembly-language programs or large assembly-language programs. Whereas the last chapter gave sufficient detail for the reader to understand the assembler output of a C compiler and to embed a limited amount of assembly language code in a C procedure, this chapter provides additional tools and greater depth to enable you to write large assembly-language programs using a relocatable, conditional, or macro assembler, and to join together separately assembled programs using a linker program.

This chapter is optional. Current economics leads to writing programs in a high-level language like C or C++ rather than in assembly language. Most programmers will not need to write a lot of programs in assembly language nor to write large programs in assembly language. Such readers can skip this chapter without losing any background needed for the rest of this book.

The first section introduces the complementary ideas of cross assembler and downloader. The next section describes the pair of ideas of relocatable assembler and linker program. Section 5.3 discusses how conditional assembly is used. The next section shows the power of macros in assembly-language programs. A final section recommends good documentation standards for programs written in assembly language.

Upon completion of this chapter, the reader should understand the tools needed for writing a large number of assembly-language programs or large assembly-language programs. He or she should have little difficulty writing assembly-language programs in the order of a couple of hundred lines long.

5.1 Cross Assemblers and Downloaders

In this section we introduce a close cousin of the assembler, the *cross-assembler,* which, like the assembler, converts sequences of (ASCII) characters into machine instructions. For the most part, this section's material is descriptive, almost philosophical, rather than precise and practical. It is important general knowledge, and it is included here because the reader should understand what he or she is doing when using a personal computer to assemble a program for a microcontroller.

The cross-assembler is a special kind of assembler. A true assembler is a program that runs on a computer that generates machine code for that same computer. It is common for a microcontroller to be too limited to be able to assemble code for itself, particularly for a microcontroller that is used in a laboratory for a university course. Such microcontrollers may not have enough memory, or a disk capable of holding the assembler program or the assembly language to be input to this program, or a printer capable of printing the listing. It is common to have a PC available with such user-friendly characteristics as Windows, large hard disks, and editors with which the user is already familiar. Such a computer, called the *host computer,* assembles programs for the microcontroller, which is called the *target machine*. The cross-assembler is written to run on the host machine and output target machine code.

The powerful host machine can handle the assembly-language program. An editor is a program on the host that helps you write the program. Editors can be used to write any kind of (ASCII) character data. The cross-assembler is used to generate the machine code for the target machine. The host machine's printer is used to print the listing, and the host machine's disk is used to hold the (ASCII) assembly-language program, the program listing, and the machine code output that is to be put into the target machine.

If your personal computer has a cross-assembler, it will usually also have a *downloader*. The downloader is a program running on the personal computer that takes the object code of the cross-assembler from the personal computer's disk or its primary memory and writes it into the target microcontroller's memory. A *monitor program* in the microcontroller receives data from the downloader; it generally does not need to be loaded because it is stored once and for all time in the microcontroller's ROM.

In a microcontroller in which it is desired to maximize the available RAM, one might have a small program, called a *bootstrap,* whose only purpose is to load the downloader into its RAM. After the program is put in by the loader, the memory space occupied by the downloader could be used by the program for data storage. The bootstrap, now occupying only a small amount of memory space, is generally in ROM.

It is also possible to connect host computers in a laboratory by means of a high-speed link to implement a *local area network*. The computers in this network are called *servers*. A printer attached to this network, with a computer to support it, is a *print server*. A computer that you experiment with is called a *workstation*. With each workstation connected to the print server, each can print a listing when it needs to do so, provided that the printer is not already in use. One printer can serve about a dozen workstations. This is economical and efficient. The downloader in a workstation sends the code generated by a cross-assembler to a target microcontroller. This kind of distributed processing, using local area networks, is one way to use a cross-assembler.

The program's machine code will be broken up into several *records,* to constrain the record's length within a suitable size. For each record, the downloader provides a starting address for the first byte and the number of bytes, as shown in Figure 5.1.

A byte called the *checksum* is usually included with each block, as indicated in Figure 5.1. This byte is formed by adding up all the bytes used to describe that particular block, ignoring carries. After loading, the checksum is computed and compared with the one supplied by the assembler. If any pattern of single errors in the columns has occurred, the two checksums will always be different, and an error message can be generated by the monitor.

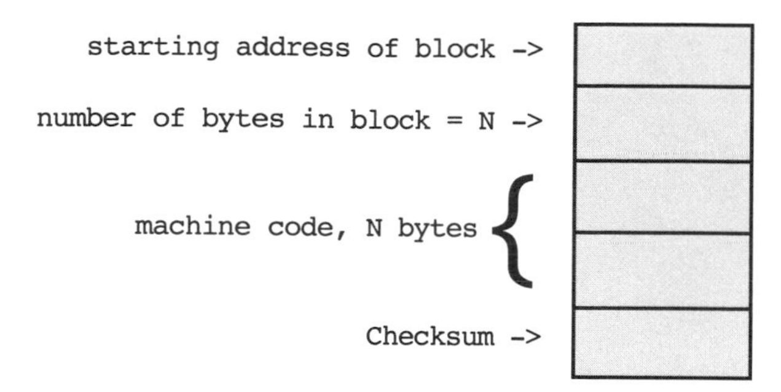

Figure 5.1. Loader Record

S-records are generally used by Motorola and other microcontroller vendors to store loader records, because ASCII character strings are easy to store on disks and send through a personal computer's communication software to the target microcontroller. S-records are essentially character strings made of several fields that identify the record type, record length, memory address, code/data, and checksum. Each byte of binary data is encoded as a 2-character hexadecimal number; the first character represents the high-order four bits, and the second represents the low-order four bits of the byte. An S-record's fields are shown in Figure 5.2, and Table 5.1 explains the content of each of the S-record's fields. Each record may be terminated with a carriage return, line feed, or both. Additionally, an S-record may have an initial field to accommodate other data such as line numbers generated by some time-sharing systems. Simple downloaders use only three of eight types of S-records: S0, S1, and S9 records. The S0 record generally contains header information. All data before the first S1 record is ignored. Thereafter, all code/data records must be of type S1, until the S9 record terminates data transfer.

An S0 record, which is a header record, is ignored by the downloader, but the cross-assembler may write in it any descriptive information identifying the following block of S-records. The S0 record address field is normally zeroes. An S1 record contains code/data and the 2-byte starting address at which the code/data is to reside. An S9 record terminates S1 records. Its address field may optionally contain the 2-byte address of the first instruction to be executed. If an S9 record doesn't specify an address, the first entry point specification encountered in the object module input is used. There is no code/data field.

A short program, whose listing is shown in Figure 5.3a, generates the S-record in Figure 5.3b. This simple program has only one S1 and the terminal S9 records, which are explained following the figure. Generally, programs have several S1 records.

The S1 code/data record begins with the ASCII characters S1, indicating a code/data record of length `$16` bytes, to be loaded at the 2-byte address `$0820`. The next 20 character pairs are the ASCII bytes of the actual program code/data. In this assembly language example, the hexadecimal opcodes of the program are written in sequence in the code/data fields of the S1 records. The first byte is `B6`, and the second byte is `08`. Compare this S-record string in Figure 5.3b with the listing in Figure 5.3a.

TYPE	RECORD LENGTH	ADDRESS	CODE/DATA	CHECKSUM

Figure 5.2. An S-Record

Table 5.1. S-Record Fields

Field	Printable Characters	Contents
Type	2	S-Record type - S0, S1, etc.
Record length	2	The count of the character pairs in the record, excluding the type and record length.
Address	4, 6, or 8	The 2-, 3-, or 4-byte address at which the data field is to be loaded into memory.
Code/data	0-2n	From 0 to n bytes of executable code, memory-loadable data, or descriptive information. For compatibility with teletypewriters some programs may limit the number of bytes to as few as 28 (56 printable characters in the S-Record)
Checksum	2	The least significant byte of the one's complement of the sum of the values represented by the pairs of characters making up the record length, address, and the code/data fields.

The second record begins with ASCII characters S9, indicating a termination record. Its length is 3. It is followed by a 4-character 2-byte address field, which is zeroes, and $FC, which is the checksum of the S9 record.

```
0000                    ORG   $820
0820 B60811             LDAA  $811
0823 BB0813             ADDA  $813
0826 7A0815             STAA  $815
0829 B60810             LDAA  $810
082C B90812             ADCA  $812
082F 7A0814             STAA  $814
0832 00                 BGND
```

a. Listing

```
S1160820B60811BB08137A0815B60810B908127A0814004E
S9030000FC
```

b. S-record

Figure 5.3. A Program

5.2 Relocatable Assemblers and Loaders

The downloader described in the previous section is sometimes called an *absolute loader,* because the location of the program is completely specified by the assembler that writes S-records. There is a limitation with an absolute loader. Suppose that you have written a program that uses a standard subroutine from a collection provided by Motorola or a third party. You have to copy it physically into your program, perhaps changing the ORG so that it does not overlap with your program, and then assemble the whole package. What you might like to do instead is to be able to assemble your program separately and then have the loader take your machine code, together with the machine code of any standard subroutines already assembled, and load them all together in consecutive RAM locations so that memory bytes would not be wasted.

To achieve merging of programs that are assembled separately, assemble each program segment as if it began at location zero. Each such program segment is called a *relocatable section*. A *linker* program determines the sizes of each section and allocates nonoverlapping areas of target memory to each. Each section's addresses are then adjusted by adding the section's beginning address to the address within the section.

Assembler directives, listed in Table 5.2, are used to name sections and identify labels that are declared in other files or are to be made available for use in other files. See Figure 5.4. A relocatable section begins with a `SECTION` directive. In each section, certain symbols would have to be declared in `XDEF` and `XREF` directives.

Table 5.2. Relocation Directives

Name	Example	Explanation
SECTION	label SECTION	Begins a section whose name is "label"
XDEF	XDEF x,y,z	Declares x,y,z visible in other sections
XREF	XREF x,y,z	Declares x,y,z variables are from outside
XREFB	XREFB x,y,z	Declares external variables x,y,z are 8-bit variables, use Page 0 addressing

```
.data:  SECTION
        XDEF    N,V         ; Make N, V externally visible
N:      DS      1           ; Declare a global variable
V:      DS      50          ; Declare a global vector
```

a. A data section in file Program1.asm.

```
.text:  SECTION
        XDEF    FUN         ; Make FUN externally visible
*
FUN:    RTS                 ; Should be replaced by full sub.
```

b. A program section in file Program2.asm

Figure 5.4. Sections in Different Source Code Files

```
.pgm:    SECTION
         XDEF     ENTRY          ; Make externally visible
         XREF     FUN,N          ; Use externally defined names
*
ENTRY:   LDS      #$BFF          ; First initialize stack pointer
         LDAA     N              ; Access external variable
         JSR      FUN            ; Call external subroutine
```

c. Another program section in file Program3.asm

Figure 5.4. Continued

For example, `JSR FUN` might occur in a section `.pgm` in one file, but the code for subroutine `FUN` is in a different section `.text` in another file. `FUN` is declared, in each file where it is used, in an `XREF` directive. In the file where the `FUN` subroutine is written, label `FUN` would be declared in an `XDEF` directive. Labels used by more than one file are similarly declared where defined (with `XDEF`) and where used (with `XREF`).

The `SECTION` directive's label is the name of the section, which is used in the *parameter file* that controls linking. Figure 5.5 illustrates a parameter file that directs the linker to put certain sections in certain parts of memory. If two or more sections have the same name, the linker appends them together, forming a larger section that is allocated in memory as a whole. A `SECTION` directive may have the word `SHORT` as an operand; this means that the section is on page zero and may use page-zero addressing. The linker requires, as a minimum, sections with names .data and .text.

The parameter file simply tells the linker what files are output from it and input to it. This parameter file uses C syntax for comments and constants. The parameter file declares *memory segments,* which are areas of memory available for assembly-language sections, and then declares what sections are put in which segments. Finally, parameter file statements tell where execution begins and what to put in the target's interrupt vectors, including the reset vector. Interrupt and reset vectors are discussed in Chapter 11.

```
LINK   Program.abs      /* the linker will write a file called prog.abs */
NAMES                   /* list all assembler output files input to the linker */
    Program1.o Program2.o Program3.o
END          /* several assembler files can be read: end with END */

SEGMENTS     /* list all segments, read/write permission & address ranges */
    ROM = READ_ONLY   0x800 TO 0x9FF /* a memory segment */
    RAM = READ_WRITE  0xA00 TO 0xBFF /* a memory segment */
END          /* several segments can be declared: end with END */

PLACEMENT  /* list all sections, in which segments they are to be put */
    .text,.pgm INTO ROM   /* puts these sections into segment ROM */
    .data INTO RAM        /* puts a section called .data into segment RAM */
END                 /* several segments can be filled with sections: end with END */

INIT         ENTRY      /* label of first instruction to be executed */
VECTOR ADDRESS 0xFFFE ENTRY /* puts label ENTRY at 0xFFFE */
```

Figure 5.5. A Parameter File

The assembler begins program segments with either an ORG statement, introduced in the previous chapter, or a SECTION statement, introduced above. Another way of looking at this: SECTION is an alternative to ORG. Sections beginning with ORG are called *absolute sections*; each section runs until the next ORG, SECTION, or END directive or the end of the file. The programmer is responsible for ensuring that absolute sections, and memory segments, declared in the parameter file, do not overlap.

The *linking loader,* in a manner similar to a two-pass assembler, takes the parameter file and the list of external and entry symbols declared in XREF and XDEF declarations; it calculates the addresses needed at load time, and inserts these into the machine code in a .o file that was generated by the assembler. In this example, the 2-byte relative address for FUN in the instruction above is determined and then inserts the code for the instruction. The linker output file, a .abs file, is ready to be downloaded into a target microcontroller.

5.3 Conditional Assemblers

A near cousin of the assembler, that we examined in the last chapter, is the *conditional assembler.* A conditional assembler allows the use of conditional directives such as IFEQ and ENDC, as listed in Table 5.3. For example, the segment

```
IFEQ    MODE
LDAA    #1
STAA    LOC1
ENDC
```

inserted in an assembly language program causes the assembler to include instructions

```
LDAA    #1
STAA    LOC1
```

in the program if the value of MODE is equal to zero. If the value of MODE is not equal to zero, assembler directives in lines from the IFEQ directive to the ENDC directive, except ENDC and ELSE, are ignored and do not generate machine code. The label MODE is usually defined, often at the beginning of the program, through an EQU directive, say MODE EQU 1. There are often several conditional directives such as IFEQ MODE throughout the program, for a single directive such as MODE EQU 1. The single EQU directive uniformly governs all of these conditional directives. This way, a directive at the beginning of the program can control the assembly of several program segments throughout the program.

The conditional statement argument can be a more complex expression. There are other conditional directives—IFNE, IFGE, IFGT, IFLE, IFLT, and IFNE—that can be used instead of IFEQ, and the ELSE statement can cause code to be assembled if the condition is false, so code immediately after the conditional is not generated. And conditional expressions can be nested. For instance, the program segment

```
IFGT    OFFSET-1
LEAX    OFFSET,X
ELSE
IFEQ    OFFSET-1
INX
ENDC    ; matches IFEQ
ENDC    ; matches IFGT
```

tests the predefined symbol OFFSET. If OFFSET is greater than one, a LEAX instruction is used; if it is one, a shorter INX instruction is used, and if it is zero, no code is generated.

The conditional IFDEF directive is often used to be sure that a symbolic name is defined before it is used in an assembler line, such as another conditional directive, to avoid generating an error message. The IFNDEF directive is particularly useful in INCLUDE files described in a later section on documentation.

The IFC and IFNC conditional directives are able to test strings of characters to see if they are exactly matching. The former assembles lines to the ENDC or ELSE directive if the strings match; the latter assembles the lines if the strings don't match. It finds special use in macros, discussed in the next section where an example is given.

One of the principal uses of conditional assembly directives is for debugging programs. For example, a number of program segments following each IFDEF MODE up to the matching ENDC can be inserted or deleted from a program by just inserting the EQU directive defining MODE. All of these directives allow the programmer to uniformly control how the program is converted to object code at assembly time.

Another significant use of conditional assembly is the maintenance of programs for several different target microcomputers. Some code can be conditionally assembled for certain target microcontrollers, but not for other microcontrollers.

Table 5.3. Conditional Directives

Name	Name	Explanation
IF	IF <expression>	Insert if expression is TRUE
IFEQ	IFEQ <expression>	Insert if expression is FALSE (zero)
IFNE	IFNE <expression>	Insert if expression is TRUE (not zero)
IFGE	IFGE <expression>	Insert if expression ≥ 0
IFGT	IFGT <expression>	Insert if expression > 0
IFLE	IFLE <expression>	Insert if expression ≤ 0
IFLT	IFLT <expression>	Insert if expression < 0
IFDEF	IFDEF <symbol>	Insert if symbol is defined
IFNDEF	IFNDEF <symbol>	Insert if symbol is not defined
IFC	IFC str1 str2	Insert if string str1 equals string str2
IFNC	IFNC str1 str2	Insert if string str1 is not = string str2
ELSE		Alternative of a conditional block
ENDIF		End of conditional block

5.4 Macro Assemblers

A *macro assembler* is able to generate a program segment, which is defined by a *macro,* when the name of the macro appears as an opcode in a program. The macro assembler is still capable of regular assembler functioning, generating a machine instruction for each line of assembly language code; but like a compiler, it can generate many machine instructions from one line of source code. Its instruction set can be expanded to include new mnemonics, which generate these program segments of machine code. The following discussion of how a macro works will show how this can be done.

A frequently used program segment can be written just once, in the macro definition at the beginning of a program. For example, the macro

```
AAX:    MACRO
        EXG     A,B
        ABX
        EXG     A,B
        ENDM
```

allows the programmer to use the single mnemonic AAX to generate the sequence

```
        EXG     A,B
        ABX
        EXG     A,B
```

The assembler will insert this sequence each time the macro `AAX` is written in the assembler source code. If the mnemonic `AAX` is used ten times in the program, the three instructions above will be inserted into the program each time the mnemonic `AAX` is used. The advantage is clear. The programmer almost has a new instruction that adds the unsigned contents of A to X, and he or she can use it like the real instruction `ABX` that actually exists in the machine. The general form of a macro is

label MACRO
instructions
ENDM

Here the symbolic name "label" that appears in the label field of the directive is the name of the macro. It must not be the same as an instruction mnemonic or an assembler directive. The phrases `MACRO` and `ENDM` are assembler directives indicating the start and the end of the macro. See Table 5.4. Both appear in the operation field of the instruction.

Table 5.4. Macro Directive

Name	Name	Explanation
MACRO	label MACRO	Begins a macro, whose name is "label"
ENDM	ENDM	Ends a macro
MEXIT	MEXIT	Exit a macro
SET	label SET expression	Sets the label to have the value of expression. Label may be redefined.

Table 5.5. Macro Arguments

Symbol	Explanation
\0	Size parameter attached to macro name
\1	First parameter in the list
\2	Second parameter in the list
...	(numbers and letters \3, \4, ... \Y)
\Z	Last parameter in the list (35 max)
\@	Macro invocation number

Every time the programmer uses this macro, he or she writes

```
label    name    parameter,parameter,...,parameter
```

where `name` is the name of the macro, which is placed in the operation field, to which an optional size `.B`, `.W`, or `.L` may be appended. The designation `label` at the beginning of the line is treated just like a label in front of a conventional assembly line.

The parameters can be inserted into the body of the macro using the two-character symbols in Table 5.5. As an example, the macro

```
MOVE:   MACRO
        LDD     \1
        STD     \2
        ENDM
```

will move the two bytes at parameter location \1 to parameter location \2, like the `MOVW` instruction, but the macro is faster and shorter where the second parameter is a page-zero address. When used in the program, say as

```
        MOVE    Z+3,M                                   (1)
```

the two bytes at locations Z + 3, Z + 4 will be moved to locations M, M + 1. In this example, all the usual rules for choosing between direct and page-zero addressing would apply. Additionally, if the actual parameters involve an index mode of addressing that uses a comma, the actual parameters must be enclosed within parentheses as in `MOVE (3,X),Y` for the sequence

```
        LDD     3,X
        STD     Y
```

As implied in the example above, when a macro is used, the actual parameters are inserted in a one-to-one correspondence with the order in (1).

If "goto" labels are to be used within macros, then, because the macro may be used more than once during the program, assembler-generated labels must be used. The symbol character pair "\@" means an underbar followed by the macro invocation number, which is initially zero and is incremented each time a macro is called. When the first macro is expanded \@ generates the label _00000, when the second macro is expanded, \@ generates _00001, and so on. Throughout the first macro, \@ generates _00000 even though this macro may call another macro expansion, whenever it is used, before or after other macros are called and expanded. This generated symbol can be concatenated to other

```
ADD:    MACRO
        LDX     \1
        LDAB    \2
        CLRA
\@:     ADDA    1,X+
        DBNE    B,\@
        ENDM
```

Figure 5.6. Loop Macro to Add Consecutive Values

letters to generate unique labels inside the macro itself. This capability is especially useful if a macro expansion has two or more "goto" labels in it; for without it, unambiguous labels could not be generated. Using the macro invocation number, each macro expansion generates labels that are different from the labels generated from the same macro that are expanded at a different time. For example, the macro in Figure 5.6, when implemented by `ADD #M,N`, adds the contents of the `N` bytes beginning in location `M`, putting the result in accumulator `A`.

A macro definition can use a macro defined earlier, or even itself (recursively). For macro assemblers that have conditional assembly, conditional directives within their definition can be used to control the expansion. The actual parameters of the macro can then be tested with these directives to determine how the macro is expanded. In particular, the `IFC` and `IFNC` directives can be used to compare a macro parameter, considered as a string, against any constant string. If a parameter is missing, it is a null string "\0." We can compare a given parameter, such as the second parameter, considered as a string denoted "`\2`," to a null string "\0." When the strings are equal, the second parameter is missing, so this condition can terminate the macro expansion.

The example in Figure 5.7 illustrates the use of recursion, conditional assembly, and early exiting of macros using `MEXIT`. You might want to use the `ADDA` instruction with a series of arguments, to add several bytes to accumulator A. If you wish, you can use a text editor to make copies of the `ADDA` instruction with different arguments. However, you can define a macro whose name is `ADDQ`, in which the body of the macro expands into one or more `ADDA` directives to implement the same effect. This `ADDQ` macro uses recursion to add one parameter at a time, up to eight parameters in this simple example, stopping when a parameter is missing (null string). When a null (missing) parameter is encountered, the macro "exits" by executing `MEXIT`, thereby not generating any more expansion or code.

```
ADDQ:   MACRO
        IFNC "\1",""
        ADDA \1
        ENDC
        IFC "\2",""
        MEXIT
        ENDC
        ADDQ  \2,\3,\4,\5,\6,\7,\8
        ENDM
```

Figure 5.7. Recursive Macro to Add up to Eight Values

5.5 Documentation

A significant part of writing large assembly-language programs is the effort to make the program understandable to a programmer who has to debug it. Some tools to do this are provided by state-of-the-art assemblers. Table 5.6 shows assembler directives that significantly clarify assembler listings. The LIST and NOLIST directives can clean up a listing by not printing parts of the program that are neither interesting nor informative. CLIST and MLIST directives control whether conditional expressions and macros are fully listed. Other directives, TITLE, LLEN, NOPAGE, PLEN, and TABS, control the format of a listing, and PAGE and SPC tidy up a listing.

The INCLUDE directive permits other files to be inserted into a source code file. The assembler saves the location after the INCLUDE directive, opens the file that is named in the directive's operand, reads all of that file, and then returns to the saved location to resume reading the file in which the INCLUDE directive appeared. Included files can include other files but cannot eventually include themselves (the INCLUDE directive is not recursive). It is common to include a file defining a program's constants, in EQU directives, in each file that uses these constants. However, in such files, so that they are included only once, the contents of an INCLUDE file might be written as follows:

```
IFNDEF     F1
     . . . (an entire file, except the three statements shown here)
F1: EQU    0  ; it really doesn't matter what F1 is set to here
     ENDC
```

The conditional statement will include the contents of a file the first time it is named as the parameter of an INCLUDE statement. Subsequent INCLUDE statements will fail to assemble the contents of the file because the symbolic name F1 will be defined after the first time it is INCLUDEd.

Table 5.6. Listing Directives

Name	Explanation
CLIST	Conditional assembly listed or not
END	Stop reading source file
INCLUDE	Include a file named as the argument
LIST	Following statements are listed
LLEN	Specifies line length
MLIST	Macros listed or not
NOLIST	Following statements are not listed
NOPAGE	Disable paging in the listing
PAGE	Insert page break
PLEN	Specifies page length
SPC	Insert empty line
TABS	Specify tab spacing
TITLE	Put directive's operand on each page

It should be noted that, in the relocatable assembler, the `INCLUDE` directive allows breaking up a large program into smaller files that are easier to understand, debug, and maintain. You should break up large files into smaller ones, each of which implements a conceptually uncluttered collection of operations, and put each such collection into its own section. Sections reused in many application programs can be saved in files collectively called a *library*. When a later program needs the same function, it can `INCLUDE` a file from the library that has the function already debugged and assembled. Your application program can `INCLUDE` prefabricated files from a library and parts of the program now being written, which are in different files, to break up the larger program into smaller sections, each of which is easier to understand, debug, and use. While further development of tools to do this are incorporated into object-oriented programming, the basic ideas can be incorporated into state-of-the-art assembly-language programs as well.

We continue this section with a list of coding techniques that make your programs or subroutines more readable.

1. Use meaningful labels that are as short as possible.

2. One should try to write code so that the program flow is progressively forward except for loops. This means that one should avoid using unnecessary unconditional jumps or branches that break up the forward flow of the program.

3. Keep program segments short, probably 20 instructions or less. If not commented in some way, at least use a line beginning with "*" or ";" to break up the code into program segments for the reader.

4. Including lucid, meaningful comments is the best way to make your program clear. Either a comment line or a comment with an instruction is acceptable, but it is probably best to use complete lines of comments sparingly. Do not make the sophomore mistake of "garbaging up" your program with useless comments, such as repeating in the comment what simple instructions do.

5. The comments at the beginning of the subroutine should be clear enough that the body of the subroutine need not be read to understand how to use it. Always give a typical calling sequence if parameters are being passed on the stack, after the call or in a table. Indicate whether the data themselves are passed (call-by-value) or the address of the data is passed (call-by-name). In addition for short subroutines, indicate whether registers remain unchanged in a subroutine. Also explain what the labels mean if they are used as variables.

```
CMPY    #1170
BEQ     L1
BMI     L2
CMPB    S1
BEQ     L3
CMPB    S2
BEQ     L4
BRA     T1
```

Figure 5.8. Example Control Sequence Program

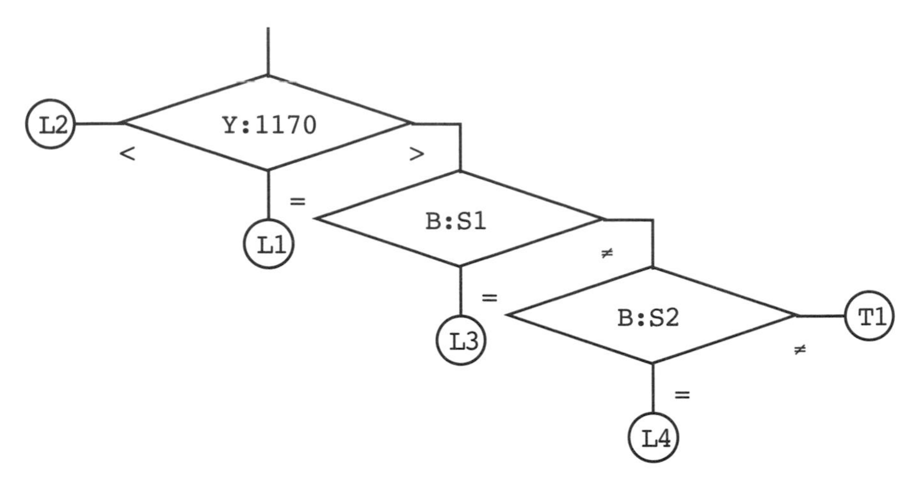

Figure 5.9. Flowchart for Example Control Sequence

Finally, one should consider attaching a flowchart with each subroutine as part of the documentation. Flowcharts are particularly useful for detailing complex control sequences. For example, the rather impenetrable sequence in Figure 5.8 becomes more fathomable when displayed as part of the subroutine flowchart shown in Figure 5.9.

5.6 Summary

In this chapter, we learned that a more powerful assembler can help you write much larger programs than you would be able to write by hand coding in machine code or by using a simplified assembler. Relocation, conditional assembly, and macro expansion permit the programmer to significantly expand his or her techniques. However, high-level languages provide alternative means to these and other ends, so that high-level language programming has superceded the more powerful assembly-language mechanisms described in this chapter.

Do You Know These Terms?

See the End of Chapter 1 for Instructions.

cross-assembler
host computer
target machine
downloader
monitor program
bootstrap
local area network
servers
print server
workstation
records
checksum
S-records
absolute loader
relocatable section
linker
parameter file
memory segment
absolute section
linking loader
conditional assembler
macro assembler
macro
library

PROBLEMS

1. Give the S1 and S9 records for the program in Figure 1.5.

2. Give the program source code (in the style of Figure 1.5) for the following S-record: `S10D0800FC0852FD0854137C08564E`.

3. Write a shortest program segment to translate an ASCII S1 record located in a 32-character vector SRECORD, to write its data into SRAM.

4. Write a parameter file for the 'B32. Its SRAM, EEPROM, and flash memory are to be the segments, with the same names; the sections are .data, .text, and .pgm; segment SRAM contains section .data; segment EEPROM contains section .text; and segment flash contains section .pgm. The starting address, named BEGIN, is to be put in $FFFE. The input file is to be progB32.o, and the output file is to be called progB32.abs.

5. Write a parameter file for the 'A4. Its SRAM and EEPROM are to be the segments, with the same names; the sections are .data and .text; segment SRAM contains section .data; and segment EEPROM contains section .text. The starting address, named START, is to be put in $FFFE. The input file is to be progA4.o, and the output file is to be called progA4.abs.

6. Write a parameter file for an expanded bus 'A4. Its internal SRAM, extended memory SRAM at $7000 to $7FFF, internal EEPROM at $4000 to $4FFF, and external ROM at $8000 to $FFFF, are to be the segments with the names ISRAM, ESRAM, EEPROM, and ROM; and the sections are .data , .edata, .text, and .pgm. Segment ISRAM contains section .data, segment ESRAM contains section .edata, segment EEPROM contains section .text, and segment ROM contains section .pgm. The starting address, named BEGIN, is to be put in $FFFE. The input files are to be camcorder1.o, camcorder2.o, camcorder3.o, and camcorder4.o, and the output file is to be called camcorder.abs.

7. Write a relocatable assembler program that uses fuzzy logic, that has a section .text that just calls fuzzy logic subroutines FUZZY and ADJUST one after another without arguments, and a section .pgm that has in it fuzzy logic subroutines FUZZY and ADJUST, which just have RTS instructions in them. Comment on the use of a relocatable assembler to break long programs into more manageable parts.

8. Write a relocatable assembler program that has a section .text that just calls subroutines OUTCH, OUTS, OUTDEC, and OUTHEX one after another. The argument in Accumulator A, for OUTCH, OUTDEC, and OUTHEX, is $41. The argument for OUTS, passed in index register X, is the address of string STRING1. A section .pgm has in it subroutines OUTCH, OUTS, OUTDEC, and OUTHEX, which just have RTS instructions in them, and the string STRING1, which is "Well done\r". Comment on the use of a relocatable assembler to break long programs into more manageable parts.

9. A program is to have symbolic names A4, B32, and MACHINE. Write the EQU directives to set A4 to 1, B32 to 2, and MACHINE to B32. Write a program segment that will store accumulator A to location $0 if MACHINE is B32 but will store accumulator A into location $1 if MACHINE is A4. Comment on the use of conditional expressions to handle programs that will be assembled to run in different environments.

10. A program is to have symbolic names LITE, FULL, and FEATURES. Write the EQU directives to set LITE to 1, B32 to 2, and FEATURES to FULL. Write a program segment that will execute a subroutine SUB if FEATURES is FULL. Comment on the use of conditional expressions to handle programs that will be assembled for different levels of features for different-priced markets.

11. A program is to have symbolic names TRUE, FALSE, and DEBUG. Write the EQU directives to set TRUE to 1, FALSE to 0, and DEBUG to TRUE. Write a program segment that will execute the 6812 background instruction if DEBUG is TRUE. Comment on the use of conditional expressions in debugging.

12. Write a shortest macro BITCOUNT that computes the number of bits equal to 1 in accumulator D, putting the number in accumulator B.

13. Write shortest macros for the following nonexistent 6812 instructions:

(a) NEGD (b) ADX (c) INCD (d) ASRD

14. Write a shortest macro NEGT that replaces the number stored at location A with its two's complement whose size is indicated by the size parameter appended to the macro name NEGT. This macro should change no registers. The call's form should be NEGT.B A, NEGT.W A, or NEGT.L A. Why can't NEG be the name of this macro?

15. Write a shortest macro INCR that increments the number stored at the address given as the macro's parameter A, whose size is indicated by the size parameter appended to the macro name INCR. The call's form should be INCR.B A, INCR.W A, or INCR.L A. No registers should be changed by the macro. Use the macro invocation counter rather than the location counter "*." Why can't we use the name INC for this macro?

16. Write a shortest macro MOVE, as in §5.4, so that no registers are changed except the CC register, which should be changed exactly like a load instruction.

17. Write a shortest macro XCHG that will exchange N bytes between locations L and M. No registers should be changed by the macro. The call should be of the form XCHG L,M,N. If N is missing, assume it is 2. A typical use would be

```
XCHG    L,M,N
```

18. Write a shortest macro for each of CLEAR, SET, and TEST, that will clear, set, and test the ith bit in the byte at location L. (Bits are labeled right to left in each byte beginning with 0 on the right.) For example, CLEAR L,5 will clear bit #5 in location L.

19. Write a shortest macro MARK that uses symbolic names given in Problem 11, which writes the characters in its parameter if DEBUG is TRUE. The macro expansion should put the character string argument inside the macro expansion and branch around it. A character is printed by calling subroutine OUTCH with the character in Accumulator A. Comment on the use of both conditional expressions and macros in debugging.

20. Write the assembly-language directives needed to print macro and conditional expansions in the listing. Consult HiWare documentation for the directive formats.

21. Write macros LISTON and LISTOFF that use a symbolic name LISTLEVEL, initially zero, as a count. LISTON increments this level, LISTOFF decrements this level, and if the level changes from 1 to 0, the assembler listing is turned off, while if the level changes from 0 to 1, the assembler listing is turned on.

Adapt912 is Technological Arts' version of Motorola's 912EVB evaluation board for the 68HC912B32 microcontroller. Offering the same modular hardware design as other Adapt12 products, Adapt912 includes Motorola's DBug-12 in on-chip Flash. This gives it the versatility to function as a standalone development system, a BDM Pod, or even a finished application (when combined with the user's circuitry on a companion Adapt12 PRO1 Prototyping card).

6

Assembly Language Subroutines

Subroutines are fantastic tools that will exercise your creativity. Have you ever wished you had an instruction that executed a floating-point multiply? The 6812 does not have such powerful instructions, but you can write a subroutine to execute the floating-point multiplication operation. The instruction that calls the subroutine now behaves pretty much like the instruction that you wish you had. Subroutines can call other subroutines as you build larger instructions out of simpler ones. In a sense, your final program is just a single instruction built out of simpler instructions. This idea leads to a methodology of writing programs called *top-down design*. Thus, creative new instructions are usually implemented as subroutines where the code is written only once. In fact, macros are commonly used just to call subroutines. In this chapter, we concentrate on the use of subroutines to implement larger instructions and to introduce programming methodologies.

To preview some of the ideas of this chapter, consider the following simple subroutine, which adds the contents of the X register to accumulator D.

```
SUB:    PSHX              ; Push copy of X onto stack
        ADDD   2,SP+      ; Add copy into D; pop copy off stack
        RTS
```

It can be called by the instruction

```
        BSR    SUB
```

Recall from Chapter 2 that the `BSR` instruction, besides branching to location `SUB`, pushes the return address onto the hardware stack, low byte first, while the instruction `RTS` at the end of the subroutine pulls the top two bytes of the stack into the program counter, high byte first. See Figure 6.1. In this figure, H:L denotes the return address and the contents of X is denoted XH:XL. Notice particularly that the instruction

```
        ADDD   2,SP+
```

in the subroutine above not only adds the copy of the contents of X into D but also pops the copy off the stack so that the return address will be pulled into the program counter by the `RTS` instruction.

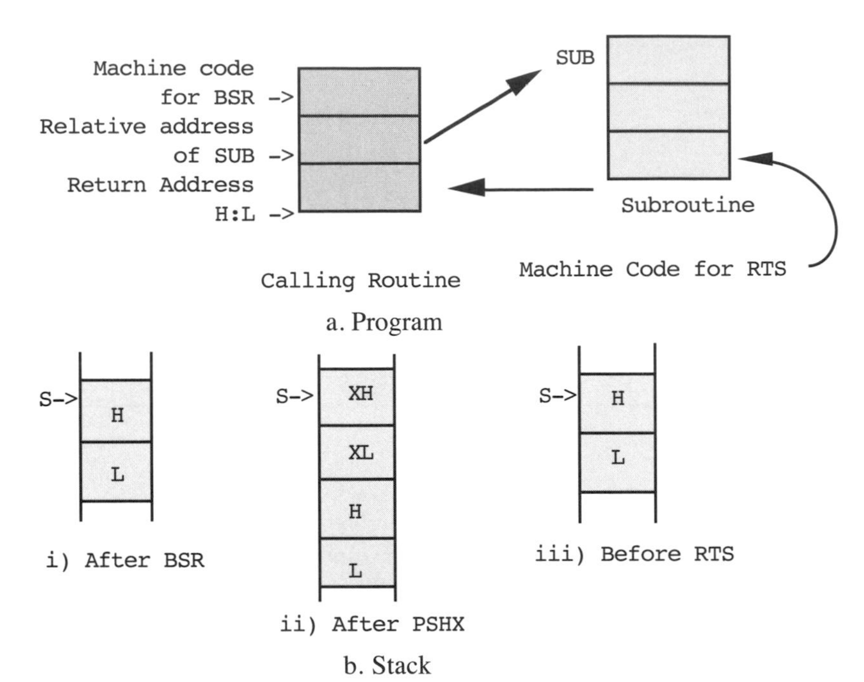

Figure 6.1. Subroutine Calling and Returning

The BSR and RTS instructions are *calling and returning mechanisms*, and the PSHX and ADDD instructions are the *program segment* in this subroutine. The X and D registers at the time the BSR is executed contain the *input parameters*, and the D register at the time that the RTS is executed contains the *output parameter*. We *pass* these parameters into the subroutine at the beginning of its execution and out of the subroutine at the end of its execution. The value pushed on the stack by the PSHX instruction becomes a *local variable* of this subroutine, a variable used only by this subroutine.

In this chapter, we discuss the mechanics of writing subroutines and, to a much lesser extent, the creative issue of what should be a subroutine and what should not. Echoing the theme of Chapter 1, we want to teach you how to correctly implement your good ideas, so you'll know how to use the subroutine as a tool to carry out these ideas.

This chapter is divided into sections that correspond to each capability that you need to write good subroutines. We first examine the storage of local variables. This discussion gives us an opportunity to become more familiar with the stack, so that the later sections are easier to present. We next discuss the passing of parameters and then consider calling by value, reference, and name. We then discuss the techniques for calling subroutines and returning from them. Finally, we present a few examples that tie together the various concepts that have been presented in the earlier sections and present some conclusions and recommendations for further reading.

Upon completion of this chapter, you should be able to pick the correct methods to call a subroutine, use local variables in it, and pass parameters to and from it. You should know how to test and document a subroutine and test the routine that calls the subroutine. With these capabilities, you should be ready to exercise your imagination creating your own subroutines for larger programs.

6.1 Local Variables

In this section we offer a mechanism that will help you write much clearer programs. The main contribution to clarity is modularization of a program into smaller segments. Think small, and the relationships within your program will be clearer. The problem with programs that are not so modularized is that a change to correct a bug in one part of the program may cause a bug in another part. If this propagation of errors explodes cancerously, your program becomes useless. The way to squelch this propagation is to divide the program into segments and to control the interaction between the segments with standard techniques. The main source of uncontrolled interaction is the storage of data for different segments. In this section, we introduce you to the tools to break up a program into segments and to store efficiently the data needed by these segments.

A *program segment* is a sequence of instructions that are stored one after another in memory as part of your program. An *entry point* of the segment is any instruction of the segment that can be executed after some instruction which is not in the segment. An *exit point* of the segment is any instruction that can be executed just before some instruction that is not in the segment. Figure 6.2a shows the flowchart of a program segment with multiple entry and exit points. For simplicity, however, you may think of a segment as having one entry point, which is the first instruction in the segment, and one exit point, which is the last instruction in the segment. See Figure 6.2b.

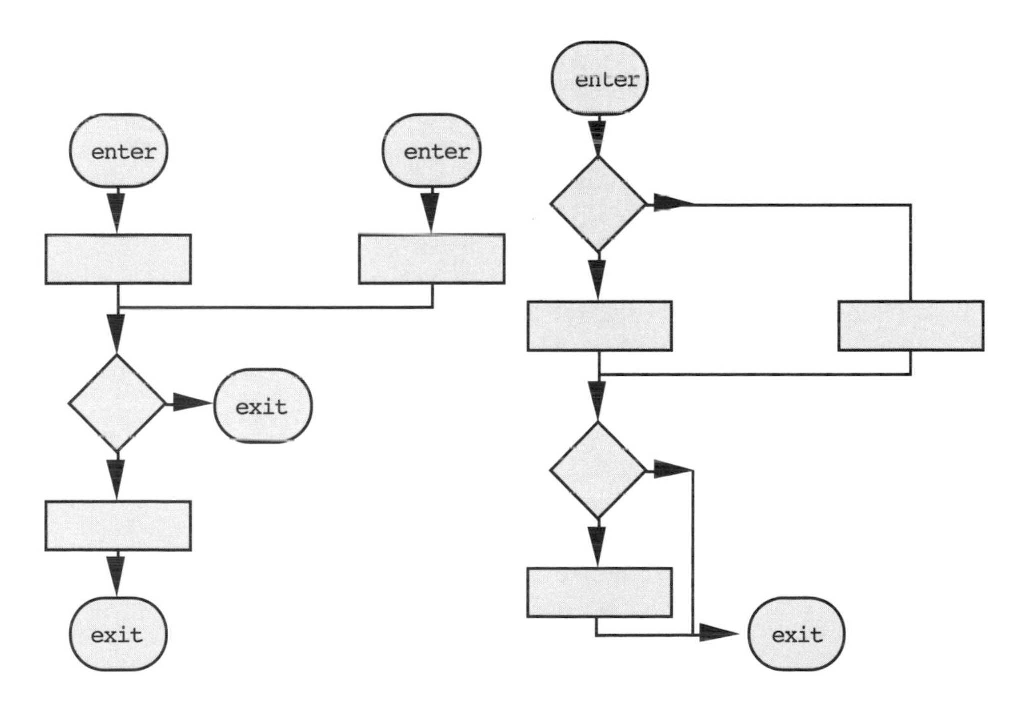

a. Multiple Entry and Multiple Exit b. Single Entry, Single Exit

Figure 6.2. Program Segment with a Plurality of Entry and Exit Points

For the purpose of this discussion, we assume that the program segment has information passed into it by its *input parameters* (or *input arguments*) and that the results of its computation are passed out through its *output parameters* (or *output arguments*). Any variables used in the segment for its computation but not used elsewhere, and which are not parameters are called *local variables*. The program segment can be thought of as a box with the parameters and variables that relate to it. Some of the parameters may be *global variables*—variables that are, or can be, used by all program segments. If you have learned about sequential machines, the input parameters are like the inputs to the sequential machine, the local variables are like the internal states of the machine, and the output parameters are like the outputs of the machine. As a sequential machine can describe a hardware module without the details about how the module is built, one can think about the input and output parameters and local variables of a program segment without knowing how it is written in assembly language. Indeed, one can think of them before the segment is written.

Consider the following example. Suppose that we have two vectors `V` and `W`, each having two l-byte elements (or components) `V(1),V(2)` and `W(1),W(2)`. We want to compute the inner product

```
V(1)*W(1) + V(2)*W(2)                                    (1)
```

which is often used in physics. Input parameters are the components of V and W, and the one output is the dot product result. The components of V and W are assumed to be l-byte unsigned numbers; for instance, assume that V1 is 1, V2 is 2, W1 is 3, and W2 is 4. The 2-byte dot product is placed in accumulator D at the end of the segment, but it becomes clear that we will have to store one of the products in expression (1) somewhere while we are computing the second product in accumulator D. This 2-byte term that we save is a local variable for the program segment we are considering. It may also be convenient to place copies of the vectors V and W somewhere for easy access during the calculation. These copies could also be considered local variables. We will continue this example below to show how local variables are stored in a program segment.

We first consider the lazy practice of saving local variables as if they were global variables, stored in memory and accessed with page-zero or direct addressing. One technique might use the same global location over and over again. Another technique might be to use differently named global variables for each local variable. While both techniques are undesirable, we will illustrate them in the examples below.

A single global variable, or a small number of global variables, might be used to hold all local variables. For example, assuming that the directive `TEMP DS 2` is in the program, one could use the two locations `TEMP` and `TEMP+1` to store one of the 2-byte local variables for the dot product segment. Figure 6.4 illustrates this practice. Six bytes are needed for various temporary storage. We first allocate six bytes using the declaration `TEMP DS 6`. Then the values of these variables are initialized by `MOVB` instructions. Next, these local variables are used in the program segment that calculates the inner product. The algorithm to compute the inner product is clearly illustrated by comments.

Using this approach can lead to the propagation of errors between segments discussed earlier. This can be seen by looking at the "coat hanger" diagram of Figure 6.3. A horizontal line represents a program segment, and a break in it represents a call to a subroutine. The diagonal lines represent the subroutine call and its return. Figure 6.3

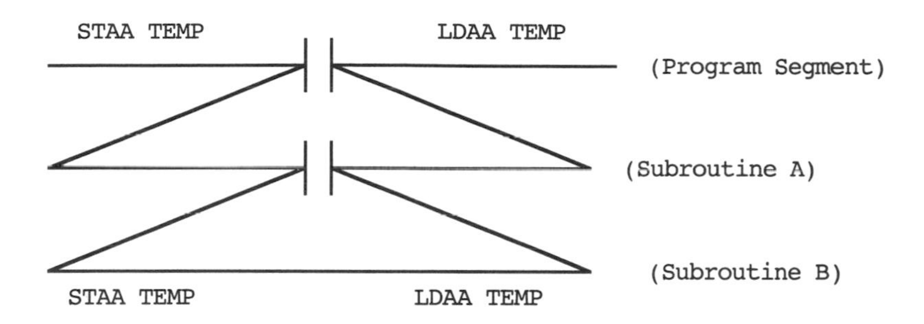

Figure 6.3. Changing a Global Variable before It Has Been Completely Used

illustrates a program segment using TEMP to store a variable to be recalled later. Before that value is recalled, however, TEMP has been changed by subroutine B, which is called by subroutine A, which itself is called by the program segment. This case is difficult to debug because each subroutine will work correctly when tested individually but will not work when one is called, either directly or indirectly through other subroutines, from within the other. This technique also confuses documentation, specifically the meaning of the local variable TEMP, generally making the program less clear.

With the other technique, the local variables will be put in different memory locations, having different symbolic names. See Figure 6.5. This approach is superior to the last approach, because differently named local variables, stored in different locations, will not interfere with the data stored in other locations. The names can be chosen to denote their meaning, reducing the need for comments. However, memory is taken up by these local variables of various program segments, even though they are hardly ever used. In a single-chip 'A4 or 'B32, only 1K bytes of SRAM are available. Using all these bytes for rarely used local variables leaves less room for the program's truly global data.

```
TEMP: DS    6               ; Allocate 6 bytes of memory for temporary variables
enter: MOVB#1,TEMP          ; Allocate and initialize V(1)
       MOVB #2,TEMP+1       ; Allocate and initialize V(2)
       MOVB #3,TEMP+2       ; Allocate and initialize W(1)
       MOVB #4,TEMP+3       ; Allocate and initialize W(2)
       LDAA TEMP            ; V(l) into A
       LDAB TEMP+2          ; W(1) into B
       MUL                  ; The value of first term is now in D
       STD  TEMP+4          ; Store first term in TERM
       LDAA TEMP+1          ; V(2) into A
       LDAB TEMP+3          ; W(2) into B
       MUL                  ; Calculate second term
       ADDD TEMP+4          ; Add in TERM; dot product is now in D
```

Figure 6.4. Inner Product Utilizing a Global Variable such as TEMP (a Bad Example)

```
V1:    DS     1       ; Allocate a byte of memory just for V1
V2:    DS     1       ; Allocate a byte of memory just for V2
W1:    DS     1       ; Allocate a byte of memory just for W1
W2:    DS     1       ; Allocate a byte of memory just for W2
TERM:  DS.W   1       ; Allocate two bytes of memory just for TERM
enter: MOVB   #1,V1   ; Allocate and initialize V(1)
       MOVB   #2,V2   ; Allocate and initialize V(2)
       MOVB   #3,W1   ; Allocate and initialize W(1)
       MOVB   #4,W2   ; Allocate and initialize W(2)
       LDAA   V1      ; V(l) into A
       LDAB   W1      ; W(1) into B
       MUL            ; The value of first term is now in D
       STD    TERM    ; Store first term in TERM
       LDAA   V2      ; V(2) into A
       LDAB   W2      ; W(2) into B
       MUL            ; Calculate second term
       ADDD   TERM    ; Add in TERM; dot product is now in D
```

Figure 6.5. Inner Product Utilizing Different Global Variables (a Bad Example)

Rather than storing a subroutine's local variables in global variables, put them in either registers or the hardware stack. Figure 6.6 illustrates how registers can be used to store local variables; this is basically what we did throughout the previous chapters. Because the 6812 has few registers, the stack holds most local variables, which will be called *stacked local variables*. We review relevant stack index addressing and stack-oriented instructions presented in Chapters 2 and 3.

Recall from Chapter 3 that index addressing using the stack pointer can access data in the stack and can push or pull data from the stack. We will use a general and simple rule for balancing the stack so that the segment will be balanced. Push all the stacked local variables of the segment on the stack at the entry point, and pull all of the local variables off the stack at the exit point. Do not push or pull words from the stack anywhere else in the program segment except for two- or three-line segments used to implement missing instructions. While an experienced programmer can easily handle exceptions to this rule, this rule is quite sound and general, so we recommend it to you. In following sections, our program segments will be balanced unless otherwise noted and we will usually follow this rule, only occasionally keeping local variables in registers.

```
LDAA   #1     ; V(1) into A
LDX    #3     ; V(2) into low byte of X
LDAB   #2     ; W(1) into B
LDY    #4     ; W(2) into low byte of Y
MUL           ; First term is now in D
EXG    D,Y    ; Store first term in Y, get W(2) in B
EXG    A,X    ; V(2) into A
MUL           ; Calculate second term
LEAY   D,Y    ; Add terms, to get result in Y
```

Figure 6.6. Inner Product Utilizing Registers

```
enter:  MOVB   #4,1,-SP   ; Allocate and initialize W(2)
        MOVB   #3,1,-SP   ; Allocate and initialize W(1)
        MOVB   #2,1,-SP   ; Allocate and initialize V(2)
        MOVB   #1,1,-SP   ; Allocate and initialize V(1)
        LEAS   -2,SP      ; Allocate room for term
        LDAA   2,SP       ; V(l) into A
        LDAB   4,SP       ; W(1) into B
        MUL               ; The value of first term is now in D
        STD    0,SP       ; Store first term in TERM
        LDAA   3,SP       ; V(2) into A
        LDAB   5,SP       ; W(2) into B
        MUL               ; Calculate second term
        ADDD   0,SP       ; Add in TERM; dot product is now in D
        LEAS   6,SP       ; Deallocate local variables; balance stack
```

Figure 6.7. Inner Product Program Segment Utilizing Local Variables on the Stack

We now look more closely at this rule to show how local variables can be bound, allocated, deallocated, and accessed using the stack. *Binding* means assigning an address (the actual address used to store the variable) to a symbolic name for a variable. *Allocation* of a variable means making room for that variable in memory, and *deallocation* means removing the room for that variable. *Accessing* is the process of finding that variable which, for stacked local variables, will be on the stack. An *input parameter* supplies a value to be used by the program segment. While input parameters are usually not changed by the program segment, local variables and output parameters that are not also input parameters generally need to be *initialized* in the program segment. That is, before any instruction reads a value from them, a known value must be written in them. A stacked local variable or output parameter is usually initialized immediately after the entry point of the program segment.

Stacked local variables are bound in two steps. Before the program is run, the symbolic address is converted to a number that is an offset that is used in index addressing with the stack pointer. The binding is completed when the program is running. There, the value in the stack pointer SP, at the time the instruction is executed, is added to the offset to calculate the actual location of the variable, completing the binding of the symbolic address to the actual address used for the variable. This two-step binding is the key to reentrant and recursive subroutines, as discussed in Chapter 3.

We will allocate stacked local variables in a balanced program segment using `LEAS` with negative offset and instructions like `PSHX`, which make room for the stacked local variables. We will deallocate the variables using `LEAS` with positive offset and instructions like `PULX`, which remove the room for the local variables. We will bind and access stacked local variables in a couple of ways to illustrate some alternative techniques that you may find useful. First, to access local variables, we use explicit offsets from the stack pointer, reminiscent of the programming that you did in the first three chapters. Then we use `EQU`, `ORG`, and `DS` directives to bind local variable names to offsets for the stack pointer. This allows symbolic names to be used to access these local variables, taking advantage of the assembler to make our program clearer.

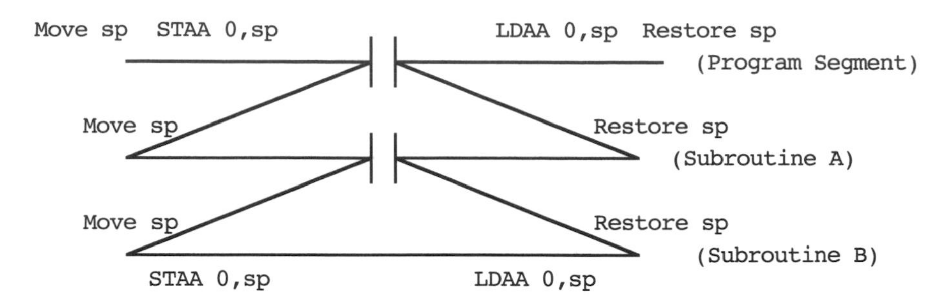

Figure 6.8. Nested Subroutines Using Local Variables Stored on the Stack

Let's now look at our dot product example in Figure 6.7, where we will initialize the copies of V(1), V(2), W(1), and W(2) to have values 1, 2, 3 and 4, respectively. The first term of the dot product shown in formula (1), which will also be placed on the stack, will be denoted `TERM`. Notice how the simple rule for balancing the stack is used in this segment. If the stack pointer were changed in the interior of the segment, offsets for local variables would change, making it difficult to keep track of them. As it is now, we have to determine the offsets from the stack pointer for each local variable. The local variable `TERM` occupies the top two bytes, the local variables V(1) and V(2) occupy the next two bytes, and the local variables W(1) and W(2) occupy the next two bytes.

Figure 6.8 illustrates how the use of the stack avoids the aforementioned problem with global variables. Because the stack pointer is moved to allocate room for local variables, the temporary variables for the outermost program are stored in memory locations different from those that store local variables of an inner subroutine like B.

The advantage of using the stack can be seen when two subroutines are called one after another, as illustrated in Figure 6.9. The first subroutine moves the stack pointer to allocate room for the local variable, and the local variable is stored with an offset of 0 in that room. Upon completion of this subroutine, the stack pointer is restored, deallocating stacked local variables. The second subroutine moves the stack pointer to allocate room for its local variable, and the local variable is stored with an offset of 0 in that room. Upon completion of this subroutine, the stack pointer is restored, deallocating stacked local variables. Note that the same physical memory words are used for local variables in the first subroutine that was called, as are used for local variables in the second subroutine that was called. However, if the second subroutine were called from within the first subroutine, as in Figure 6.8, the stack pointer would have been moved, so that the second subroutine would not erase the data used by the first subroutine. Using the stack for local variables conserves SRAM utilization and prevents accidental erasure of local variables.

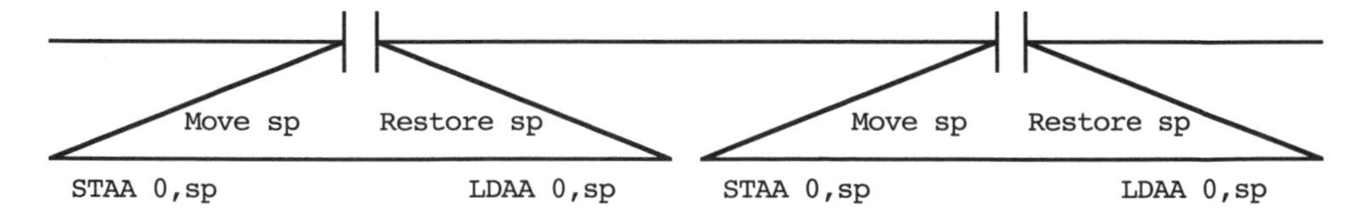

Figure 6.9. Local Variables Stored on the Stack, for Successive Subroutines

```
TERM:   EQU    0
V1:     EQU    2
V2:     EQU    3
W1:     EQU    4
W2:     EQU    5
NBYTES: EQU    6
        LEAS   -NBYTES,SP   ; Allocate all local variables
        MOVB   #1,V1,SP     ; Initialize V(l)
        MOVB   #2,V2,SP     ; Initialize V(2)
        MOVB   #3,W1,SP     ; Initialize W(l)
        MOVB   #4,W2,SP     ; Initialize W(2)
        LDAA   V1,SP        ; V1 into A
        LDAB   W1,SP        ; W1 into B
        MUL                 ; First term is now in D
        STD    TERM,SP      ; Store first term in TERM
        LDAA   V2,SP        ; V2 into A
        LDAB   W2,SP        ; W2 into B
        MUL                 ; Calculate second term
        ADDD   TERM,SP      ; Add in TERM; dot product in D
        LEAS   NBYTES,SP    ; Deallocate locals; balance stack
```

Figure 6.10. Using Symbolic Names for Stacked Local Variables

A problem with the stack approach is that remembering that a variable is at `2,SP` (or is it at `5,SP`?) is error prone, compared to giving names to variables (see Figure 6.7). We need to give names to the relative locations on the stack; that is, we need to bind the variables. We discuss two ways to bind names to locations on the stack: the `EQU` and `DS` assembler directives. See Figure 6.10.

Since `TERM` is on top of the stack, the `EQU` assembler directive can bind the value of the name `TERM` to the location `0,SP;` then `TERM,SP` can access it. Note that you bind the name of the container `TERM` to `0,SP` and not the contents of `TERM` to `0` with the `EQU` directive. That is, we use 0 wherever we see `TERM,` and we use it in calculating the address with `TERM` as an offset in index addressing with the SP register. Similarly, we can bind `V1` to `2,SP` and `W2` to `5,SP`. Also, the offsets used in the `LEAS` instructions to allocate and deallocate the local variables can be set when they are defined. Initialization is changed a bit to make `NBYTES` easier to use, but the effect is the same. The program segment to calculate the dot product can now be rewritten as in Figure 6.10. (We write the changed parts in boldface, here and in later examples, to focus your attention on them.) Note that each statement is quite readable without comments. However, good comments are generally still valuable.

As described above, this technique requires the programmer to calculate the values for the various labels and calculate the value for `NBYTES.` Adding or deleting a variable requires new calculations of these values. Figure 6.11 shows how this can be avoided. With this use of the `EQU` directive, each new stacked local variable is defined in terms of the local variable just previously defined plus the number of bytes for that local variable. Insertions or deletions of a stacked local variable in a segment now requires changing only two lines, a convenience if the number of local variables gets large.

```
TERM:   EQU    0
V1:     EQU    TERM+2
V2:     EQU    V1+1
W1:     EQU    V2+1
W2:     EQU    W1+1
NBYTES: EQU    W2+1
```

Figure 6.11. Defining Symbolic Names for Stacked Local Variables by Sizes

Another technique, shown in Figure 6.12, uses the `DS` directive to play a trick on the assembler. The technique uses the `DS` directive to bind the stacked local variables partially with the stack pointer `SP`, using the location counter and the `ORG` directive to modify the location counter. Recall that `ALPHA DS 2` will normally allocate two bytes for the variable `ALPHA`. The location counter is used to bind addresses to labels like `ALPHA` as the assembler generates machine code. The location counter and hence the address bound to the label `ALPHA` correspond to the memory location where the word associated with the label `ALPHA` is to be put. A `DS` statement, with a label, binds the current value of the location counter to the label (as the name of the container, not the contents) and adds the number in the `DS` statement to the location counter. This will bind a higher address to the next label, allocating the desired number of words to the label of the current directive. Note that `ALPHA EQU *` will bind the current location counter to the label `ALPHA` but not affect the location counter. Also, recall that the `ORG` directive can set the location counter to any value. These can be used as shown in Figure 6.12.

You can reset the location counter to zero many times, and you should do this before each group of `DS` directives that are used to define local storage for each program segment. These `DS` statements should appear first in your program segment. Each set should be preceded by a directive such as `LCSAVE DS 0` to save the location counter using `LCSAVE` and an `ORG 0` directive to set the location counter to 0; and each set should be followed by a directive such as `ORG LCSAVE` to set the origin back to the saved value to begin generating machine code for your program segment. The last directive in Figure 6.12, `ORG LCSAVE`, can be replaced by `DS LCSAVE-*`, which avoids the use of the `ORG` statement. The `DS` directive adds its operand `LCSAVE-*` to the location counter, so this directive loads `LCSAVE` into the location counter.

```
LCSAVE: EQU *          ; Save current location counter
        ORG 0          ; Set the location counter to zero
TERM:   DS  2          ; First term of dot product
V1:     DS  1          ; Copy of input vector element V(1)
V2:     DS  1          ; Copy of input vector element V(2)
W1:     DS  1          ; Copy of input vector element W(1)
W2:     DS  1          ; Copy of input vector element W(2)
NBYTES: EQU *          ; Number of bytes of local variables
        ORG LCSAVE     ; Restore location counter
```

Figure 6.12. Declaring Symbolic Names for Local Variables Using DS Directives

```
VV:     EQU  0                        ; Input vector V(1),V(2)
WW:     EQU  2                        ; Input vector W(1),W(2)
SIZEA:  EQU  4
*
STARTA: LEAS -SIZEA,SP                ; Start of segment A
        MOVW #$102,VV,SP              ; Initialize both bytes of VV
        MOVW #$304,WW,SP              ; Initialize both bytes of WW
*
TERM:   EQU  0
SIZEB:  EQU  2
*
STARTB: LEAS -SIZEB,SP                ; Start of segment B
        LDAA VV+SIZEB,SP              ; V(1) into A
        LDAB WW+SIZEB,SP              ; W(1) into B
        MUL                           ; First term is now in D
        STD  TERM,SP                  ; Store first term in TERM
        LDAA VV+1+SIZEB,SP            ; V(2) into A
        LDAB WW+1+S1ZEB,SP            ; W(2) into B
        MUL                           ; Calculate second term
        ADDD TERM,SP                  ; Add in TERM; dot product in D
ENDB:   LEAS SIZEB,SP                 ; End of segment B; balance stack
*
ENDA:   LEAS SIZEA,SP                 ; End of segment A; balance stack
```

Figure 6.13. Declaring Symbolic Names for Extended Local Access

It is easy to insert or delete variables without making mistakes, using this technique, because the same line has the variable name and its length, as contrasted with the last technique using the EQU directive. The number of bytes needed to store the local variables is also automatically calculated with this technique. You do, however, have to write three more lines of code with this technique. You have to invent some kind of convention in naming variables to avoid this problem, but that is not too difficult to do.

In Chapter 3, we introduced nested local variables. Suppose that a program segment B is nested in, that is, entirely contained in, a program segment A. An instruction inside B may need to access a local variable that is allocated and bound for all of segment A. There are two techniques that can be used to access the variable in B that is defined for A. These are the extended local access and stack marker access techniques described below.

The idea of the *extended local access* technique assumes that there is a way to fix the location of the desired variable over one or more allocations of stacked local variables. See Figure 6.13. In this version, the outer segment A copies vectors into the stack where the inner segment B calculates the dot product. The dot product is placed in D by segment B and left there by A.

The stack marker technique uses an index register to provide a reference to local variables of outer segments. See Figure 6.14. Just before a program segment allocates its variables, the old value of the stack pointer is transferred to a register. Just after the local variables are allocated, the value in this register is put into a stacked local variable for the inner segment. It is called a *stack marker* because it marks the location of the stack that was used for the local variables of the outer program segment. It is always in a known position on the stack (in this case, on the very top of the stack), so it is easy to find. See Figure 6.14, where the inner program segment can access the local variables of the outer segment by loading the stack marker into any index register and using index addressing to get the variable. Note that the stack marker is deallocated together with the other stacked local variables at the end of the program segment.

```
MARKA:   EQU  0                ; Stack mark for segment A
VV:      EQU  2                ; Input vector
WW:      EQU  4                ; Input vector
SIZEA:   EQU  6
*
STARTA:  TFR  SP,X             ; Start for segment A
         LEAS -SIZEA,SP
         STX  MARKA,SP
         MOVW #$102,VV,SP      ; Initialize both bytes of VV
         MOVW #$304,WW,SP      ; Initialize both bytes of WW
*
MARKB:   EQU  0                ; Stack mark for segment B
TERM:    EQU  2
SIZEB:   EQU  4
*
STARTB:  TFR  SP,X
         LEAS -SIZEB,SP
         STX  MARKB,SP
         LDAA VV,X             ; V(1) into A
         LDAB WW,X             ; W(1) into B
         MUL                   ; First term is now in D
         STD  TERM,SP          ; Store first term in TERM
         LDAA VV+1,X           ; V(2) into A
         LDAB WW+1,X           ; W(2) into B
         MUL                   ; Calculate second term
         ADDD TERM,SP          ; Add in TERM; dot product in D
ENDB:    LEAS SIZEB,SP         ; End of segment B
*
ENDA:    LEAS SIZEA,SP         ; End of segment A
```

Figure 6.14. Accessing Stacked Local Variables Using a Stack Marker

Either the extended local access or the stack marker access mechanisms can be used in cases where program segments are further nested. Consider program segment C, with SIZEC stacked local variables, which is nested in segment B and needs to load accumulator A with the value of SA, a stacked local variable of segment A. Using extended local access, as in the first example, the following instruction will accomplish the access.

```
         LDAA    SIZEC+SIZEB+SA,SP
```

Using the stack marker of the second example, the following instructions will access the variable.

```
LDX     0,SP      ; Get to segment B
LDX     0,X       ; Get to segment A
LDAA    SA,X      ; Access local variable SA
```

The extended local access mechanism appears to be a bit simpler for smaller assembly language programs because it takes fewer instructions or directives. The stack marker mechanism seems to be used in some compilers because the compiler program has less to "remember" using this approach than using the other, where the compiler has to keep track of the sizes of each allocation of stacked local variables, particularly if a subroutine is called by many program segments that have different numbers of local variables. It is not unreasonable to expect a large program to have 20 levels of nesting. Because the stack marker to the next outer segment is always on top of the stack, the compiler does not have to remember where it is. In fact, the labels for stack markers are not really necessary for access at all, their only real use being for allocation and deallocation of the local variables. With two good mechanisms, you will find it easy to use one of them to handle nested program segments to many levels.

This section introduced the idea of a local variable and the techniques for storing local variables on the 6812. We demonstrated that local variables should not be stored in global variables, whether using the same name (`TEMP`) for each or giving each variable a unique name. Local variables can be stored on the stack in any program segment. They are especially easy to use in the 6812, because the `LEAS` instructions are able to allocate and deallocate them, the index addressing mode using the stack pointer is useful in accessing them, and the `EQU` or `DS` directives are very useful in binding the symbolic names to offsets to the stack pointer. These techniques can be used within subroutines, as we discuss in the remainder of this chapter. They can also be used with program segments that are within macros or those that are written as part of a larger program. The nested program segments can be readily handled too, using either the extended local access or stack marker technique to access local variables of outer program segments.

6.2 Passing Parameters

We now examine how parameters are passed between the subroutine and the *calling routine*. We do this because an assembly-language programmer will have frequent occasions to use subroutines written by others. These subroutines may come from other programmers that are part of a large programming project, or they may be subroutines that are taken from already documented software, such as assembly-language subroutines from a C support package. They may also come from a collection of subroutines supplied by the manufacturer in a user's library. In any case, it is necessary to understand the different ways in which parameters are passed to subroutines, if only to be able to use the subroutines correctly in your own programs or perhaps modify them for your own specific applications.

Before we begin, however, we reiterate that these techniques are quite similar to those used in Section 6.1 to store local variables. However, these techniques are used between subroutines, while the latter were used entirely within a subroutine.

```
* SUBROUTINE DOT PRODUCT
DOTPRD:  MUL                  ; First term is now in D
         EXG    D,Y           ; Store first term in Y, get W(2) in B
         EXG    A,X           ; V(2) into A
         MUL                  ; Calculate second term
         LEAY   D,Y           ; Add terms, to get result in Y
         RTS                  ; Return to the calling program
```

a. A subroutine

```
         LDAA   #2            ; Copy of V(1) into A
         LDX    #7            ; Copy of V(2) into low byte of X
         LDAB   #6            ; Copy of W(1) into B
         LDY    #3            ; Copy of W(2) into low byte of Y
         BSR    DOTPRD        ; Call the subroutine
         STY    DTPD          ; Store dot product in DTPD
```

b. A calling sequence

```
         LDAA   LV,SP         ; Copy of V(1) into A
         LDX    LW-1,SP       ; Copy of V(2) into low byte of X
         LDAB   LV+1,SP       ; Copy of W(1) into B
         LDY    LW,SP         ; Copy of W(2) into low byte of Y
         BSR    DOTPRD        ; Call the subroutine
         STY    DTPD          ; Store dot product in DTPD
```

c. Another calling sequence

Figure 6.15. A Subroutine with Parameters in Registers

In this section we examine six methods used to pass parameters to a subroutine. We illustrate each method with the dot product from Section 6.1. We first consider the simplest method, which is to pass parameters in registers as we did in our earlier examples. Then the passing of parameters by global variables is discussed and discouraged. We then consider passing parameters on the stack and after the call, which are the most common methods used by high-level languages. We then discuss the technique of passing parameters using a table, which is widely used in operating system subroutines.

The first method is that of passing parameters through registers, which is preferred when the number of parameters is small. We will also use this method to illustrate the idea of a calling sequence. See Figure 6.15. Suppose that the calling routine, the program segment calling the subroutine, puts a copy of the vector V into registers A and X and a copy of the vector W into registers B and Y, where, as before, the low byte of a 16-bit register contains the 8-bit element. Both components of each vector are 1-byte unsigned numbers. Assume that the dot product is returned in accumulator D, a subroutine `DOTPRD` that performs the calculation is shown in Figure 6.15a. The instructions in the calling routine shown in Figure 6.15b cause the subroutine to be executed for vector V equal to (2, 6) and vector W equal to (7, 3). Notice that the values in A and B have been changed by the subroutine, because an output parameter is being returned in D. The sequence of instructions in the calling routine that handles the placement of the input and output parameters is termed the *calling sequence*. In our calling sequence we have, for convenience, assumed that constant input parameters are given to `DOTPRD` while the output parameter in D is copied into the global variable `DTPD`. These constants and the global variable could just as easily have been stacked local variables for the calling routine.

To emphasize that a calling sequence is in no way unique, suppose that the calling routine has vectors that are pairs of 8-bit local variables on the stack, labeled `LV` and `LW`, as offsets to the stack pointer. To compute the dot product of `LV` and `LW`, execute the calling sequence in Figure 6.15c.

We have, for simplicity, omitted the binding, allocation, and deallocation of the local variables of the calling routine. The point of this second example is to stress that any calling sequence for the subroutine `DOTPRD` must load copies of the vectors for which it wants the dot product into X and Y and then call `DOTPRD`. It must then get the dot product from D to do whatever it needs to do with it. From a different point of view, if you were to write your own version of `DOTPRD`, but one that passed parameters in exactly the same way, your version could not directly access the global variable `LV` used in the calling sequence in Figure 6.15. If it did, it would not work for the calling sequence in Figure 6.14c.

Parameters are passed through registers for most small subroutines. The main limitation with this method of passing parameters is that there are only two 16-bit registers (you do not pass parameters in the stack pointer SP register itself), two 8-bit registers, and a few condition code bits. Although this limits the ability of the 6812 to pass parameters through registers, you will, nevertheless, find that many, if not most, of your smaller subroutines will use this simple technique.

The next technique we discuss is that of passing parameters through global variables. We include it because it is used in small microcomputers like the 6805, but we discourage you from using it in larger machines like the 6812. It is easy to make mistakes with this technique, so much so that most experienced programmers avoid this method of passing parameters when other techniques are available. Figure 6.16 shows a coat hanger diagram that illustrates how incorrect results can occur when parameters are passed with global variables. Notice in particular how subroutine B writes over the value of the global variable passed by the calling routine to subroutine A, so when subroutine A performs the load instruction, it may not have the calling routine's value.

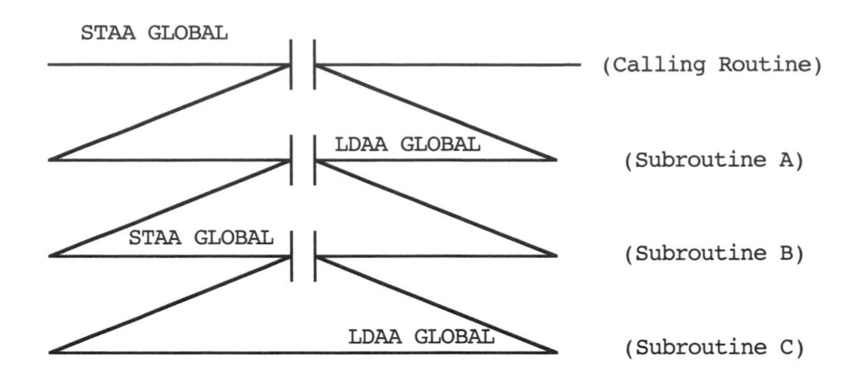

Figure 6.16. Change a Global Parameter before Its Subroutine Has Used It

```
*         SUBROUTINE DOTPD – LOCAL VARIABLES
TERM:     EQU     0              ; First term of the dot product
NBYTES:   EQU     2
*
DOTPRD:   LEAS    -NBYTES,SP     ; Allocate local variables
          LDAA    V1             ; First component of V into A
          LDAB    W1             ; First component of W into B
          MUL                    ; First term of dot product into D
          STD     TERM,SP        ; Save first term
          LDAA    V2             ; Second component of V into A
          LDAB    W2             ; Second component of W into B
          MUL                    ; Second term of dot product into D
          ADDD    2,SP+          ; Dot product into D, Deallocate loc var
          STD     DTPD           ; Place dot product
          RTS
```

Figure 6.17. A Subroutine with Parameters in Global Variables

In assembly language, global variables are defined through a `DS` directive that is usually written at the beginning of the program. These variables are often stored on page zero on smaller microcontrollers so that direct page addressing may be used to access them. However in the 6812, page zero is used for I/O ports. Assuming that the directives are written somewhere in the program, the subroutine in Figure 6.17 does the previous calculation, passing the parameters through these locations. Note that we use local variables in this subroutine, as discussed in Section 6.1.

The subroutine in Figure 6.17 uses global variables `V1`, `V2`, `W1`, `W2`, and `DTPD` to pass parameters to the subroutine and from it. If the calling routine wants to compute the dot product of its local variables `LV` and `LW`, which each store a pair of 2-element 1-byte vectors, putting the result in `LDP`, the calling sequence in Figure 6.18 could be used. Notice that the calling routine's local variables are copied into global variables `V1`, `V2`, `W1`, and `W2` before execution and copied out of the global variable `DTPD` after execution. Any other calling sequence for this version of `DOTPRD` must also copy the vectors of which it wants to compute the dot product, call the subroutine, and get the dot product result from DTPD. Note also that

```
ADDD   2,SP+  ; Dot product into D, also deallocate local variable
```

rendered the last LEAS instruction of the subroutine unnecessary.

```
        MOVB   LV,SP,V1     ; Copy V(1)
        MOVB   LV+1,SP,V2   ; Copy V(2)
        MOVB   LW,SP,W1     ; Copy W(1)
        MOVB   LW+1,SP,W2   ; Copy W(2)
        BSR    DOTPRD
        MOVW   DTPD,LDP,SP  ; Place result in local variable LDP
```

Figure 6.18. Calling a Subroutine for Figure 6.17

```
aLOCV:  EQU    0                 ; Input parameter copy of the vector V
aLOCW:  EQU    2                 ; Input parameter copy of the vector W
aLOCDP: EQU    4                 ; Output parameter copy of dot product
PSIZE:  EQU    6                 ; Number of bytes for parameters
*
        LEAS   -PSIZE,SP         ; Allocate space for parameters
        MOVW   V,aLOCV,SP        ; Initialize parameter LOCV,SP
        MOVW   W,aLOCW,SP        ; Initialize parameter LOCW,SP
        BSR    DOTPRD
        MOVW   aLOCDP,SP,DTPD    ; Place output in global variable
        LEAS   PSIZE,SP          ; Deallocate space for parameters
```

Figure 6.19. Calling a Subroutine with Parameters on the Stack for Figure 6.21

We now consider a very general and powerful method of passing parameters on the stack. We illustrate the main idea, interpreting it as another use of local variables, as well as the technique that makes and erases "holes" in the stack, and we consider variations of this technique that are useful for very small computers and for larger microcontrollers like the 68332.

Input and output parameters can be passed as if they were local variables of the program segment that consists of the calling sequence that allocates and initializes. The local variables are allocated and initialized around the subroutine call. In this mode the parameters are put on the stack before the BSR or JSR. For our particular dot product example, the calling sequence might look like Figure 6.19.

For simplicity, we have assumed that input parameter values come from global variables V and W, and the output parameter is placed in the global variable DTPD. All of these global variables could, however, just as well have been local variables of the calling routine. The idea is exactly the same. The stack is as shown in Figure 6.20 as execution progresses. The dot product subroutine is now as shown in Figure 6.21.

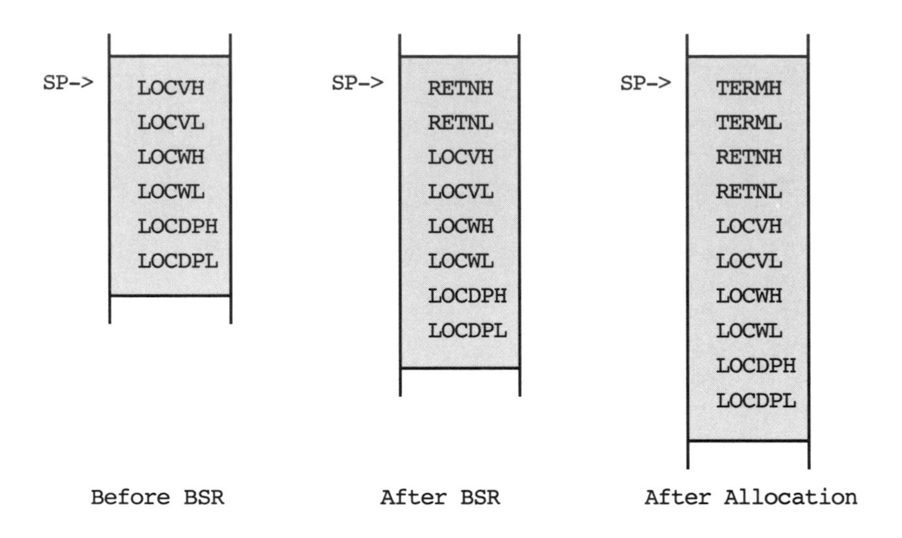

Figure 6.20. Location of Parameters Passed on the Hardware Stack

```
* SUBROUTINE DOTPRD – LOCAL VARIABLES
TERM:    EQU    0           ; First term of the dot product
NBYTES:  EQU    2
* PARAMETERS
RETN:    EQU    2           ; Return address
LOCV:    EQU    4
LOCW:    EQU    6
LOCDP:   EQU    8
DOTPRD:  LEAS   -NBYTES,SP  ; Allocation for local variables
         LDAA   LOCV,SP
         LDAB   LOCW,SP
         MUL
         STD    TERM,SP     ; Copy first term to local variables
         LDAA   LOCV+1,SP
         LDAB   LOCW+1,SP
         MUL
         ADDD   TERM,SP     ; Dot product into D
         STD    LOCDP,SP    ; Place dot product in output parameter
         LEAS   NBYTES,SP   ; Deallocate local variables
         RTS
```

Figure 6.21. A Subroutine with Parameters on the Stack

```
* SUBROUTINE DOTPRD – LOCAL VARIABLES
TERM:     EQU    0                ; First term of the dot product
NBYTES:   EQU    2
* PARAMETERS
LOCV:     EQU    0
LOCW:     EQU    2
LOCDP:    EQU    4
PSIZE:    EQU    6
*
DOTPRD:   LEAS   -NBYTES,SP ; Allocation for local variables
          LDAA   LOCV+NBYTES+2,SP
          LDAB   LOCW+NBYTES+2,SP
          MUL
          STD    TERM,SP  ; First term to local variables
          LDAA   LOCV+NBYTES+2+1,SP
          LDAB   LOCW+NBYTES+2+1,SP
          MUL
          ADDD   TERM,SP       ; Dot product into D
          STD    LOCDP+NBYTES+2,SP  ; Dot product to output parameter
          LEAS   NBYTES,SP     ; Deallocate local variables
          RTS
```

Figure 6.22. Revised Subroutine with Local Variables and Parameters on the Stack

```
          LEAS   -PSIZE,SP     ; Allocate space for parameters
          MOVW   V,LOCV,SP     ; Put copy in parameter location
          MOVW   W,LOCW,SP     ; Put copy in parameter location
          BSR    DOTPRD
          MOVW   LOCDP,SP,DTPD  ; Put in global location
          LEAS   PSIZE,SP      ; Deallocate space for parameters
```

Figure 6.23. Calling a Subroutine with Parameters on the Stack for Figure 6.22

Notice several things about the way this version of the subroutine is written. We do not need the local variables to hold copies of vectors V and W as we did in the earlier versions because the copies are already on the stack as parameters where we can access them using the extended local access technique described in Section 6.1. Because the number of parameters and local variables is small and because each is equal to two bytes, we can easily calculate the stack offsets ourselves, particularly if we use a dummy parameter `RETN` for the return address. Notice particularly how we have redefined the labels `LOCV`, `LOCW`, and `LOCDP` in the subroutine with the `EQU` directive to avoid adding an additional offset of 4 to each parameter to account for the number of bytes in the return address and the local variables. Suppose now that we write the subroutine as shown in Figure 6.22. When `EQU` is used in this way, the additional offset of `NBYTES+2` is needed to access the parameters to account for the local variables and the return address. No `EQU` directives are needed in the calling sequence, however, because `EQU` is a global definition; that is, the labels `LOCV`, `LOCW`, `LOCDP`, and `PSIZE` are fixed, respectively, at 0, 2, 4, and 6 throughout the program. The calling sequence for this case is shown in Figure 6.23.

Putting the additional offset of NBYTES+2 in the subroutine, which is written only once, makes the calling sequence, which may be written many times, more straightforward. However, keeping the EQU directives with the subroutine as shown will force 2-byte offsets for parameter accesses in all of the calling sequences placed before the subroutine. For this reason, the EQU tables for subroutines might be placed at the beginning of a program to improve the static efficiency of the calling sequences. One could also force 1-byte offsets by using "<" before the expressions that access the parameters, as in LDAA <LOCW+NBYTES+2,SP. This, however, would still not get the 5-bit offset for those local variables that would be accessed with offset expressions in the range from zero to fifteen.

The reader should recognize that in this method of passing parameters, the calling sequence is just another instance of a balanced program segment, with local variables that are the parameters to the subroutine. Variables are copied from the calling routine to the parameter locations and back so that the subroutine will have a precise place to find them. Compare this to the earlier example where parameters were passed by global variables, and review the discussion after that example.

There is another way to think of this technique, which some students have found to be more concrete. You can think of "holes in the stack." To pass an argument out of the subroutine, such as LOCDP, you create room for it in the calling routine. The instruction LEAS -PSIZE,SP creates a hole for this output parameter, among other things, and the subroutine puts some data in this hole. Conversely, the input parameters LOCV and LOCW are used up in the subroutine and leave holes after the subroutine is completed. These holes are erased by the instruction LEAS PSIZE,SP.

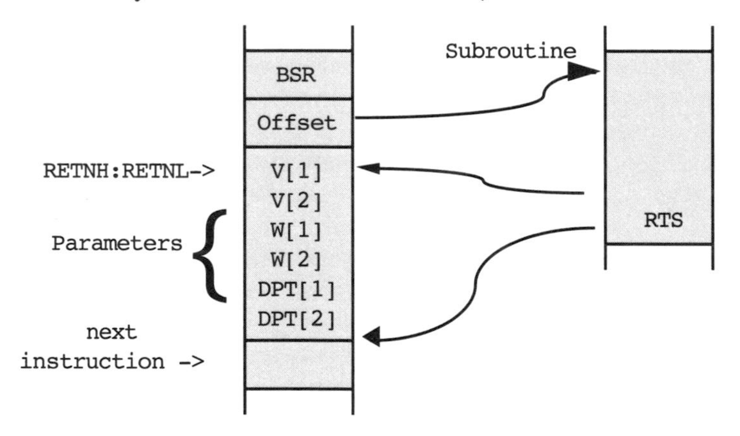

Figure 6.24. Parameters Passed after the Call

Some readers may appreciate the generality of the idea of parameters as local variables of the calling sequence, whereas others may prefer the more concrete technique of providing and removing holes for input and output parameters. They are the same.

The reason that the stack mode of passing arguments is recommended is that it is very general. Because registers are quite limited in number and are useful for other functions, it is hard to pass many parameters through registers. You can pass as many arguments to or from a subroutine as you will ever need using the stack. Compilers often use this technique. It is easier to use a completely general method in a compiler,

```
*          SUBROUTINE DOTPRD  – LOCAL VARIABLES
TERM:      EQU      0              ; First term of the dot product
NBYTES:    EQU      2
*          PARAMETERS
PARV:      EQU      0              ; Copy of vector V
PARW:      EQU      2              ; Copy of vector W
PARDP:     EQU      4              ; Dot product of V and W
PSIZE:     EQU      6
*
DOTPRD:    PULX                    ; Return address into X
           LEAS     -NBYTES,SP     ; Allocation for local variables
           LDAA     PARV,X
           LDAB     PARW,X
           MUL
           STD      TERM,SP        ; Copy first term into local variable
           LDAA     PARV+1,X
           LDAB     PARW+1,X
           MUL
           ADDD     TERM,SP        ; Dot product into D
           STD      PARDP,X        ; Place dot product in out parameter
           LEAS     NBYTES,SP      ; Deallocate local variables
           JMP      PSIZE,X
```

Figure 6.25. A Subroutine with Parameters after the Call, which Pulls the Return

```
PARV:    EQU    0
PARW:    EQU    2
PARDP:   EQU    4
*
         MOVW   V,PARV+L,PCR               ; Copy of V into parameter list
         MOVW   W,PARW+L,PCR               ; Copy of W into parameter list
         BSR    DOTPRD
L:       DS     6
         MOVW   PARDP+L,PCR,DTPD   ; Copy result into DTPD
```

Figure 6.26. A Subroutine Calling Sequence for Figure 6.25

rather than a kludge of special methods that are restricted to limited sizes or applications. The compiler has less to worry about and is smaller because less code in it is needed to handle the different cases. This means that many subroutines that you write for high-level languages such as C may require you to pass arguments by the conventions that it uses. Moreover, if you want to use a subroutine already written for such a language, it will pass arguments that way. It is a good idea to understand thoroughly the stack mode of passing parameters.

```
*        SUBROUTINE DOTPRD
*        LOCAL VARIABLES
TERM:    EQU    0            ; First term of the dot product
NBYTES:  EQU    2
*
*        PARAMETERS
*
PARV:    EQU    0            ; Copy of vector V
PARW:    EQU    2            ; Copy of vector W
PARDP:   EQU    4            ; Dot product of V and W
*
DOTPRD:  LDX    0,SP         ; Return address into X
         LEAS   -NBYTES,SP   ; Allocation for local variables
         LDAA   PARV+2,X
         LDAB   PARW+2,X
         MUL
         STD    TERM,SP      ; Copy first term into local variable
         LDAA   PARV+1+2,X
         LDAB   PARW+1+2,X
         MUL
         ADDD   TERM,SP      ; Dot product into D
         STD    PARDP+2,X    ; Place dot product in out parameter
         LEAS   NBYTES,SP    ; Deallocate local variables
         RTS
```

Figure 6.27. A Subroutine with Parameters after the Call, which Uses RTS

```
entry:  MOVW V,PARV+L,PCR       ; Copy of V into parameter list
        MOVW W,PARW+L,PCR       ; Copy of W into parameter list
        BSR  DOTPRD
        BRA  L1
*
L:      DS   6
*
L1:     MOVW PARDP+L,PCR,DTPD   ; Copy parameter list into DTPD
```

Figure 6.28. A Subroutine Call with Parameters after the Call for Figure 6.27

We now consider another common method of passing arguments in which they are put after the BSR or equivalent instruction. This way, they look rather like addresses that are put in the instruction just after the op code. Two variations of this technique are discussed below.

```
*      SUBROUTINE DOTPRD
*              LOCAL VARIABLES
*
TERM:    EQU     0              ; First term of the dot product
NBYTES:  EQU     2
*
*              PARAMETERS
*
PARV:    EQU     0              ; Copy of vector V
PARW:    EQU     2              ; Copy of vector W
PARDP:   EQU     4              ; Dot product of V and W
*
DOTPRD:  PULX                   ; Return address into X
         LEAS    -NBYTES,SP     ; Allocation for local variables
         LDAA    [PARV,X]
         LDAB    [PARW,X]
         MUL
         STD     TERM,SP        ; Copy first term into local variable
         LDY     PARV,X
         LDAA    1,Y
         LDY     PARW,X
         LDAB    1,Y
         MUL
         ADDD    TERM,SP        ; Dot product into D
         STD     [PARDP,X]      ; Place dot product in out parameter
         LEAS    NBYTES,SP      ; Deallocate local variables
         JMP     0,X
```

Figure 6.29. A Subroutine With In-Line Parameters that Are Addresses

```
BSR    DOTPRD
DC.W   ADDRV          ; Address for V
DC.W   ADDRW          ; Address for W
DC.W   ADDRDP         ; Address for dot product
```

Figure 6.30. An In-Line Argument List of Addresses for Figure 6.29

In the first alternative, the parameters are placed after the `BSR` or `JSR` instructions in what is called an *in-line argument list*. Looking at Figure 6.24, we see that the return address, which is pushed onto the hardware stack, points to where the parameters are located. When parameters are passed this way, sometimes referred to as *after the call,* the subroutine has to increment the return address appropriately to jump over the parameter list. If this is not done, the MPU would, after returning from the subroutine, try to execute the parameters as though they were instructions. For our dot product example, assume that the parameter list appears as shown in Figure 6.24. Notice that the subroutine must skip over the six bytes in the parameter list when it returns to avoid "executing the parameters." The subroutine shown in Figure 6.25 does this.

Notice that the return address is pulled into X before the local variables are allocated and that the labels for the parameters are now used as offsets with X. In particular, the label `PSIZE` used in the `JMP` instruction automatically allows the proper return. If we make the assumption that the global variables `V`, `W`, and `DTPD` are moved to and from the parameter list, a calling sequence would look as in Figure 6.26.

One should note that this technique generally takes more bytes of code than doing the correction within the subroutine because each call requires an additional two bytes.

A second alternative permits the return from the subroutine to be simply an RTS instruction without modifying the return address saved by the BSR instruction. To account for the BRA instruction, the labels for the parameters have to be increased by 2 so that the subroutine is now written as in Figure 6.27. The trick is to put a BRA instruction in front of the argument list to branch around it, as shown in Figure 6.28.

The typical use of passing parameters after the call assumes that all of the arguments are either constant addresses or constant values. They are often just addresses. The addresses are not usually modified, although the data at the addresses can be modified. In particular, one will always be modifying the data at the addresses where output parameters are placed. Calling sequences for this situation are particularly simple, as we show in Figure 6.30 for our dot product example in Figure 6.29.

Because programs are usually in ROM in microprocessor applications, parameters passed after the call must be constants or constant addresses. At a place in the program, these addresses are constants. At another place, the addresses would be different constants.

Passing parameters in an in-line argument list is often used in FORTRAN programs. A FORTRAN compiler passes the addresses of parameters, such as the parameters in the example above. Like the stack method, this method is general enough for FORTRAN, and it is easy to implement in the compiler. In assembly language routines, this method has the appeal that it looks like an "instruction," with the opcode replaced by the calling instruction and addressing modes replaced by the argument list.

```
        BSR     SWITCH
        DC.W    L0,L1,L2,L3
L0:                     ; program segment for case 0
        BRA     L4
*
L1:                     ; program segment for case 1
        BRA     L4
*
L2:                     ; program segment for case 2
        BRA     L4
*
L3:                     ; program segment for case 3
L4:     BRA     *
```

a. A calling sequence with an in-line argument list of jump addresses

```
SWITCH: CLRA
        LSLD
        PULX
        JMP   [D,X]
```

b. The subroutine

Figure 6.31. Implementation of a C or C++ Switch—Case Statement

Before we look at other argument passing techniques using our running inner product subroutine example, we illustrate a common subroutine used in C and C++ compilers to implement the *case* statement by means of an in-line argument list of addresses. Figure 6.31 illustrates how, when a variable between 0 and 3 is in accumulator B, the program can jump to label L0 if the variable is 0, to L1 if the variable is 1, to L2 if the variable is 2, and to L3 if the variable is 3. For each case, an address is put in the in-line argument list; the SWITCH subroutine reads one of these arguments into the PC, as selected by the value in accumulator B. Note that this technique is more efficient than a decision tree (Figure 2.14) when the same variable is tested for numbers, which happen to be consecutive, going to a program segment that the number indicates. However, this example is neither position independent nor does it use 8-bit in-line arguments to further improve efficiency. The reader is encouraged to improve this technique to do this.

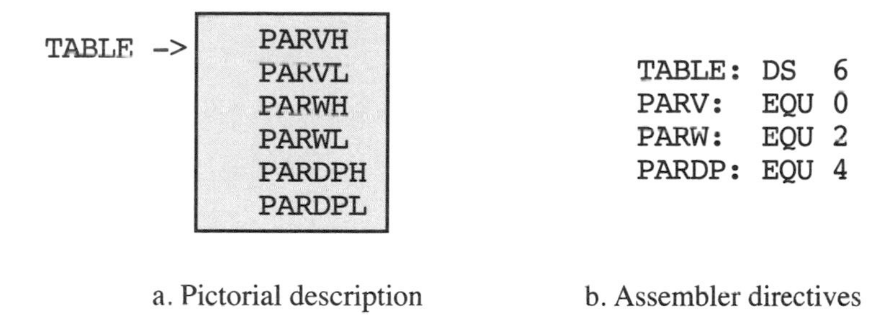

a. Pictorial description b. Assembler directives

Figure 6.32. Parameters in a Table

We now consider the technique of passing parameters via a table. The argument list, which is in-line when parameters are passed after the call, can be a table stored anywhere in memory. This technique is quite similar to passing parameters after the call. For our example, suppose that one uses a table whose address is passed in X and that looks like Figure 6.32, where, as before, the suffixes H and L stand for the high and low bytes of the 2-byte parameters PARV, PARW, and PARDP. The subroutine shown in Figure 6.33a is called by a sequence shown in Figure 6.33b.

Passing parameters by a table is often used to control a floppy disk in a way that is transparent to the user. The number of parameters needed to control a disk can be very large; therefore the table can serve as a place to keep all the parameters, so only the address of the table is sent to each subroutine that deals with the floppy disk.

In this section, we considered ways to pass arguments to and from subroutines. The register technique is best for small subroutines that have just a few arguments. The stack technique is best for larger subroutines because it is the most general. The in-line argument list that passes parameters after the call is used in FORTRAN subroutines, and the table technique is commonly used in operating system subroutines. The technique that passes parameters in global variables was covered for completeness and is useful in very simple micrcontrollers such as the 6805, but it is discouraged in the 6812.

```
*       SUBROUTINE DOTPRD
*       PARAMETERS
*
PARV:   EQU  0          ; Copy of vector V
PARW:   EQU  2          ; Copy of vector W
PARDP:  EQU  4          ; Dot product of V and W
*
*       LOCAL VARIABLES
*
TERM:   EQU  0
NBYTES: EQU  2
DOTPRD: LEAS -NBYTES,SP ; Allocation for local variables
        LDAA PARV,X
        LDAB PARW,X
        MUL             ; First term of DP into D
        STD  TERM,SP    ; Store in local variable
        LDAA PARV+1,X
        LDAB PARW+1,X
        MUL             ; Second term into D
        ADDD TERM,SP
        STD  PARDP,X    ; Place dot product
        LEAS NBYTES,SP  ; Deallocate local variables
        RTS
```

a. The subroutine

```
LDX  #TABLE
MOVW V,PARV,X       ; Place copy of V into parameter
MOVW W,PARW,X       ; Place copy of W into parameter
BSR  DOTPRD         ; Call Subroutine
MOVW PARDP,X,DTPD   ; Copy result into global variable
```

b. Calling sequence

Figure 6.33. Calling Sequence for Passing Arguments in a Table

6.3 Passing Arguments by Value, Reference, and Name

Computer science students, as opposed to electrical engineering students, study the passing of parameters in high-level language subroutines on a different level than that used in the preceding section. We include this section to explain that level to you. On the one hand, this level is very important if, say, you are writing or using a subroutine that is used by a high-level language program and that subroutine has to conform to the properties discussed below. On the other hand, the differentiation between some of the characteristics discussed below is rather blurry in assembly language programs.

The most important characteristic of a parameter is whether you pass the value of the parameter to the subroutine, or the address of the parameter to the subroutine. In the example used throughout §6.2, values of the vectors and the dot product usually were passed to and from the subroutine rather than the addresses of these arguments. (The one exception was in the discussion of passing parameters after the call, where constant addresses were passed in the argument list.) If the value of a parameter is passed, we say the parameter is passed or *called by value*. Output parameters passed by value are also said to be passed or *called by result*. If the address of the parameter is passed, the parameter is passed by reference or by name as we describe below.

The passing of parameters by value is completely general but could be time consuming. Consider a simple example where the string `STRING` of ASCII characters, terminated by an ASCII carriage return ($0D), is passed one character at a time into the subroutine, and the string is up to 100 characters long. Clearly, a lot of time would be used copying the characters into parameter locations. Were there enough registers, we would have to load 100 bytes into registers before calling the subroutine. A hundred bytes would be moved to global memory using the global technique, or 100 bytes would be pushed on the stack using the stack technique. Obviously, it is more efficient to pass an address rather than these values. For this case, the address would be the address of the first character of the string, or the label `STRING` itself. The parameter is *called by reference*, or is *called by name*. There is a slight difference between call by reference and call by name, but in assembly language, this difference is not worth splitting hairs about. We will refer to this technique, where the address of the data is passed as an argument, as call by name. Figures 6.29 and 6.31 are examples of subroutines that pass parameters by name. The other figures in this chapter called input parameters by value and called output parameters by result.

Call by name is useful when the parameters themselves are subroutines as, for example, in a subroutine that integrates the function `FUN` between 0 and 1. In this case, the function `FUN` could be supplied as an argument to an integration subroutine and it is reevaluated each time the subroutine calculates a new point of `FUN`. For example, the call to `FUN` may supply a starting point and an increment delta. The nth call to the subroutine returns the value of function `FUN`(x) at x = starting point + (n – 1) * delta. This continues until the calling routine changes the value of the starting point. Each call to `FUN` inside the integration subroutine returns a different value for `FUN`. This, then, is an example of call by name, because the address (of a subroutine) is passed.

In this section, we have described the types of information that are passed about parameters. The most important distinction is whether a value is passed (a call by value or call by result) or whether an address is passed (a call by name or a call by reference).

6.4 Calling and Returning Mechanisms

The 6812 provides several mechanisms to call a subroutine and return from it. The standard subroutine uses `BSR`, or the equivalent instruction, to call the subroutine and `RTS`, or the equivalent instruction, to return from it. We have used this mechanism in the earlier sections of this chapter. However, there are the `SWI` and `RTI` instructions that can be used for the software interrupt handler; the `LDX #RETURN` instruction can be used to call and return from a program segment that is very much like a subroutine. Such a calling and returning mechanism is used in the Motorola 500 series of RISC microcomputers discussed in Chapter 12. In the spirit of showing design alternatives, we will survey these techniques, pointing out advantages and disadvantages of each approach.

As we proceed, we will discuss several related important topics. We will look at hardware interrupts and at the fork and join mechanisms used in timesharing. These important topics are best covered in this section on calling and returning mechanisms.

There are alternatives to the most commonly used `BSR/RTS` mechanisms. We first discuss alternative returning mechanisms and then alternative calling mechanisms.

The alternative to the returning mechanism can be used to greatly improve the clarity of your programs. A significant problem in many programs that call subroutines occurs when the subroutine alters the contents of a register, but the writer of the calling routine does not think it does. The calling routine puts some number in that register, before calling the subroutine, and expects to find it there after the subroutine returns to this calling routine. It will not be there. Two simple solutions to this problem come to mind. One is to assume all subroutines will change any register, so the calling routine will have to save and restore any values in registers that it wants to save. The other is to assume that the subroutine will save and restore all registers except those used for output parameters. The latter solution is generally more statically efficient because any operation done by a subroutine requires instructions that are stored just once, where the subroutine is stored, whereas any operation done in the calling routine requires instructions to be stored at each place that the subroutine is called.

```
SUB:    TFR   X,D
        LEAY  D,Y
        RTS
```

Figure 6.34. Simple Subroutine

```
SUB:    PSHD          ; Save D
        TFR   X,D
        LEAY  D,Y
        PULD          ; Restore D
        RTS           ; Return
```

Figure 6.35. A Subroutine Saving and Restoring Registers

Suppose that a subroutine is called ten times. Then the former solution needs ten pairs of program segments to save and restore the register values. The latter solution requires only one pair of segments to save and restore the registers.

Consider the simple example in Figure 6.34 to add the contents of the register X to that of register Y without altering any register except Y. If we use `TFR X,D,` followed by `LEAY D,Y`, accumulator D is changed. However, as shown in Figure 6.35, if we save D first and restore it after the addition is done by the `LEAY` instruction (which doesn't affect condition codes), we don't affect other registers when we add X to Y. This technique for saving registers can be used to save all the registers used in a subroutine that do not return a result of the subroutine.

The extended local access of stacked local variables can be used with this technique. Consider the body of the subroutine after the PSHC instruction and before the PULC instruction as a program segment. The local variables are allocated just after the PSHC instruction that saves registers and are deallocated just before the PULC instruction that returns to the calling routine. You now look at the saved register values as stacked local variables of an outer program segment that includes the PSHD, PSHC, PULC, and PULD instruction. Using extended local access, you can read these registers and write into them, too. This allows you to output data in a register, even if the data are saved.

```
REGCC:  EQU     0
REGD:   EQU     1
REGX:   EQU     3
REGY:   EQU     5
*
SUB:    PSHY                    ; Save Y
        PSHX                    ; Save X
        PSHD                    ; Save D
        PSHC                    ; Save CC
        LDD     REGX,SP         ; Get to caller's X value
        ADDD    REGY,SP         ; Add to caller's Y value
        STD     REGY,SP         ; Result to caller's Y value
        PULC                    ; Restore CC
        PULD                    ; Restore  D
        PULX                    ; Restore  X
        PULY                    ; Restore  Y
        RTS                     ; Return
```

Figure 6.36. Saving and Restoring All the Registers

```
SUB     LBRA SUB0
        LBRA SUB1
        LBRA SUB2
SUB0:   ; perform initialization
        RTS
SUB1:   ; perform output
        RTS
SUB2:   ; perform termination
        RTS
```

Figure 6.37. A Subroutine with Multiple Entry Points

As an example of this, look at the preceding example again, now using saved registers as stacked local variables of an outer program segment. See Figure 6.36. This idea was expanded earlier to cover the passing of arguments on the stack. The basic idea is that you just have to know where the data are, relative to the current stack pointer SP, in order to access the data. Thus, access to the saved registers, the caller's stacked local variables, or the caller's saved registers is as easy as access to local variables. You are really using the extended local access technique regardless of the variations.

Although BSR is the most efficient instruction to call subroutines, its 8-bit offset limits it to call subroutines within –128 to +127 locations of the instruction after that BSR instruction. The JSR instruction with program relative addressing can be used for subroutines that are outside that range. Both instructions are position independent. This means that if the BSR or JSR instruction and the subroutine itself are in the same ROM, the subroutine will be called correctly wherever the ROM is placed in memory. The JSR instruction using direct or indirect addressing does not have this property and, because of this, should be avoided except for one case.

The one exception is where the subroutine is in a fixed place in memory. A monitor or debugging program is a program used to help you step through a program to find errors in it. These programs are often in ROM at fixed addresses in memory to allow the reset mechanism (that is executed when power is first turned on) to work correctly. Calling a subroutine of the monitor program from outside of the monitor program should be done with a JSR instruction, using direct or indirect addressing, because, wherever the calling routine is assigned in memory, the subroutine location is fixed in absolute memory, not relative to the address of the calling routine. The simple statement that all subroutines must be called using BSR or JSR with program relative addressing to make the program position independent is wrong in this case.

A variation of the calling mechanism uses computed addresses, computed in an index register, say X, and loaded from there into the program counter using JSR 0,X. One place where this calling mechanism is used is to call a subroutine with many entry points. In this situation, we need to have a standard means to call these different entry points that does not change if the subroutine is modified. If this is not done, a subroutine modification to fix a bug may cause a change in the calling routine because the entry point it used to jump to is now at a different location. This is another example where fixing an error in one part of a program can propagate to other parts of the program, making the software very difficult to maintain. The solution is to have standard places to call that contain the appropriate jumps to the correct entry points.

```
        ASLB
        ASLB                ; Multiply contents of B by 4
        LDX     #SUB
        JSR     B,X         ; Jump to ith entry point
```

Figure 6.38. Calling the ith Subroutine for Figure 6.37

The standard places are at the beginning of the subroutine, so that their locations will not be affected by changes in the subroutine. These places contain LBRA instructions that jump to the proper entry point. The LBRA instructions are always the same length regardless of whether or not the entry point may be reached by a BRA.

When LBRA instructions are used, the standard places to jump are always some multiple of three bytes from the beginning of the subroutine. An example will make this clearer. Suppose a subroutine SUB to interface to a printer has three entry points, SUB0, SUB1, and SUB2. We call SUB0 before we use the printer to initialize it, we call SUB1 to output a character to it, and we call SUB2 after we use the printer to turn it off. The layout of the subroutine is shown in Figure 6.37. If one wants to call subroutine SUB at its ith entry point, i = 0 to initialize it, i = 1 to output to it, and i = 2 to terminate its use, then, assuming that the value of i is in accumulator B, the calling sequence in Figure 6.38 can be used.

Notice that the machine code for the calling sequence stays the same regardless of internal changes to subroutine SUB. That is, if SUB1 were to increase in size due to modifications of the code, the calling program in Figure 6.38 is not changed at all. This technique limits the interaction between program segments so that changes in one segment do not propagate to other segments. So if a bug is fixed in the subroutine and the calling routine is in a different ROM or a different part of EEPROM, it won't have

```
SUB:   DC.W    SUB0
       DC.W    SUB1
       DC.W    SUB2
```

Figure 6.39. A Jump Vector

to be changed when the subroutine size changes. A variation of this technique uses indirect addressing and addresses instead of LBRA instructions, because fewer bytes are used with DC.Ws than LBRA instructions. However, this does not yield a position independent subroutine. For example, if the layout of SUB has the LBRA instructions replaced by the program segment in Figure 6.39, then its calling sequence is that shown in Figure 6.40. Although we described this technique and its variation as useful for a subroutine with several entry points, either works equally well for distinct subroutines.

```
ASLD              ; Multiply contents of D by 2
LDX     #SUB
JSR     [D,X]     ; Jump to ith entry point
```

Figure 6.40. Calling the ith Subroutine of a Jump Vector

```
     LEAS   -2,SP    ; Make space for return address
     PSHY            ; Save Y above return
     PSHX            ; Save X above that
     PSHA            ; Save accumulator A
     PSHB            ; Save accumulator B
     PSHC            ; Save condition codes
     LEAX   RET,PCR  ; Get return address ( position independent)
     STX    7,SP     ; Place return address
     LDX    $FFF6    ; Get SWI handler address
     JMP    0,X      ; Go to handler
RET: (next instruction)
```

Figure 6.41. Emulation of SWI

Now we consider some variations of subroutines. A *handler* is really just a subroutine that "handles" an interrupt. The *software interrupt* instruction SWI pushes all the registers onto the hardware stack, except SP, and then loads the program counter with the contents of locations $FFF6 and $FFF7. The sequence in Figure 6.41 produces the same effect as the SWI instruction except for minor changes in the CC and X registers. The subroutine at the address contained in $FFF6 and $FFF7 is called the *SWI handler*. A handler must end in an RTI instruction rather than an RTS instruction because all registers are pushed onto the stack with the SWI instruction. The RTI instruction at the end of an SWI handler does the same thing as the code in Figure 6.42.

SWI differs from BSR in that the address of the handler is obtained indirectly through the implied address $FFF6. This makes the SWI instruction shorter, in this case one byte long. An SWI handler can be made to perform a function, such as our ubiquitous dot product. See Figure 6.43. The initialization of the high-address "vector" need be done only once, before the first SWI call is made, as shown in Figure 6.44a. Then, each time it's called, insert the SWI instruction in your calling program, as shown in Figure 6.44b. Note that in the calling routine, we pass arguments in registers; but inside the handler, we access these arguments using stack techniques.

When you are debugging a program, you can use a program called a *debugger* or a *monitor* to help you debug the program that you are writing. You may want to display the values in some of the registers or memory locations or to insert some data into some of the registers or memory locations. As used with most debug programs, the SWI handler is a routine in the debug program that displays a prompt character, such as "*," on the terminal and waits for the user to give commands to display or change the values in registers or memory locations. This can be used to display or change any amount of data or even to modify the program. This SWI instruction is called a *breakpoint*.

A typical monitor program inserts a breakpoint at the start of an instruction by replacing the opcode byte with an SWI instruction. The address of the replaced opcode byte, as well as the byte itself, are kept in a part of RAM that the monitor uses for this purpose. The program now runs until it encounters the SWI breakpoint. Then the registers might be displayed together with a prompt for further commands to examine or change memory or register contents. It is indispensable that the SWI instruction be one byte long to be used as a breakpoint. If you tried to put breakpoints in the program with a JSR instruction, you would have to remove three bytes. If your program had a branch in it to the second byte being removed, unfathomable things might begin to happen! The problem in your program would be even harder to find now. However, if the single-word SWI instruction is used, this cannot happen, and the SWI handler call can be made to help you debug the program. One limitation of breakpoints is that the program being debugged must be in RAM. It is not possible to replace an opcode in ROM with an SWI instruction. Programs already in ROM are therefore more difficult to debug.

```
        PULC          ; Restore condition codes
        PULB          ; Restore accumulator B
        PULA          ; Restore accumulator A
        PULX          ; Restore X
        PULY          ; Restore Y
        RTS           ; Restore PC
```

Figure 6.42. Emulation of RTI

```
* original variables
RESULT: DS.W      1
*
* SUBROUTINE DOTPRD
* LOCAL VARIABLES
*
TERM:   EQU       0
NBYTES: EQU       2
* Saved registers
*
REGCC:  EQU       0+NBYTES    ; saved condition code register
REGB:   EQU       1+NBYTES    ; saved accumulator B -- note "backwards"
REGA:   EQU       2+NBYTES    ; saved accumulator A -- note "backwards"
REGX:   EQU       3+NBYTES    ; saved index register X
REGY:   EQU       5+NBYTES    ; saved index register Y
*
DOTPRD: LEAS      -NBYTES,SP
        LDAA      REGA,SP
        LDAB      REGB,SP
        MUL
        STD       TERM,SP     ; First term to local variables
        LDAA      REGX+1,SP
        LDAB      REGY+1,SP
        MUL
        ADDD      TERM,SP     ; Dot product into D
        STAA      REGA,SP     ; Dot product high byte to output parameter
        STAB      REGB,SP     ; Dot product low byte to output parameter
        LEAS      NBYTES,SP   ; Deallocate local variables
        RTI
```

Figure 6.43. An SWI Handler

```
entry:  MOVW     #DOTPRD,$FFF6
```

a. Initialization (done just once)

```
        LDAA     #1
        LDAB     #2
        LDX      #3
        LDY      #4
        SWI
        STD      RESULT  ; Put in global location
        BRA      *
```

b. Calling sequence (done each time the operation is to be performed).

Figure 6.44. Calling an `SWI` Handler

Early in this book, we said that the `SWI` instruction can be used at the end of each program. This instruction is really a breakpoint. You cannot turn off a microcomputer at the end of a program, but you can return to the debug program in order to examine the results of your program. The "halt" instruction is just a return to the debug program.

The `TRAP` instructions do essentially the same thing as the SWI instruction except that the program counter is loaded from different consecutive addresses ($FFF8). These instructions also happen to be two words long rather than one. The handlers, whose addresses are put there, are used in lieu of subroutines for very commonly used operations needing to save all of the registers. For example, operating system subroutines frequently use these instructions.

```
* TRAP HANDLER
* saved registers
REGCC:    EQU      0  ; saved condition code register
REGB:     EQU      1  ; saved accumulator B
REGA:     EQU      2  ; saved accumulator A
REGX:     EQU      3  ; saved index register X
REGY:     EQU      5  ; saved index register Y
REGPC:    EQU      7  ; saved PC
JUMPVECTOR: DC.W F0,F1,F2,F3
*
TRAP:     CLRA
          LDX      REGPC,SP
          LDAB     -1,X
          SUBB     #$30
          LSLD
          LDX      #JUMPVECTOR
          JMP      [D,X]
F0:       RTI     ; do inner product
F1:       RTI     ; do quadradic
F2:       RTI     ; do temperature conversion
F3:       RTI     ; do parallel resistor
```

Figure 6.45. A Trap Handler

```
entry:   MOVW     #TRAP,$FFF8
         LDAA     #1
         LDAB     #2
         LDX      #3
         LDY      #4
         DC.W     $1831 ; TRAP #$31
         LDX      #5
         LDY      #6
         DC.W     $1833 ; TRAP #$33
         bra      *
```

Figure 6.46. Calling a Trap Handler

A trap handler, shown in Figure 6.45, can execute the program segment to compute the inner product, if the instruction whose opcode is $1830 is executed; can execute the program segment to compute the quadratic formula, if the instruction whose opcode is $1831 is executed; can execute the program segment to compute the temperature conversion, if the instruction whose opcode is $1832 is executed; and can execute the program segment to compute the parallel resistor calculation, if the instruction whose opcode is $1833 is executed. Each of the program segments in the trap handler will look like the `SWI` handler shown in Figure 6.43. Calling routines can generate these nonstandard opcodes using `DC` directives, as in Figure 6.46. This program passes arguments in the registers for the quadradic formula and then for the parallel resistor formula.

The `TRAP` instructions can be used to *emulate* other instructions. Emulation means getting exactly the same result but perhaps taking more time. This technique is often used in minicomputers that are sold in different models that have different costs and speeds. The faster, more expensive model may have an instruction such as floating point add that is implemented on the slower, cheaper model as `TRAP`. The same program can be run in both the cheaper and more expensive models. When such an opcode is encountered in more expensive models, it results in the execution of the instruction. In cheaper models it results in calling a handler to emulate the instruction. In a sense, the instruction `TRAP` is a wildcard instruction because it behaves a bit like an instruction but is really a call to a software handlers. The 68000 family uses these types of instructions, which they call f-line instructions, to emulate some instructions in the cheaper 68332 that are implemented in hardware in the 68020 and 68881.

We digress for a moment to discuss *hardware interrupts*. These are handlers that are called up by an I/O device when that device wants service. An I/O hardware interrupt can occur at any time, and the main program will stop as if the next instruction were an `SWI` instruction. When the handler has finished, the `RTI` instruction is executed. This causes the main program to resume exactly where it left off. One exception, the *reset* interrupt, occurs when the reset pin on 6812 has a low signal put on it. The program counter is then loaded with the contents of locations $FFFE and $FFFF. The reset pin can be put in a circuit such that the pin is put low whenever the power is turned on, thus allowing the microprocessor to start running its program stored in ROM. The hardware interrupt is not all that magical. It is merely a handler that is called by I/O hardware by putting the appropriate signal on some pin of the 6812. This hardware is outside the direct control of the program that you are writing. Interrupts are further considered in Chapter 11.

In this section we have considered the subroutine and its alternatives. The subroutine is most often called by the `BSR` or `JSR` instruction, but it is occasionally called using the `JSR` instruction with direct or index addressing to achieve position independence. The `SWI` and `RTI` instructions call and return from handlers, which are like subroutines but are also like user-defined machine instructions. The hardware interrupt is a handler that is initiated by a hardware signal rather than a program call. With these tools, you are ready to modularize your programs into subroutines or equivalent program segments, so that each subroutine is more compact and easier to understand and so that interactions between these subroutines are carefully controlled to prevent unnecessary propagation of errors.

6.5 Summary

This chapter introduced the subroutine. It was dissected into parts to show alternative techniques for each part. Stacked local variables were studied along with alternatives for accessing local variables of nested program segments. Calling and returning mechanisms were studied, and the principal one, using `BSR` and `RTS`, was shown to have several alternatives, including ones that save the registers and restore them efficiently. Alternatives such as `SWI` and `TRAP` were studied. We then turned to the techniques that are used to pass arguments. The register, global, stack, in-line argument list, before the subroutine entry point, and table techniques were considered.

You should now be able to write subroutines, call them, and pass arguments to them in an effective manner. You should be prepared to use them in the following chapters to manipulate data structures, perform arithmetic, and interface to I/O hardware. Moreover, you should know how to approach a problem using top-down design or how to test a module with a driver. The need for the last two techniques is not apparent to the student who writes a lot of small programs and never faces the problem of writing a large program. When he or she does face that problem without the tools that we have introduced, inefficiencies and chaos generally will be the result. We introduced these techniques early and suggest that you use them whenever you can do so, even if they are not absolutely needed.

Do You Know These Terms?

See the end of chapter 1 for instructions.

top-down design
calling and returning mechanism
program segment
input parameter
output parameter
pass parameters
local variable
program segment
entry point
exit point
input parameter
input argument
output parameter
output argument
local variable
global variable
stacked local variable
binding
allocation
deallocation
accessing
reset
initialized
extended local access
stack marker
calling routine
calling sequence
in-line argument list
after the call
case
call by value
call by result
call by reference
call by name
handler
software interrupt
SWI handler
debugger
monitor
breakpoint
emulate
hardware interrupts
reset

PROBLEMS

1. Write a program segment that evaluates the quadratic function $ax^2 + bx + c$, where signed 16-bit arguments a, b, c, and x are stored on the stack and are initialized to 1, 2, 3, and 4, respectively, by pushing 4, then 3, then 2, and then 1, in the manner of Figure 6.6, and the output is stored on the stack in a "hole" created by a `LEAS -2,SP` instruction before the segment begins. In order to demonstrate local variables, as part of your program segment, save 16-bit value ax^2 in a 16-bit local variable on the stack.

2. Write a shortest program segment that computes the parallel resistance of two resistors R1 and R2, where unsigned 16-bit arguments R1 and R2 are stored in local variables, which are both initialized to 100, by pushing 100 and then 100, and the result is stored on the stack in a "hole" created by a `LEAS -2,SP` instruction, in the manner of Figure 6.6. In order to demonstrate local variables, as part of your program segment, store R1 times R2 in a 32-bit local variable on the stack.

3. Write a program segment that evaluates the quadratic function $ax^2 + bx + c$, where signed 16-bit arguments a, b, c, and x are stored in local variables `PARA`, `PARB`, `PARC`, and `PARX` on the stack, which are initialized to 1, 2, 3, and 4, respectively, and the output is returned in local variable `RESULT` on the stack, in the manner of Figure 6.10. In order to demonstrate local variables, as part of your program segment, store ax^2 in a 16-bit local variable on the stack.

4. Write a shortest program segment that computes the parallel resistance of two resistors R1 and R2, where unsigned 16-bit arguments are stored in local variables named `R1` and `R2`, which are both initialized to 100 and the output is returned in register D, in the manner of Figure 6.10. In order to demonstrate local variables, as part of your program segment, store `R1` times `R2` in a 32-bit local variable on the stack.

5. Write a program segment that evaluates the quadratic function $ax^2 + bx + c$, where signed 16-bit arguments a, b, c, and x are stored in local variables `PARA`, `PARB`, `PARC`, and `PARX` on the stack as outer segment local variables, which are initialized to 1, 2, 3, and 4, respectively, and the output is returned in register D, in the manner of Figure 6.13. In order to demonstrate local variables, as part of your inner program segment, store ax^2 in a 16-bit local variable on the stack.

6. Write a shortest program segment that computes the parallel resistance of two resistors `R1` and `R2`, where unsigned 16-bit arguments are stored in outer segment local variables `R1` and `R2`, which are both intitilaized to 100, and the output is returned in register D, in the manner of Figure 6.13. To demonstrate local variables, as part of your inner program segment, store `R1` times `R2` in a 32-bit local variable on the stack.

7. Write a program segment that evaluates the quadratic function $ax^2 + bx + c$, where signed 16-bit arguments a, b, c, and x are stored in local variables `PARA`, `PARB`, `PARC`, and `PARX` on the stack as outer segment local variables, which are initialized to 1, 2, 3, and 4, respectively, and the output is returned in register D, using a stack marker in the manner of Figure 6.14. In order to demonstrate local variables, as part of your inner program segment, store ax^2 in a 16-bit local variable on the stack.

8. Write a shortest program segment that computes the parallel resistance of two resistors `R1` and `R2`, where unsigned 16-bit arguments are stored in outer segment local variables `R1` and `R2`, which are both initialized to 100, and the output is returned in register D, using a stack marker in the manner of Figure 6.14. In order to demonstrate local variables, as part of your inner program segment, store `R1` times `R2` in a 32-bit local variable on the stack.

9. Write a position-independent reentrant subroutine `QUAD` that evaluates the quadratic function $ax^2 + bx + c$, where signed 8-bit arguments a, b, c, and x are passed in registers `A`, `B`, (low byte of) Y, and (low byte of) X, and the output is passed in A. In order to demonstrate local variables, as part of your subroutine, store ax^2 in an 8-bit local variable on the stack. Write a calling sequence that loads 1 into A, 2 into B, 3 into Y, and 4 into X; calls `QUAD`; and stores the result in global variable `ANSWER`.

10. Write a shortest position-independent reentrant subroutine `PAR` that computes the parallel resistance of two resistors `R1` and `R2`, where unsigned 16-bit arguments R1 and R2 are passed in registers D and Y and the output is passed in D. In order to demonstrate local variables, as part of your subroutine, store R1 times R2 in a 32-bit local variable on the stack. Write a calling sequence that loads 100 into D and Y, calls `PAR`, and stores the result in global variable `ANSWER`.

11. Write a position-independent reentrant subroutine `QUAD` that evaluates the quadratic function $ax^2 + bx + c$, where signed 16-bit arguments a, b, c, and x are passed in global variables `PARA`, `PARB`, `PARC`, and `PARX` and the output is passed in global variable `RESULT`. In order to demonstrate local variables, as part of your subroutine, store ax^2 in a 16-bit local variable on the stack. Write a calling sequence that loads 1 into `PARA`, 2 into `PARB`, 3 into `PARC` and 4 into `PARX`; calls `QUAD`; and stores the result in global variable `ANSWER`.

12. Write a shortest position-independent reentrant subroutine `PAR` that computes the parallel resistance of two resistors `R1` and `R2`, where unsigned 16-bit arguments `R1` and `R2` are passed in global variables `R1` and `R2` and the output is passed in global variable `RESULT`. In order to demonstrate local variables, as part of your subroutine, store `R1` times `R2` in a 32-bit local variable on the stack. Write a calling sequence that loads 100 into `R1` and `R2`, calls `PAR`, and stores the result in global variable `ANSWER`.

13. Write a shortest reentrant, position-independent subroutine `SEARCH` that returns the number of times that the integer K appears in the vector Z of length N and each element is one byte. If the address of Z is in X, the value of K is in A, the value of N is in B, and the return value `NUM` is left on the stack, `SEARCH` is called as in Figure 6.47.

```
PSHA
PSHB
PSHX
CLR    1, -SP                  Hole for NUM
BSR    SEARCH
PULA                           Put NUM into A
LEAS   4,SP                    Balance stack
```

Figure 6.47. Program for Problem 13

14. Write a position-independent reentrant subroutine QUAD that evaluates the quadratic function $ax^2 + bx + c$, where signed 16-bit arguments a, b, c, and x are passed on the stack, named PARA, PARB, PARC, and PARX, and the output is passed on the stack, named RESULT. In order to demonstrate local variables, as part of your subroutine, store ax^2 in a 16-bit local variable on the stack. Write a calling sequence that pushes 1, 2, 3, and 4; calls QUAD; pulls the result from the stack; and stores the result in global variable ANSWER.

15. Write a shortest position-independent reentrant subroutine PAR that computes the parallel resistance of two resistors R1 and R2, where unsigned 16-bit arguments are passed on the stack and named R1 and R2, and the output is passed on the stack, named RESULT. In order to demonstrate local variables, as part of your subroutine, store R1 times R2 in a 32-bit local variable on the stack. Write a calling sequence that pushes 100 twice, calls PAR, pulls the result from the stack, and stores the result in global variable ANSWER.

16. Do the same thing as in Problem 13, assuming that the input parameters are passed after the call in the same order, while the parameter NUM is returned on the stack as before. Do not use a BRA instruction before the parameter list, but follow the style of Figure 6.27. Provide an example of a calling sequence.

17. Write a position-independent reentrant subroutine QUAD that evaluates the quadratic function $ax^2 + bx + c$, where signed 16-bit arguments a, b, c, and x are passed after the call, named PARA, PARB, PARC, and PARX, and the output is passed after the call, named RESULT. In order to demonstrate local variables, as part of your subroutine, store ax^2 in a 16-bit local variable on the stack. Do not use a BRA instruction before the parameter list but follow the style of Figures 6.25 and 6.26. Write a calling sequence that writes 1, 2, 3, and 4 into PARA, PARB, PARC, and PARX; calls QUAD; and moves the result to global variable ANSWER.

18. Write a shortest position-independent reentrant subroutine PAR that computes the parallel resistance of two resistors, where unsigned 16-bit arguments are passed after the call and named R1 and R2 and the output is passed after the call, named RESULT. In order to demonstrate local variables, as part of your subroutine, store R1 times R2 in a 32-bit local variable on the stack. Do not use a BRA instruction before the parameter list, but follow the style of Figures 6.25 and 6.26. Write a calling sequence that writes 100 into R1 and R2, calls PAR, and moves the result to global variable ANSWER.

19. One reason for not passing output parameters after the call is that a subroutine that calls another subroutine and has some parameters passed back to it after the call will not always be reentrant. Explain why this is so. Are there similar restrictions on input parameters?

20. Give an example of passing output parameters after the call where the program can still be stored in ROM.

21. Write a subroutine to search an N-byte vector Z until a byte is found that has the same bits in positions 0, 3, 5, and 7 as the word `MATCH`. The address of Z is passed as the first entry `AZ` in a table, the value of `MATCH` is passed below it in the table, the value of N is passed below it in the table, and the address of the first byte found, `ADDR`, is passed below it in a table. If no byte is found in Z with a match, $FFFF should be placed in `AZ`. The address of the table is in X when the subroutine is called.

22. Write a position-independent reentrant subroutine `QUAD` that evaluates the quadratic function $ax^2 + bx + c$, where signed 16-bit arguments a, b, c, and x are passed in a table, named `PARA`, `PARB`, `PARC`, and `PARX`; and the output is passed in the table, named `RESULT`. The address of the table is in X when the subroutine is called. In order to demonstrate local variables, as part of your subroutine, store ax^2 in a 16-bit local variable on the stack. Write a calling sequence that writes 1, 2, 3, and 4 into `PARA`, `PARB`, `PARC`, and `PARX`; calls `QUAD`; and moves the result to global variable `ANSWER`.

23. Write a shortest position-independent reentrant subroutine `PAR` that computes the parallel resistance of two resistors `R1` and `R2`, where unsigned 16-bit arguments are passed in a table and in elements `R1` and `R2` and the output is passed in the same table in an element named `RESULT`. In order to demonstrate local variables, as part of your subroutine, store `R1` times `R2` in a 16-bit local variable on the stack. The address of the table is in X when the subroutine is called. Write a calling sequence that writes 100 into `R1` and `R2`, calls `PAR`, and moves the result to global variable `ANSWER`.

24. Repeat Problem 14, saving and restoring all the registers that were used.

25. Repeat Problem 15, saving and restoring all the registers that were used.

26. Repeat Problem 22, saving and restoring all the registers that were used.

27. Repeat Problem 23, saving and restoring all the registers that were used.

28. How would the calling sequence of Figure 6.40 be modified if the subroutine `SUB` in Figure 6.41 replaced the `LBRA` instructions with

```
SUB     DC.W    SUB0-SUB
        DC.W    SUB1-SUB
        DC.W    SUB2-SUB
```

What are the advantages of doing this, if any?

29. A device driver is an operating system component that interfaces to an I/O device. It has subroutines Init, Read, Write, and Terminate. Subroutine Init is called when the I/O device is being prepared for use, Subroutine Read is called to read data from the I/O device, Subroutine Write is called to output data from the I/O device, and Subroutine Terminate is called when the I/O device is no longer needed. Write a program segment for the beginning of a position-independent device driver, which, for an address in index register X, `JSR 0,X` always executes the Init subroutine; `JSR 4,X` always executes the Read subroutine; `JSR 8,X` always executes the Write subroutine; and `JSR 16,X` always executes the Terminate subroutine, regardless of where the device driver is stored.

30. Write an instruction sequence that produces the same moves as the instruction `SWI` and, in addition, sets the bits in the CC register in exactly the same way.

31. Write a shortest trap handler whose opcode is \$1830, to test whether a 2-byte number N is prime. The number N should be passed by value in D and the carry bit should be returned set if N is prime. Write a program segment that loads 11 into D, executes \$1830, and branches if carry set to location `ISPRIME`.

32. Write a shortest trap handler whose opcode is \$1831 that evaluates the quadratic function $ax^2 + bx + c$, where signed 8-bit arguments a, b, c, and x are passed in registers A, B, Y, and X, and the 16-bit output is passed in A. In order to demonstrate local variables, as part of your handler, store ax^2 in an 8-bit local variable on the stack. Write a program segment that loads 1 into A, 2 into B, 3 into Y, and 4 into X; executes \$1831; and stores the result in global variable `ANSWER`.

33. Write a trap handler whose opcode is \$1832, that computes the parallel resistance of two resistors `R1` and `R2`, where unsigned 16-bit arguments `R1` and `R2` are passed in registers D and Y, and the 16-bit output is passed in D. In order to demonstrate local variables, as part of your handler, store `R1` times `R2` in a 32-bit local variable on the stack. Write a program segment that loads 100 into D and Y, executes \$1832, and stores the result in global variable `ANSWER`.

34. Write a position-independent trap handler, whose address is in \$FFF8, which branches to address `PRIME` if the trap instruction is \$1830, to address `QUAD` if the trap instruction is \$1831, and to address `PAR` if the trap instruction is \$1832.

This inexpensive Axiom PB68HC12A4 board is well suited to senior design, and other prototyping projects. Its wire-wrap pins can be reliably connected to external wire-wrap sockets and connectors.

7

Arithmetic Operations

This chapter deals with how one number crunches, using algebraic formulas, and how one writes subroutines for conversion between different bases, multiple-precision integer arithmetic, floating-point arithmetic, and fuzzy logic.

The first section describes unsigned and signed multiplication and division. Although the subroutines developed herein are in fact equivalent to 6812 instructions, they clearly illustrate the algorithms used to execute these instructions in a typical controller, and easily developed variants of these subroutines are usable on other microcontrollers that do not have these instructions.

In the next section, we develop ways to convert integers between number systems, such as binary and decimal. These techniques are needed when you input numbers from a terminal keyboard and output numbers to a terminal display. A full comparison of these techniques is made, examining all known possibilities and selecting the best, something that you should try to do in writing any of your own subroutines for any purpose.

The third section presents a technique to write program segments that evaluate algebraic formulas. The operations in the formulas are implemented with macros. Formula evaluation becomes a sequence of macro calls. You can hand assemble these macros. The actual variables used in the formula could be 32-bit integers, floating-point numbers, or any other type of variable that the subroutines have been written to handle.

Some C compilers do not support *long* and *float* data types. Section 7.4 shows how to perform 32-bit signed and unsigned integer arithmetic, which is useful to one who uses a microprocessor in a numerical control application. The fifth section deals with floating-point arithmetic. These sections will provide subroutines that enable you to perform arithmetic using these number systems.

Section 7.6 deals with fuzzy logic, for which the 6812 has special instructions. This section will give you some background so that you can describe a fuzzy logic system, and you can write assembly language subroutines to execute fuzzy logic.

After reading this chapter, you should be able to convert integers from one base to another. You should be able to write a sequence of subroutine calls to evaluate any algebraic formula and write subroutines to work out any multiprecision arithmetic operation, especially 32-bit long arithmetic. You should understand the principles of floating point arithmetic and fuzzy logic to the point that you could write subroutines for the usual floating-point and fuzzy logic operations.

7.1 Multiplication and Division

This section illustrates basic multiplication and division algorithms using the 6812 instruction set. Because there are 6812 instructions that give the same results, these subroutines are not useful for this machine. However, for other instruction sets that do not have multiply and divide instructions, these subroutines are the only ways to perform multiplication and division on these machines. Finally, these subroutines provide an understanding of how the operations can be implemented in a controller and how these operations can be extended to higher precisions where the instructions at that level of precision may be unavailable in the microcontroller's instruction set.

We first look at multiple-precision multiplication. The 6812 microcontroller has several instructions that multiply 8-bit unsigned integers and 16-bit signed and unsigned numbers. To see the advantages of the `MUL` instruction, look at the subroutine `BINMUL` of Figure 7.1, which does exactly the same multiplication as `MUL`, which takes three clock cycles. Neglecting the clock cycles for the `BSR` and `RTS` instructions, `BINMUL` takes 87 to 95 clock cycles to execute the multiplication. This illustrates that generally, hardware operations are 10 to 100 times faster than the same operations done in software.

Turning to the division of unsigned integers, the subroutine of Figure 7.2 divides the unsigned contents of B by the unsigned contents of A, returning the quotient in B and the remainder in A. To understand better the division subroutine of Figure 7.2, let's look more closely at binary division, where, for simplicity, we consider just 4-bit numbers. To divide 6 into 13, we can mimic the usual base-10 division algorithm as follows:

```
*
* BINMUL multiplies the two unsigned numbers in A and B, putting
* the product in D. Register Y is unchanged.
*
BINMUL: PSHA                 ; Save first multiplier
        CLRA                 ; Accumulator D will become the product
        LDX     #8           ; Count out 8 loops
*
LOOP:   CLC                  ; Clear carry, if not adding first number
        BITB    #1           ; If multiplier bit is 1
        BEQ     SHIFT
*
        ADDA    0,SP         ; Add first number
*
SHIFT:  RORA                 ; Shift 16 bits, feeding carry back into sum
        RORB
*
        DBNE    X,LOOP       ; Count down, repeat
*
        LEAS    1,SP         ; Balance stack
        RTS
```

Figure 7.1. 8-Bit Unsigned Multiply Subroutine

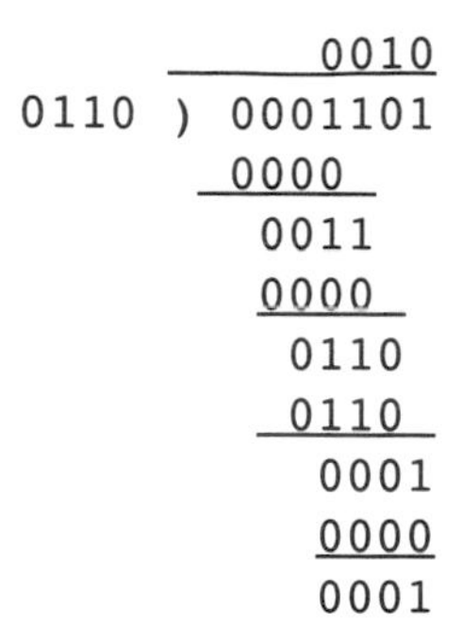

Another way of doing the bookkeeping of this last version is to first put all zeros in the nibble A and put the dividend in nibble B so that the contents of A:B look like

A	B
0000	1101

which we will think of as our initial 8-bit *state*. Next, shift the bits of A:B left, putting 1 in the rightmost bit of B if the divisor can be subtracted from A without a carry, putting in 0 otherwise. If the subtraction can be done without a carry, put the result of the subtraction in A. Repeat this shift-subtract process three more times to get the remainder in A and the quotient in B.

```
*
*       DIVIDE divides the 1-byte unsigned number in B (dividend) by the
* 1-byte unsigned number in A (divisor), putting the quotient in B and
* the remainder in A.  Register Y is unchanged.
*
DIVIDE:  PSHA                   ; Save divisor
         CLRA                   ; Expand dividend, fill with zeros
         LDX     #8             ; Initialize counter
*
LOOP:    ASLD                   ; Shift dividend and quotient left
*
         CMPA    0,SP           ; Check if subtraction will leave positive rslt
         BLO     JUMP           ; If so
*
         SUBA    0,SP           ; Subtract divisor
         INCB                   ; Insert 1 in quotient
*
JUMP:    DBNE    X,LOOP         ; Decrement counter
*
         LEAS    1,SP           ; Balance stack
         RTS                    ; Return with quotient in B, remainder in A
```

Figure 7.2. 8-Bit Unsigned Divide Subroutine

The 8-bit case is identical to the above except that A and B become bytes, our state A:B becomes sixteen bits, and there are eight shifts instead of four. The shift left on D can be done with `ASLD`, which always puts a zero in the rightmost bit of B. If the divisor can be subtracted from A without a carry, one needs to execute `INCB` and perform the subtraction from A. The complete subroutine is shown in Figure 7.2. You should be able to see how to adapt this subroutine for, say, dividing two unsigned 16-bit numbers or an unsigned 16-bit number by an 8-bit one. (When the dividend and the divisor are of different lengths, one has to check if a one has been shifted out when the state is shifted left, and if so, the divisor should be subtracted from the remainder part of the state. This prevents the remainder from being truncated.)

We now discuss briefly the multiplication and division of signed integers with a special look at the multiplication of 8-bit signed integers. Using the usual rule for the sign of the product of two signed numbers, we can extend the multiplication subroutines of this section in a straightforward way to handle signed integers. Extending the subroutines for division in this way is also straightforward after recalling that the sign of the remainder is the same as the sign of the dividend and that the sign of the quotient is positive if the signs of the dividend and divisor are equal; otherwise, it is negative.

There are also algorithms to multiply n-bit signed integers directly. We leave the details of this to more advanced treatments of arithmetic, because it is quite easy to modify any of the subroutines that we have presented for unsigned integers to work for signed integers. We illustrate the technique by showing how to modify the results of the MUL instruction to obtain an 8-bit signed multiply of $\mathcal{A}$ and $\mathcal{B}$, which are 8-bit signed integers located in registers A and B. Suppose $\mathcal{A}$ is represented as bits $a_7, \ldots, a_0$ and $\mathcal{B}$ is represented as $b_7, \ldots, b_0$. Then the numerical value of signed number $\mathcal{A}$ is,

$$\mathcal{A} = -a_7 * 2^7 + a_6 * 2^6 + \ldots + a_0 * 2^0$$

Now, by adding $a_7 * 2^8$ to both sides, thereby making $a_7 * 2^8 + -a_7 * 2^7 = a_7 * 2^7$ we get, from the previous expression:

$$(\mathcal{A} + a_7 * 2^8) = a_7 * 2^7 + a_6 * 2^6 + \ldots + a_0 * 2^0$$

(This addition of a constant is called *biasing*.) We recognize this equation's right side as an unsigned number `A`. It can be input to the MUL instruction, which actually multiplies unsigned `A` times unsigned `B`. The result in accumulator D is

$$(\mathcal{A} + a_7 * 2^8) * (\mathcal{B} + b_7 * 2^8) \quad (1)$$

which, by multiplying out the terms, is

$$\mathcal{A} * \mathcal{B} + a_7 * \mathcal{B} * 2^8 + b_7 * \mathcal{A} * 2^8 + a_7 * b_7 * 2^{16} \quad (2)$$

We see that to get the first term $\mathcal{A} * \mathcal{B}$ from (2) requires subtracting the two middle terms of (2) from accumulator D. [The rightmost term of (2), 2^{16}, does not appear in D and can be ignored.] The subroutine in Figure 7.3a makes this adjustment in D, where we note that to subtract $2^8 * \mathcal{B}$ or $2^8 * \mathcal{A}$ from accumulator D, we subtract $\mathcal{B}$ or $\mathcal{A}$ from accumulator A. See Figure 7.3. Subroutine SGNMUL behaves like an instruction that multiplies the signed contents of accumulator A with the signed contents of accumulator B, putting the signed result in D. The Z bit is not set correctly by `SGNMUL`, however.

```
*
*     SGNMUL multiplies the l-byte signed number in B by the l-byte signed number in
*     A, putting the product in accumulator D. Registers X and Y are unchanged.
*
SGNMUL:  PSHD             ; Save two bytes to be multiplied
         MUL              ; Execute unsigned multiplication
         TST    1,SP      ; If first number is negative
         BPL    L1        ; Then
         SUBA   0,SP      ; Subtract second number
L1:      TST    0,SP      ; If second number is negative
         BPL    L2        ; Then
         SUBA   1,SP      ; Subtract first number
L2:      LEAS   2,SP      ; Balance stack
         RTS              ; Return with product in accumulator D
```

a. Using MUL

```
*
*     SGNMUL multiplies the l-byte signed number in B by the l-byte signed number in
*     A, putting the product in accumulator D. Register X is unchanged.
*
SGNMUL:  SEX    A,Y       ; move one multiplier to Y, sign extending it
         SEX    B,D       ; sign extend the other multiplier
         EMULS            ; put the low-order 16-bits in D
         RTS              ; Return with product in accumulator D
```

b. Using EMULS

Figure 7.3. 8-Bit Signed Multiply Subroutine

Another approach to multiplication of signed 8-bit numbers is to use signed 16-bit multiplication available in the `EMULS` instruction. See Figure 7.3b. This method is far better on the 6812, because the `EMULS` instruction is available, but the former method is useful on other machines and shows how signed multiplication is derived from unsigned multiplication having the same precision. A modification of it is used to multiply 32-bit signed numbers, using a procedure to multiply 32-bit unsigned numbers.

7.2 Integer Conversion

A microcomputer frequently communicates with the outside world through ASCII characters. For example, subroutines `INCH` and `OUTCH` allow the MPU to communicate with a terminal, and this communication is done with ASCII characters. When numbers are being transferred between the MPU and a terminal, they are almost always decimal numbers. For example, one may input the number 3275 from the terminal keyboard, using the subroutine INCH, and store these four ASCII decimal digits in a buffer. After the digits are input, the contents of the buffer would be

$33
$32
$37
$35

While each decimal digit can be converted to binary by subtracting $30 from its ASCII representation, the number 3275 still has to be converted to binary (e.g., $0CCB) or some other representation if any numerical computation is to be done with it. One also has to convert a binary number into an equivalent ASCII decimal sequence if this number is to be displayed on the terminal screen. For example, suppose that the result of some arithmetic computation is placed in accumulator D, say, $0CCB. The equivalent decimal number, in this case 3275, must be found and each digit converted to ASCII before the result can be displayed on the terminal screen. We focus on the ways of doing these conversions in this section.

One possibility is to do all of the arithmetic computations with binary-coded decimal (BCD) numbers, where two decimal digits are stored per byte. For example, the BCD representation of 3275 in memory would be

$32
$75

Going between the ASCII decimal representation of a number to or from equivalent BCD representation is quite simple, involving only shifts and the AND operation. With the 6812, it is a simple matter to add BCD numbers; use ADDA or ADCA with the DAA instruction. Subtraction of BCD numbers on the 6812 must be handled differently from the decimal adjust approach because the subtract instructions do not correctly set the half-carry bit H in the CC register. (See the problems at the end of the chapter.) For some applications, addition and subtraction may be all that is needed, so that one may prefer to use just BCD addition and subtraction. There are many other situations, however, that require more complex calculations, particularly applications involving control or scientific algorithms. For these, the ASCII decimal numbers are converted to binary because binary multiplication and division are much more efficient than multiplication and division with BCD numbers. Thus we convert the input ASCII decimal numbers to binary when we are preparing to multiply and divide efficiently. However, depending on the MPU and the application, BCD arithmetic may be adequate so that the conversion routines below are not needed.

We consider unsigned integer conversion first, discussing the general idea and then giving conversion examples between decimal and binary representations. A brief discussion of conversion of signed integers concludes this section. The conversion of numbers with a fractional part is taken up in a later section.

An unsigned integer N less than b^m has a unique representation

$$N = c_{m-1} * b^{m-1} + c_{m-2} * b^{m-2} + \dots + c_1 * b^1 + c_0 \tag{3}$$

where $0 \le c_i < b$ for $0 \le i < m$. The sequence $c_{m-1}, \ldots c_0$ is called an *m-digit base-b representation* of N. We are interested in going from the representation of N in one base to its representation in another base. There are two common schemes for this conversion that are based on multiplication and two schemes that are based on division. Although one of the division schemes is taught in introductory logic design courses, and you are likely to select it because you know it well, it does not turn out to be the most efficient to implement in a microcomputer. We look at all the schemes in general and then give examples of each to find the most promising one.

The two multiplication schemes simply carry out (3), doing the arithmetic in the base that we want the answer in. There are two ways to do this. Either evaluate expression (3) as it appears or else nest the terms as shown.

$$N = (\ldots (0 + c_{m-1}) * b + c_{m-2}) * b + \ldots + c_1) * b + c_0 \quad (4)$$

The other two schemes involve division. Notice from (3) that if you divide N by b, the remainder is c_0. Dividing the quotient by b again yields c_1, and so forth, until one of the quotients becomes 0. In particular, if one has a base-r representation of N and wants to go to a base-b representation, division of N by b is done in base-r arithmetic. You are

```
*
*       SUBROUTINE CVDTB puts the unsigned equivalent of five ASCII decimal digits
*       pointed to by X into D.
*
SCRATCH: ds.b   6                       ; Scratch area for product and multiplier
K:       dc.w   10000,1000,100,10,1     ; Coefficient Vector
*
CVDTB:   LDAB   #6                      ; Clear scratch area
         LDY    #SCRATCH
C1:      CLR    1,Y+
         DBNE   B,C1
*
         LDAB   #5                      ; Five terms to be evaluated
         LDY    #K                      ; Constants in vector K, Y = multiplicand address
*
C2:      LDAA   1,X+                    ; Next ASCII digit into A
         PSHX                           ; Save pointer for next character
         SUBA   #$30                    ; ASCII to binary
         STAA   SCRATCH+5               ; Save in last byte in scratch
         LDX    #SCRATCH+4              ; Get address of multiplier
         EMACS  SCRATCH                 ; Multiply and accumulate
         LEAY   2,Y                     ; Next multiplicand address
         PULX                           ; Restore pointer for next character
         DBNE   B,C2                    ; Count down and loop
         LDD    SCRATCH+2               ; Get number
         RTS                            ; Return to Caller
```

Figure 7.4. Conversion from Decimal to Binary by Multiplication by Powers of 10

probably familiar with this technique because it is easy to go from a decimal representation to any base with it using a calculator, because the calculator does decimal arithmetic. The other scheme using division simply divides the number N, assumed to be less than b^m, by b^{m-1}, so that the quotient is the most significant digit c_{m-1}. Dividing the remainder by b^{m-2} produces the next most significant digit, and so on. We will also consider these schemes below. Which of these four schemes is best for microcomputers? We now look more closely at each for conversion between decimal and binary bases.

Consider first the multiplication scheme that evaluates formula (3) directly. Suppose that $N_4, \ldots, N_0$ represent five decimal digits stored in a buffer pointed to by X. Assume that these five decimal digits have been put in from the terminal using INCH so that each digit is in ASCII. Then

$$N_4 * 10^4 + N_3 * 10^3 + \ldots + N_0 * 10^0 \tag{5}$$

is the integer that we want to convert to binary. To carry out (5) we can store constants

```
K       dc.w     10000,1000,100,10,1
```

and then multiply N_4 times (K):(K + 1), N_3 times (K + 2):(K + 3), and so on, adding up the results and putting the sum in D. The subroutine shown in Figure 7.4 does just that, indicating an overflow by returning the carry bit equal to 1. The multiplication scheme in Figure 7.4 takes advantage of the fact that the assembler can convert 10^4 through 10^9 into equivalent 16-bit binary numbers using the dc.w directive.

Looking at the second multiplication conversion scheme applied to our present example, we rewrite the decimal expansion formula (3) as (6).

```
*         CVDIB converts the five ASCII decimal digits, stored at the location
*         contained in X, into an unsigned 16-bit number stored in D.
*
CVDIB:    CLRA                    ; Generate 16-bit zero
          CLRB                    ; Which becomes the result
          LEAY    5,X             ; Get address of end of string
          PSHY                    ; Save it on stack
*
C2:       LDY     #10             ; Multiply previous by 10
          EMUL                    ; Multiply D * Y
          PSHD                    ; Low 16 bits of product to stack
          LDAB    1,X+            ; Next ASCII digit into B
          SUBB    #$30            ; ASCII to binary
          CLRA                    ; Extend to 16 bits
          ADDD    2,SP+           ; Add previous result
          CPX     0,SP            ; At end of ASCII string?
          BNE     C2              ; No, repeat
          PULY                    ; Balance the stack
          RTS                     ; Return to caller
```

Figure 7.5. Conversion from Decimal to Binary by Multiplication by 10

$$[([(0 + N_4) * 10 + N_3]* 10 + N_2) * 10 + N_1] * 10 + N_0 \quad (6)$$

Doing our calculations iteratively from the inner pair of parentheses, we can get the same result as before without storing any constants. A subroutine that does this is shown in Figure 7.5. In this subroutine, there is only one local variable, except for a program sequence to save the low 16 bits of the sum, so that the binding process can be omitted. If we compare the two subroutines, we see that the second one has the clear edge in terms of lines of code or static efficiency, particularly if you consider the ten bytes used by the dc.w directive in the first subroutine. Furthermore, if one wanted to convert a 7-digit decimal number to a 32-bit binary number, the difference in the static efficiencies of these two multiplication techniques would become even more pronounced. For example, a 40-byte table of constants would be needed for the analog of the subroutine of Figure 7.4, an increase of 30 bytes that is not needed by the corresponding analog of Figure 7.5's subroutine. Finally, each subroutine can take the ASCII decimal digits directly from the terminal, using subroutine INCH, effectively accessing the digits as a character sequence rather than a vector. (See the problems at the end of the chapter.)

```
*
*        CVBTD converts unsigned binary number in D into five ASCII decimal
*        digits at the location passed in X. For each power of 10, subtract it as many
*        times as possible from D without causing a carry. The number obtained,
*        after ASCII conversion, is the ASCII decimal coefficient of that power of 10.
*
K:       dc.w    10000,1000,100,10,1
*
CVBTD:   LEAY    K+10,PCR        ; End of power-of-tens
         PSHY                    ; Save a local variable for CPY
         LEAY    K,PCR           ; Point Y to power-of-tens
*
CVB1:    MOVB    #$30,1,X+       ; Generate ASCII '0' in character position
CVB2:    SUBD    0,Y             ; Try removing one power-of-ten
         BLO     CVB3            ; If unsuccessful, quit
         INC     -1,X            ; If successful, up the ASCII character
         BRA     CVB2            ; Repeat trial
*
CVB3:    ADDD    2,Y+            ; Restore D, point Y to next constant
*
         CPY     0,SP            ; At end of constants?
         BNE     CVB1            ; If not, repeat for next power-of-ten
*
         PULY                    ; Deallocate local variables
         RTS                     ; Return
```

Figure 7.6. Conversion from Binary to Decimal by Successive Subtraction

```
*
* CVBTD converts the unsigned contents of D into an equivalent 5-digit ASCII decimal
* number at the location passed in X. Registers A, B, CC, and X are changed.
*
CVBTD:  LEAY   K+12,PCR     ; End address of power-of-tens
        PSHY                ; Save a local variable for CPY
        LEAY   K,PCR        ; Point Y to power-of-tens
        PSHY                ; Save for LDY
        PSHX                ; Save output string pointer
L1:     LDY    2,SP         ; Get address of powers-of-ten
        LDX    2,Y+         ; Get a power of ten, move pointer
        CPY    4,SP         ; At end of string?
        BEQ    L2           ; Yes, exit
        STY    2,SP         ; Save address of powers-of-ten
        LDY    #0           ; High 16-bits of dividend (low 16-bits in D)
        EDIV                ; D is remainder Y is quotient
        XGDY                ; Put quotient in D
        ADDB   #$30         ; ASCII conversion
        PULX                ; Restore output string pointer
        STAB   1,X+         ; Store character, move pointer
        PSHX                ; Save output string pointer
        XGDY                ; Remainder to D
        BRA    L1           ; Continue
L2:     LEAS   6,SP         ; Balance stack
        RTS                 ; Exit
```

Figure 7.7 Conversion from Binary to Decimal by Division by Powers of Ten

The remaining two techniques use division. The second division technique that divides by different numbers each time, getting the most significant digit first, suffers from the same malady as the first multiplication scheme. A lot of numbers have to be stored in a table, and that reduces static efficiency. The other division scheme, getting the least significant digits first, is the one most commonly taught in introductory courses on logic design. Although it is better than the last division scheme, it is going to be less useful on the 6812 than the multiplication schemes, because this microcomputer has no decimal divide instruction, and the divide routine will take up memory and be slow. Thus the best scheme for conversion from decimal to binary is the multiplication scheme that uses the nesting formula (4) to avoid the need to store all the powers of 10.

Let's apply the division techniques to converting a binary number into ASCII decimal digits to be output to a terminal. In particular, suppose that we want to convert the unsigned 16-bit number in D into five ASCII decimal digits.

Consider the division scheme that generates the most significant digit first. We could again have a table of constants in the subroutine with the directive

```
K   dc.w 10000,1000,100,10,1
```

```
* CVBTD converts the unsigned contents of D into an equivalent 5-digit ASCII decimal
* number ending at the location passed in X. Register Y is unchanged.
*
CVBTD:  PSHX              ; Save beginning address of string
        LEAX    5,X       ; End address of string
        PSHX              ; Save on stack
L1:     LDX     #10       ; Divide by 10
        LDY     #0        ; High 16-bits of dividend (low 16-bits in D)
        EDIV              ; D is remainder Y is quotient
        ADDB    #$30      ; ASCII conversion
        PULX              ; Restore output string pointer
        STAB    1,-X      ; Store character, move pointer
        PSHX              ; Save output string pointer
        XGDY              ; Remainder to D
        CPX     2,SP      ; At end of string?
        BNE     L1        ; Continue
        LEAS    4,SP      ; Balance stack
        RTS               ; Exit
```

a. Using a Loop

```
*       SUBROUTINE CVBTD converts unsigned binary number in D into five
*       ASCII decimal digits ending at the location passed in X, using recursion.
*
CVBTD:  PSHX              ; Save string pointer
        LDY     #0        ; High dividend
        LDX     #10       ; Divisor
        EDIV              ; Unsigned (Y:D) / X -> Y, remainder to D
        ADDB    #$30      ; Convert remainder to ASCII
        PULX              ; Get string pointer
        STAB    1,-X      ; Store in string
        TFR     Y,D       ; Put quotient in D
        TBEQ    D,L1      ; If zero, just exit
        BSR     CVBTD     ; Convert quotient to decimal (recursively)
L1:     RTS               ; Return
```

b. Using Recursion

Figure 7.8 Conversion from Binary to Decimal by Division by 10

Looking at the expansion (3) we see that N_4 is the quotient of the division of (D) by (K):(K + 1). If we divide the remainder by (K + 2):(K + 3), we get N_3 for the quotient, and so forth. These quotients are all in binary, so that a conversion to ASCII is also necessary. The subroutine `CVBTD` of Figure 7.6 essentially uses this technique, except that the division is carried out by subtracting the largest possible multiple of each power of ten which does not result in a carry.

An improved technique shown in Figure 7.7 uses division, rather than successive subtraction, to speed up the algorithm, but it has the same difficulties as the one in Figure 7.3; larger numbers require a larger table of constants. We now look at the other division scheme. For example, if the contents of D is divided by 10, the remainder is the binary expansion of N_0. Dividing this quotient by 10 yields a remainder equal to the binary expansion of N_1, and so on. See Figure 7.8a. This approach can be implemented by a recursive algorithm. See Figure 7.8b.

Suppose that we wanted to convert a 16-bit unsigned binary integer in D to the equivalent decimal number using the best multiplication technique, in particular, the one that used (4) instead of (3). Write the binary expansion of the number $b_{15}, \ldots, b_0$ in decimal notation as

$$[([(0 * 2 + b_{15}) * 2 + b_{14}] * 2 + b_{13}) * 2 + \ldots + b_1] * 2 + b_0 \qquad (7)$$

and assume that the equivalent five-decimal digits are to be placed at the address passed in X, with one decimal digit stored per byte in binary. (We can convert each digit to ASCII later, if necessary.) After initializing the result to 0, we iteratively build the equivalent decimal number from the innermost parentheses above by repeating the following steps:

```
* CVBTD converts the 16-bit unsigned number in D to an equivalent 6-digit BCD
*  number stored in 3 bytes at the address passed in by the calling routine in X.
*
CVBTD:  PSHD                    ; Save number to be converted
        CLR     0,X
        CLR     1,X
        CLR     2,X             ; Initialize decimal number
        MOVB    #16,1,-SP       ; Push count
*
CBD1:   ASL     2,SP
        ROL     1,SP            ; Put next bit in C
        LDAB    #2              ; 3-byte decimal addition
*
CBD2:   LDAA    B,X             ; Get ith byte
        ADCA    B,X             ; Adding a number to itself doubles it
        DAA                     ; Double in decimal
        STAA    B,X             ; Put back ith byte
*
        DECB
        BPL     CBD2            ; Count down and loop 3 bytes
        DEC     0,SP
        BNE     CBD1            ; Count down and loop 16 bits
*
        LEAS    3,SP            ; Balance stack
        RTS
```

Figure 7.9. Conversion from Binary to Decimal by Decimal Multiplication

1. Multiply the five decimal digits pointed to by X by 2.
2. Add 1 to the five decimal digits pointed to by X if $b_n = 1$.

Recall, however, that the preceding operations must be done in decimal. For this, it is more convenient to output our digits in BCD rather than one decimal digit per byte because we can take advantage of the `DAA` instruction. In particular, notice that in the subroutine of Figure 7.9, the carry bit in loop `CBD2` equals b_n the first time through, while the next two times through, the loop completes the doubling in decimal of the BCD contents of the three bytes pointed to by X. The conversion of the BCD number to ASCII, if desired, is straightforward. (See the problems at the end of the chapter.)

Conversion of signed numbers is straightforward once the foregoing techniques are understood. This conversion can be done strictly by the formula that defines the signed number, as done above for unsigned numbers, or the signed number can be expressed as a sign and a magnitude, and the magnitude can be converted as before because it is an unsigned number. The idea of conversion is quite general. You can convert to base 12 or from base 12 using any of these ideas. You can convert between the time in a week expressed in days of the week, hours, minutes, and seconds and the time of the week in seconds represented in binary. This type of conversion is similar to going between a binary sector number and a disk track sector number, something that becomes important if you are involved in writing disk controller programs. Conversion can be met in unexpected places, and with the techniques of this section, you should be able to handle any integer conversion problem.

An important aspect of this discussion is the manner in which a software engineer approaches any problem in general and a numeric problem in particular. The general rule is to learn about all the algorithms that can be used to solve the problem. In algebraically specified problems such as conversion between number systems, the algorithms are described by different formulas. The software engineer researches all the formulas that have been applied to the problem. For example, we found that formulas (3) and (4) apply to number system conversions. Then the different programming styles, such as storing coefficients in a table, generating the required constants by a subroutine, and using loops or recursion can be attempted to derive different programs. The shortest, fastest, or clearest program is then selected for the application.

7.3 From Formulas to Subroutine Calls

In following sections, we develop techniques to write subroutines to carry out the usual arithmetic operations on long integer and floating-point numbers, namely, +, –, *, and /. We could, with further study, write a whole collection of subroutines that would carry out the functions we would meet in engineering, such as sin(x), cos(x), log(x), $(x)^2$, y^z, and so on. In a given application, we might have to write a program segment to evaluate a complicated formula involving these operations, for example,

$$z = (x + 2 * y) / (\sin(u) + w) \tag{8}$$

This section examines a general technique to write such a program segment.

For simplicity, we will assume that all operands are long integers or single-precision floating point numbers (four bytes) and that the subroutines are written so that the values of the operands are pushed on the stack before the subroutine call with only the result on top of the stack after the return. (See Figure 7.10, and notice particularly that the subroutine for an operation with two operands leaves only the result on top of the stack.) To place these parameters on the stack, we will here assume that a macro `PUSH` has been written that pushes the 4-byte number at address `ADDR` onto the stack, low byte first, with `PUSH ADDR.` Similarly, we will assume that a macro `PULL` has been written so that

```
PULL ADDR
```

will pull a 4-byte number off the stack, high byte first, and place it at the address `ADDR`. Variables x, y, u, w, and z are at addresses X, Y, `U`, `W`, and `Z`, respectively, and the constant 2 is at address `K2`. This can be implemented with the directives in (9). We can evaluate (8) with the program segment (10) , where the symbolic names of the variables are in (9). In this segment, `FPMUL`, `FPADD`, and `FPDIV` are subroutines to multiply, add, and divide floating-point numbers, while `SIN` is a subroutine to calculate the sine of a floating-point number. The movement of the stack is shown after each operation in Figure 7.11.

```
X:  DS.L 1
Y:  DS.L 1
U:  DS.L 1
W:  DS.L 1
Z:  DS.L 1
K2: DS.L 1
```
(9)

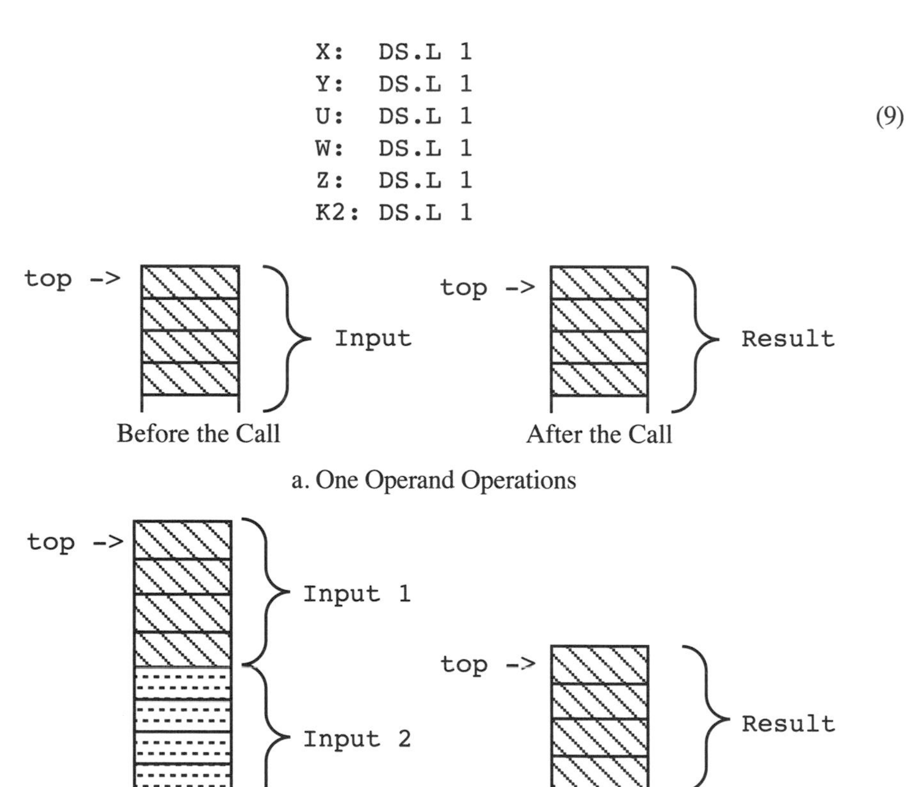

Figure 7.10. Passing Parameters on the Stack

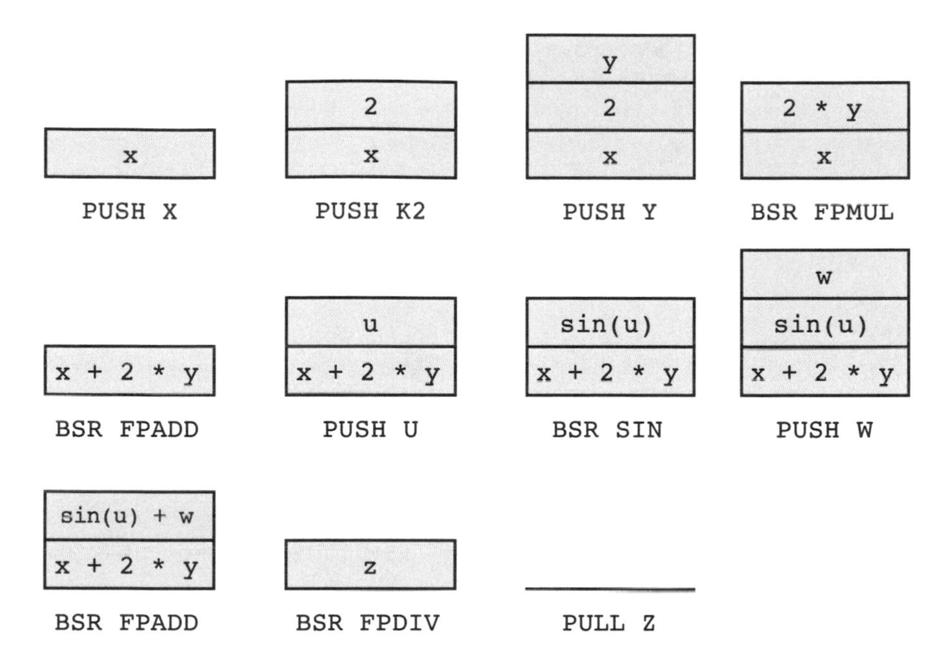

Figure 7.11. Stack Movement when Evaluating (8)

```
PUSH    X
PUSH    K2
PUSH    Y
BSR     FPMUL
BSR     FPADD
PUSH    U                (10)
BSR     SIN
PUSH    W
BSR     FPADD
BSR     FPDIV
PULL    Z
```

How does one write a program segment to evaluate formula or expression (8)? The method comes from the work of the Polish logician Jan Lucasiewicz, who investigated ways of writing expressions without using parentheses. With his technique, referred to as *Polish notation,* one would write (8) as

$$z = x\ 2\ y\ *\ +\ w\ \sin\ (u)\ +\ / \tag{11}$$

Notice that when reading (11) from left to right, each variable name generates a PUSH in (10) and each operation generates a subroutine call. Going between (10) and (11) is easy. We soon show a simple algorithm to write a segment for any expression.

The way to display a formula by parsing it into its basic operations uses a binary tree as we will use again in §10.4. Moreover, the technique to write a program segment to evaluate the formula from its tree representation will be virtually the same one that we use to scan that tree from the left in Chapter 10. Any operation with two operands, such as +, has a tree with two branches and nodes whose values are the operands for the operation. For example, the result of x + y and x * y are represented by the trees

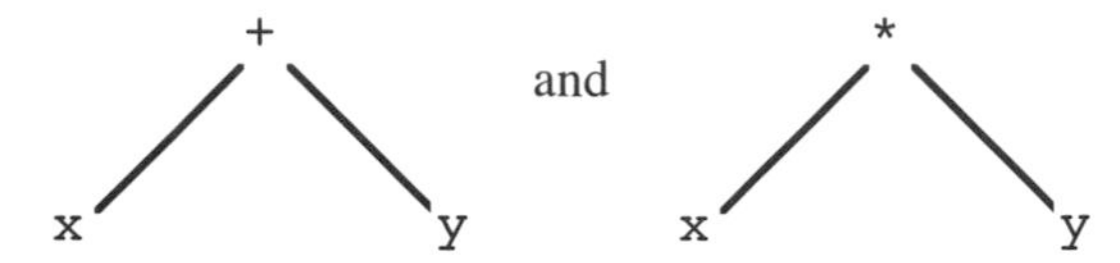

If the operation has only one operand, the operation is displayed as a tree with only one branch. For example, sin(y) and sqrt(x) (for "square root of x") have trees

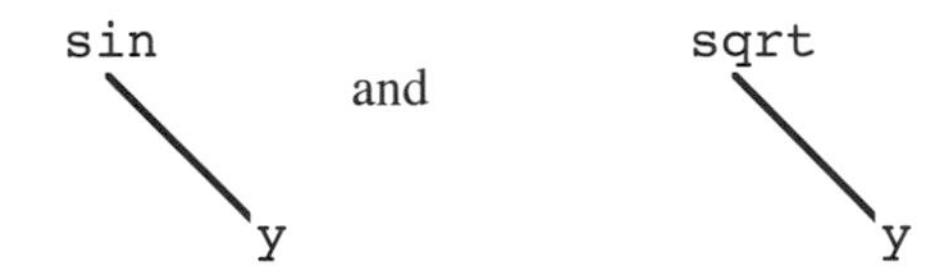

If your formula is more complicated than those shown above, you just plug in the results of one tree where the operand appears in the other tree. For instance, if you want the tree for sqrt(x + y), just substitute the tree for x + y into the node, for example,

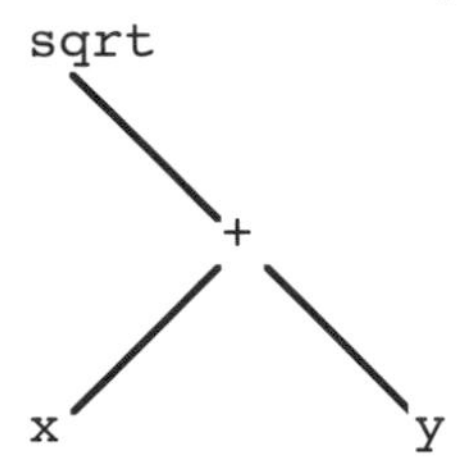

The tree for the formula (8) is shown in Figure 7.12. Notice how the formula is like a projection onto a horizontal line below the tree and how the tree can be built up from the bottom following the "plug in" technique just described.

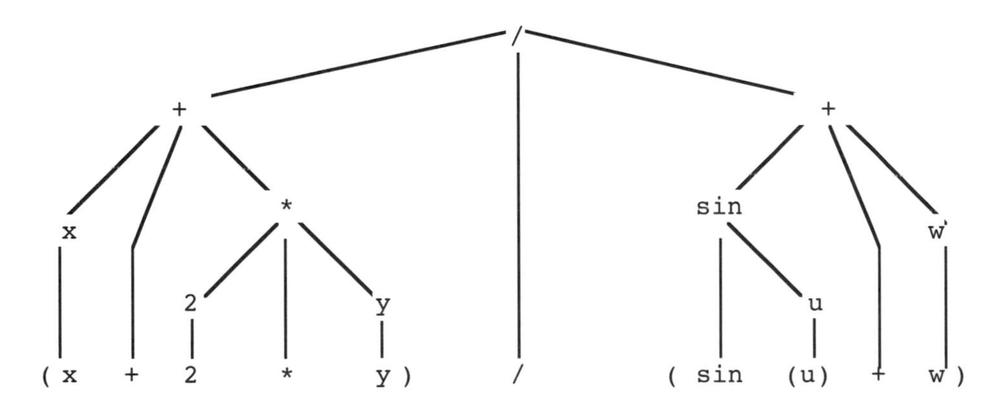

Figure 7.12. Finding a Parsing Tree for (8) from the Bottom Up

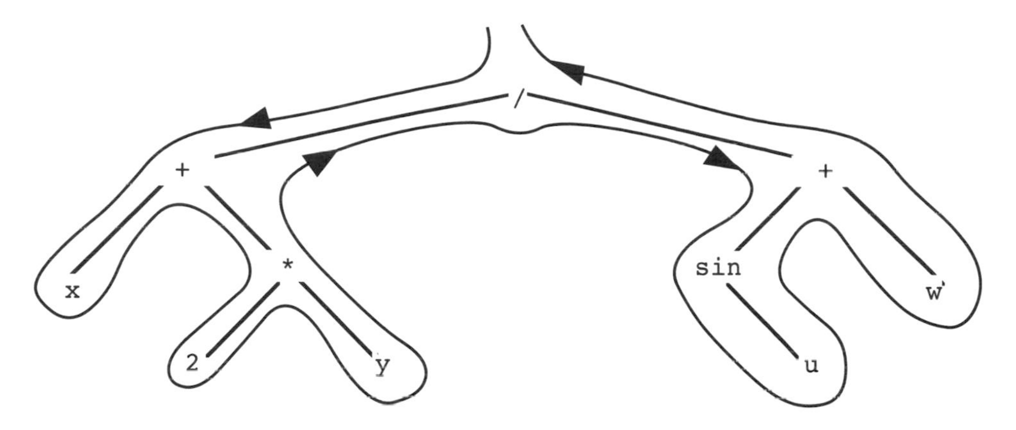

Figure 7.13. Algorithm to Write a Sequence of Subroutine Calls for (8)

The reader is invited to find these *parsing trees* for the formulas given in the problems at the end of the chapter. Note that a parsing tree is really a good way to write a formula because you can see "what plugs into what" better than if you write the formula in the normal way. In fact, some people use these trees as a way to write all of their formulas, even if they are not writing programs as you are, because it is easier to spot mistakes and to understand the expression. Once the parsing tree is found, we can write the sequence of subroutine calls in the following way. Draw a string around the tree, as shown in Figure 7.13. As we follow the string around the tree, a subroutine call or PUSH is made each time we pass a node for the last time or, equivalently, pass the node on the right. When a node with an operand is passed, we execute the macro PUSH for that operand. When a node for an operation is passed, we execute a subroutine call for that operation. Compilers use parsing trees to generate the subroutine calls to evaluate expressions in high-level languages. The problems at the end of the chapter give you an opportunity to learn how you can store parsing trees the way that a compiler might do it, using techniques from the end of Chapter 6, and how you can use such a tree to write the sequence of subroutine calls the way a compiler might.

As a second example, we consider a program evaluating the consecutive expressions

delta = delta + c
s = s + (delta * delta)

These can be described by the trees

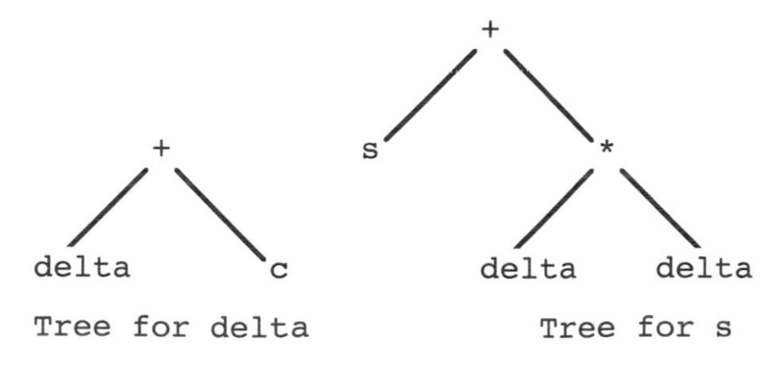

The subroutine calls for these consecutive expressions are shown in (12) assuming that the 4-byte floating-point numbers c, s, and delta are at addresses `C`, `S`, and `DELTA`, respectively.

```
PUSH    DELTA
PUSH    C
BSR     FPADD
PULL    DELTA
PUSH    S                          (12)
PUSH    DELTA
PUSH    DELTA
BSR     FPMUL
BSR     FPADD
PULL    S
```

The example above is easy to work through once you have studied the example for (8). Note, however, that a lot of pushing and pulling is done between the variable locations and the stack. Seven of the ten macros or subroutines merely move data to or from the stack, and three do arithmetic operations. It might be more efficient to use another technique for simple problems like this one. The stack method handles more complicated situations and offers a completely general technique for evaluating expressions.

This technique handles formulas of any complexity with ease and accuracy; it can be used with any system that uses a stack to hold intermediate results. This approach can be used in the 6812 such that the stack pointed to by SP is used to store the intermediate results. This approach can be used in the 6812 such that an auxilliary stack (Figure 3.11) is used to store these intermediate results. Moreover, a microcontroller such as the 68332 (§12.4) which has eight data registers analogous to the 6812's two accumulators A and B, can use this technique to assign its data registers to store these intermediate results. That is, the first item pushed is kept in data register 0, the next in data register 1, and so on. In the next section, we find that the stack described in this section is best handled by putting its topmost element in registers, like the 68332 does, and to put the remainder of the stack described in this section on the hardware stack pointed to by SP.

7.4 Long Integer Arithmetic

Multiple-precision arithmetic is very important in microcontrollers because the range of integers specified by a 16-bit word, whether signed or unsigned, is too small for many applications. Therefore, we will develop 32-bit arithmetic operations in this section. We will use the stack discussed in the last section for storage of intermediate results, to provide generally useful subroutines. However, due to the use of the register pair Y:D for multiplication, we put the top 32-bit element of the stack in register Y (high order 16 bits) and accumulator D (low order 16 bits); we call this pair of registers "register L."

It should be clear that pushing and pulling can be done by loading and storing L, which is done by loading and storing Y and D. However, to maintain the stack mechanism, pushing requires moving L into a long word on the hardware stack. The following program segment pushes a long word from location `ALPHA`.

```
        PSHD                    ; Move up low 16-bits from register L to stack
        PSHY                    ; Move up high 16-bits from register L to stack
        LDY     ALPHA           ; Get high 16-bits of register L
        LDD     ALPHA+2         ; Get low 16-bits of register L
```

This program segment's first two instructions alone will duplicate the top stack element.

```
        PSHD                    ; Duplicate low 16-bits from register L to stack
        PSHY                    ; Duplicate high 16-bits from register L to stack
```

Similarly, to maintain the stack mechanism, a long word can be pulled from the stack to fill L. The following program segment pulls a long word into location `ALPHA`.

```
        STY     ALPHA           ; Save high 16-bits of register L
        STD     ALPHA+2         ; Save low 16-bits of register L
        PULY                    ; Move down high 16-bits to register L
        PULD                    ; Move down low 16-bits to register L
```

If you use a macro assembler, these operations are easily made into macros. Otherwise, you can "hand-expand" these "macros" whenever you need these operations.

A 32-bit negate subroutine is shown in Figure 7.14. The algorithm is implemented by subtracting register L from 0, putting the result in L. This is a monadic operation, an operation on one operand. You are invited to write subroutines for other simple monadic operations. Increment and decrement are somewhat similar, and shift left and right are much simpler than this subroutine. (See the problems at the end of the chapter.)

A multiple-precision comparison is tricky in almost all microcomputers if you want to correctly set all of the condition code bits so that, in particular, all of the conditional branch instructions will work after the comparison. The subroutine of Figure 7.15 shows how this can be done for the 6812. If Z is initially set, using `ORCC #4`, entry at `CPZRO` will test register L for zero while pulling it from the stack. The first part of this subroutine suggests how subroutines for addition, subtraction, ANDing, ORing, and exclusive-ORing can be written.

```
*       SUBROUTINE NEG  negates the 32-bit number in L.
*
NEG:    PSHY                    ; Save high 16 bits
        PSHD                    ; Save low 16 bits, which are used first
        CLRA                    ; Clear accumulator D
        CLRB
        SUBD    2,SP+           ; Pull low 16 bits, subtract from zero
        TFR     D,Y             ; Save temporarily in Y
        LDD     #0              ; Clear accumulator D, without changing carry
        SBCB    1,SP            ; Subtract next-to-most-significant byte
        SBCA    2,SP+           ; Subtract most-significant byte, balance stack
        XGDY                    ; Exchange temporarily in Y with high 16-bits
        RTS                     ; Return with result in register L
```

Figure 7.14. 32-Bit Negation Subroutine

```
* SUBROUTINE COMPAR subtracts the next word on the stack from L, and pulls the
* next word into L. Condition codes reflect top - next
*
COMPAR:  SUBD   4,SP    ; Compare with low 16 bits of operand
         XGDY           ; Get high 16-bits
         SBCB   3,SP    ; Compare with next 8 bits of operand
         SBCA   2,SP    ; Compare with high 8 bits of operand
CPZRO:   TBNE   D,L1    ; If high 16 bits of result are nonzero, go to clear Z
         TBEQ   Y,L2    ; If low 16 bits of result are zero, leave Z alone
L1:      ANDCC  #$FB    ; Clear Z bit
L2:      PULX           ; Pull return address
         PULY           ; Remove operand from stack
         PULD           ; Remove operand from stack
         JMP    0,X     ; Return to caller
```

Figure 7.15. 32-Bit Compare Subroutine

Multiple-precision multiplication takes advantage of the EMUL instruction. The general 32-bit by 32-bit unsigned case is illustrated in Figure 7.16 and handled by the subroutine in Figure 7.17. A signed multiplication subroutine can be easily written that combines Figures 7.17 and 7.3. Unsigned division is shown in Figure 7.18. It can be modified to leave the remainder, rather than the quotient. For signed division, recall that the sign of the remainder is the same as the sign of the dividend and that the sign of the quotient is positive if the signs of the dividend and divisor are equal; otherwise, it is negative. A signed divide can be implemented with the unsigned divide and sign modification using the above rule. These subroutines can also be optimized. When a PUSH macro precedes an ADD subroutine, the combined operation can be done simply as seen in Figure 7.19.

We have completed our examination of multiple-precision arithmetic for both signed and unsigned integers. With the techniques developed in this section and in the examples of the earlier chapters, you should be prepared to handle any arithmetic calculation with signed or unsigned long (32-bit) integers.

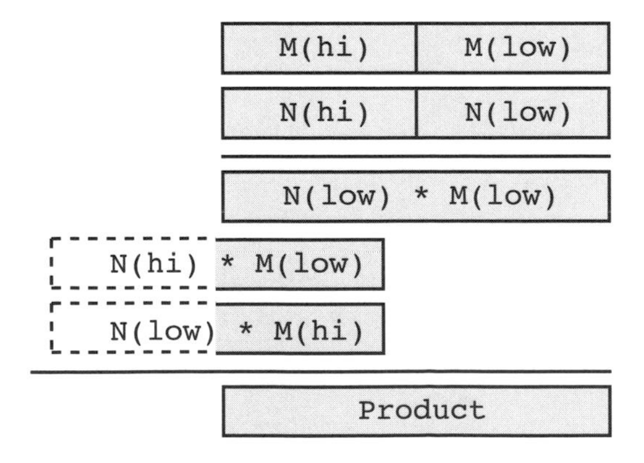

Figure 7.16. Multiplication of 32-Bit Numbers

```
*       SUBROUTINE MULT multiplies the unsigned next word on the stack with
*       L, and pulls the next word
*
LCSAVE:  EQU     *
         ORG     0
PROD:    DS.L    1
N:       DS.L    1
M:       DS.L    1

         ORG     LCSAVE
MULT:    PULX                    ; pull return address
         PSHD                    ; low part of N
         PSHY                    ; high part of N
*
         LDY     M+2-4,SP        ; note: M is operand offset in sub middle
         EMUL                    ; note - accum D is still low part of N
         PSHD                    ; low word of product
         PSHY                    ; high word of product
*
         LDD     N,SP            ; get high part of N
         LDY     M+2,SP          ; get low part of M
         EMUL
         ADDD    PROD,SP         ; add to high word
         STD     PROD,SP         ; place back
*
         LDD     N+2,SP          ; get low part of N
         LDY     M,SP            ; get high part of M
         EMUL
         ADDD    PROD,SP         ; add to high word
         TFR     D,Y             ; high 16 bits
         LDD     PROD+2,SP       ; low 16 bits
*
         LEAS    12,SP           ; remove M, N, and PROD
         JMP     0,X             ; return
```

Figure 7.17. 32-Bit by 32-Bit Unsigned Multiply Subroutine

```
LCSAVE:  EQU     *
         ORG     0
RTRN:    DS.W    1
COUNT:   DS.B    1
DVS:     DS.L    1
REM:     DS.L    1
QUOT:    DS.L    1
```

Figure 7.18. 32-Bit by 32-Bit Unsigned Divide Subroutine

```
            ORG     LCSAVE
*
*           SUBROUTINE DIV
* DIV divides the unsigned next word on the stack into L, and pulls the next word
*
DIV:        PULX                    ; unstack return address
            LEAS    -4,SP           ; room for remainder right above dividend
            PSHD                    ; save low 16 bits of divisor
            PSHY                    ; save high 16 bits of divisor
            MOVB    #32,1,-SP       ; count for 32 bits
            PSHX                    ; put back return address
            CLRA
            CLRB
            STD     REM,SP
            STD     REM+2,SP
*
DIV1:       CLC                     ; divide loop
            LDAA    #8              ; shift remainder and divisor: shift 8 bytes
            LEAX    QUOT+3,SP       ; pointer for bottom of quotient-remainder
DIV2:       ROL     1,X-
            DBNE    A,DIV2
*
            LDY     REM,SP          ; subtract from partial product
            LDD     REM+2,SP        ; (note: 4 extra bytes on stack)
*
            SUBD    DVS+2,SP
            XGDY
            SBCB    DVS+1,SP
            SBCA    DVS,SP
            XGDY
*
            BCS     DIV3            ; if borrow
            STD     REM+2,SP        ; then put it back
            STY     REM,SP
            INC     QUOT+3,SP       ; and put 1 into lsb of quotient
DIV3:       DEC     COUNT,SP        ; counter is high byte of last operand
            BNE     DIV1            ; count down - 32 bits collected
*
            PULX                    ; pull return
            LEAS    9,SP            ; balance stack - remove divisor
            PULY
            PULD                    ; pop quotient
DIVEXIT:    JMP     0,X             ; return to caller
```

Figure 7.18. Continued.

```
* ADD adds the word pointed to by X into L (combined PUSH and ADD)
ADD:    ADDD   2,X          ; add low 16 bits
        XGDY                ; put high 16 bits in D
        ADCB   1,X          ; add mid 8 bits
        ADCA   0,X          ; add high 16 bits
        XGDY                ; put high 16 bits in Y
        RTS                 ; return to caller
```

Figure 7.19. Push-and-Add Subroutine

7.5 Floating-Point Arithmetic and Conversion

We have been concerned exclusively with integers, and, as we have noted, all of the subroutines for arithmetic operations and conversion from one base to another could be extended to include signs if we wish. We have not yet considered arithmetic operations for numbers with a fractional part. For example, the 32-bit string $b_{31}, \ldots, b_0$ could be used to represent the number x, where

$$x = b_{31} * 2^{23} + \ldots + b_8 * 2^0 + \ldots + b_0 * 2^{-8} \quad (13)$$

The notation $b_{31} \ldots b_8 \cdot b_7 \ldots b_0$ is used to represent x, where the symbol "$\cdot$," called the binary point, indicates where the negative powers of 2 start. Addition and subtraction of two of these 32-bit numbers, with an arbitrary placement of the binary point for each, is straightforward except that the binary points must be aligned before addition or subtraction takes place and the specification of the exact result may require as many as 64 bits. If these numbers are being added and subtracted in a program (or multiplied and divided), the programmer must keep track of the binary point and the number of bits being used to keep the result. This process, called scaling, was used on analog computers and early digital computers. In most applications, scaling is so inconvenient to use that most programmers use other representations to get around it.

One technique, called a *fixed-point* representation, fixes the number of bits and the position of the binary point for all numbers represented. Thinking only about unsigned numbers for the moment, notice that the largest and smallest nonzero numbers that we can represent are fixed once the number of bits and the position of the binary point are fixed. For example, if we use 32 bits for the fixed-point representation and eight bits for the fractional part as in (13), the largest number that is represented by (13) is about 2^{24} and the smallest nonzero number is 2^{-8}. As one can see, if we want to do computations with either very large or very small numbers (or both), a large number of bits will be required with a fixed-point representation. What we want then is a representation that uses 32 bits but gives us a much wider range of represented numbers and, at the same time, keeps track of the position of the binary point for us just as the fixed-point representation does. This idea leads to the floating-point representation of numbers, which we discuss in this section. After discussing the floating-point representation of numbers, we examine the arithmetic of floating-point representation and the conversion between floating-point representations with different bases.

We begin our discussion of floating-point representations by considering just unsigned (nonnegative) numbers. Suppose that we use our 32 bits b_{31}, ... ,b_0 to represent the number

$$S * 2^E$$

where S, the *significand*, is of the form

$$b_{23} .b_{22} \ldots b_0$$

and 2^E, the *exponential part*, has an exponent E, which is represented by the bits b_{31}, . . . , b_{24}. If these bits are used as an 8-bit two's-complement representation of E, the range of the numbers represented with these 32 bits goes from 2^{-151} to 2^{127}, enclosing the range for the 32-bit fixed-point numbers (13) by several orders of magnitude. (To get the smallest exponent of –151, put all of the significand bits equal to 0, except b_0 for an exponent of –128–23 = –151.)

This type of representation is called a *floating-point* representation because the binary point is allowed to vary from one number to another even though the total number of bits representing each number stays the same. Although the range has increased for this method of representation, the number of points represented per unit interval with the floating-point representation is far less than the fixed-point representation that has the same range. Furthermore, the density of numbers represented per unit interval gets smaller as the numbers get larger. In fact, in our 32-bit floating-point example, there are 273 + 1 uniformly spaced points represented in the interval from 2^n to 2^{n+1} as n varies between –128 and 127.

Looking more closely at this same floating-point example, notice that some of the numbers have several representations; for instance, a significand of 1.100 . . . 0 with an exponent of 6 also equals a significand of 0.1100 . . . 0 with an exponent of 7. Additionally, a zero significand, which corresponds to the number zero, has 256 possible exponents. To eliminate this multiplicity, some form of standard representation is usually adopted. For example, with the bits b_{31}, ... , b_0 we could standardize our representation as follows. For numbers greater than or equal to 2^{-127} we could always take the representation with b_{23} equal to 1. For the most negative exponent, in this case –128, we could always take b_{23} equal to 0 so that the number zero is represented by a significand of all zeros and an exponent of –128. Doing this, the bit b_{23} can always be determined from the exponent. It is 1 for an exponent greater than –128 and 0 for an exponent of –128. Because of this, b_{23} does not have to be stored, so that, in effect, this standard representation has given us an additional bit of precision in the significand. When b_{23} is not explicitly stored in memory but is determined from the exponent in this way, it is termed a *hidden bit*.

Floating-point representations can obviously be extended to handle negative numbers by putting the significand in, say, a two's-complement representation or a signed-magnitude representation. For that matter, the exponent can also be represented in any of the various ways that include representation of negative numbers. Although it might seem natural to use a two's-complement representation for both the significand and the exponent with the 6812, one would probably not do so, preferring instead to adopt one of the standard floating-point representations.

We now consider the essential elements of the proposed IEEE standard 32-bit floating-point representation. The numbers represented are also called *single precision floating-point* numbers, and we shall refer to them here simply as *floating-point numbers*. The format is shown below.

31	30 — 23	22 — 0
s	e	f

In the drawing, s is the sign bit for the significand, and f represents the 23-bit fractional part of the significand magnitude with the hidden bit, as above, to the left of the binary point. The exponent is determined from e by a *bias* of 127, that is, an e of 127 represents an exponent of 0, an e *of* 129 represents an exponent of +2, an e of 120 represents an exponent of –7, and so on. The hidden bit is taken to be 1 unless e has the value 0. The floating-point numbers given by

$$\begin{array}{ll} (-1)^s * 2^{e-127} * 1.f & \text{for } 0 < e < 256 \\ 0 & \text{for } e = 0 \text{ and } f = 0 \end{array} \qquad (14)$$

are called *normalized*. (In the IEEE standard, an e of 255 is used to represent $\pm$ infinity together with values that are not to be interpreted as numbers but are used to signal the user that his calculation may no longer be valid.) The value of 0 for e is also used to represent *denormalized* floating-point numbers, namely,

$$(-1)^s * 2^{e-126} * 0.f \qquad \text{for } e = 0.f \neq 0$$

Denormalized floating-point numbers allow the representation of small numbers with magnitudes between 0 and 2^{-126}. In particular, notice that the exponent for the denormalized floating-point numbers is taken to be –126, rather than –127, so that the interval between 0 and 2^{-126} contains $2^{23}-1$ uniformly spaced denormalized floating-point numbers.

Although the format above might seem a little strange, it turns out to be convenient because a comparison between normalized floating-point numbers is exactly the same as a comparison between 32-bit signed-magnitude integers represented by the string s, e, f. This means that a computer implementing signed-magnitude integer arithmetic will not have to have a separate 32-bit compare for integers and floating-point numbers. In larger machines with 32-bit words, this translates into a hardware savings, while in smaller machines, like the 6812, it means that only one subroutine has to be written instead of two if signed-magnitude arithmetic for integers is to be implemented.

We now look more closely at the ingredients that floating-point algorithms must have for addition, subtraction, multiplication, and division. For simplicity, we focus our attention on these operations when the inputs are normalized floating-point numbers and the result is expressed as a normalized floating-point number.

To add or subtract two floating-point numbers, one of the representations has to be adjusted so that the exponents are equal before the significands are added or subtracted. For accuracy, this *unnormalization* always is done to the number with the smaller exponent. For example, to add the two floating-point numbers

```
      2^4 * 1.00 . . . 0
+     2^2 * 1.00 . . . 0
```

$$\begin{array}{lr} & 2^4 * 1.00\ldots0 \\ + & \underline{2^2 * 1.00\ldots0} \end{array}$$

we first unnormalize the number with the smaller exponent and then add as shown.

$$\begin{array}{lr} & 2^4 * 1.000\ldots0 \\ + & \underline{2^4 * 0.010\ldots0} \\ & 2^4 * 1.010\ldots0 \end{array}$$

(For this example and all those that follow, we give the value of the exponent in decimal and the 24-bit magnitude of the significand in binary.) Sometimes, as in adding,

$$\begin{array}{lr} & 2^4 * 1.00\ldots0 \\ + & \underline{2^4 * 1.00\ldots0} \\ & 2^4 * 10.00\ldots0 \end{array}$$

the sum will have to be *renormalized* before it is used elsewhere. In this example

$$2^5 * 1.00\ldots0$$

is the renormalization step. Notice that the unnormalization process consists of repeatedly shifting the magnitude of the significand right one bit and incrementing the exponent until the two exponents are equal. The renormalization process after addition or subtraction may also require several steps of shifting the magnitude of the significand left and decrementing the exponent. For example,

$$\begin{array}{lr} & 2^4 * 1.0010\ldots0 \\ - & \underline{2^4 * 1.0000\ldots0} \\ & 2^4 * 0.0010\ldots0 \end{array}$$

requires three left shifts of the significand magnitude and three decrements of the exponent to get the normalized result:

$$2^1 * 1.00\ldots0$$

With multiplication, the exponents are added and the significands are multiplied to get the product. For normalized numbers, the product of the significands is always less than 4, so that one renormalization step may be required. The step in this case consists of shifting the magnitude of the significand right one bit and incrementing the exponent. With division, the significands are divided and the exponents are subtracted. With normalized numbers, the quotient may require one renormalization step of shifting the magnitude of the significand left one bit and decrementing the exponent. This step is required only when the magnitude of the divisor significand is larger than the magnitude of the dividend significand. With multiplication or division it must be remembered also that the exponents are biased by 127 so that the sum or difference of the exponents must be rebiased to get the proper biased representation of the resulting exponent.

In all of the preceding examples, the calculations were exact in the sense that the operation between two normalized floating-point numbers yielded a normalized floating-point number. This will not always be the case, as we can get overflow, underflow, or a result that requires some type of rounding to get a normalized approximation to the result. For example, multiplying

$$
\begin{array}{rl}
 & 2^{56} \ * \ 1.00 \ . \ . \ . \ 0 \\
* & \underline{2^{100} \ * \ 1.00 \ . \ . \ . \ 0} \\
 & 2^{156} \ * \ 1.00 \ . \ . \ . \ 0
\end{array}
$$

yields a number that is too large to be represented in the 32-bit floating-point format. This is an example of *overflow,* a condition analogous to that encountered with integer arithmetic. Unlike integer arithmetic, however, *underflow* can occur, that is, we can get a result that is too small to be represented as a normalized floating-point number. For example,

$$
\begin{array}{rl}
 & 2^{-126} \ * \ 1.0010 \ . \ . \ . \ 0 \\
- & \underline{2^{-126} \ * \ 1.0000 \ . \ . \ . \ 0} \\
 & 2^{-126} \ * \ 0.0010 \ . \ . \ . \ 0
\end{array}
$$

yields a result that is too small to be represented as a normalized floating-point number with the 32-bit format.

The third situation is encountered when we obtain a result that is within the normalized floating-point range but is not exactly equal to one of the numbers (14). Before this result can be used further, it will have to be approximated by a normalized floating-point number. Consider the addition of the following two numbers.

$$
\begin{array}{rl}
 & 2^{2} \ * \ 1.00 \ . \ . \ . \ 00 \\
+ & \underline{2^{0} \ * \ 1.00 \ . \ . \ . \ 01} \\
 & 2^{2} \ * \ 1.01 \ . \ . \ . \ 00(01)
\end{array}
$$

(in parenthesis: least significant bits of the significand)

The exact result is expressed with 25 bits in the fractional part of the significand so that we have to decide which of the possible normalized floating-point numbers will be chosen to approximate the result. *Rounding toward plus infinity* always takes the approximate result to be the next larger normalized number to the exact result, while *rounding toward minus infinity* always takes the next smaller normalized number to approximate the exact result. *Truncation* just throws away all the bits in the exact result beyond those used in the normalized significand. Truncation rounds toward plus infinity for negative results and rounds toward minus infinity for positive results. For this reason, truncation is also called *rounding toward zero*. For most applications, however, picking the closest normalized floating-point number to the actual result is preferred. This is called *rounding to nearest*. In the case of a tie, the normalized floating-point number with the least significant bit of 0 is taken to be the approximate result. Rounding to nearest is the default type of rounding for the IEEE floating-point standard. With rounding to nearest, the magnitude of the error in the approximate result is less than or equal to the magnitude of the exact result times 2^{-24}.

One could also handle underflows in the same way that one handles rounding. For example, the result of the subtraction

```
     2-126 * 1.0110 . . . 0
-    2-126 * 1.0000 . . . 0
     2-126 * 0.0110 . . . 0
```

could be put equal to 0, and the result of the subtraction

```
     2-126 * 1.1010 . . . 0
-    2-126 * 1.0000 . . . 0
     2-126 * 0.1010 . . . 0
```

could be put equal to 2^{-126} `* 1.0000`. More frequently, all underflow results are put equal to 0 regardless of the rounding method used for the other numbers. This is termed *flushing to zero*. The use of denormalized floating-point numbers appears natural here, as it allows for a gradual underflow as opposed to, say, flushing to zero. To see the advantage of using denormalized floating-point numbers, consider the computation of the expression `(Y — X) + X`. If Y – X underflows, X will always be the computed result if flushing to zero is used. On the other hand, the computed result will always be Y if denormalized floating-point numbers are used. The references mentioned at the end of the chapter contain further discussions on the merits of using denormalized floating point numbers. Implementing all of the arithmetic functions with normalized and denormalized floating-point numbers requires additional care, particularly with multiplication and division, to ensure that the computed result is the closest represented number, normalized or denormalized, to the exact result. It should be mentioned that the IEEE standard requires that a warning be given to the user when a denormalized result occurs. The motivation for this is that one is losing precision with denormalized floating-point numbers. For example, if during the calculation of the expression `(Y — X) * Z`. If Y –X underflows, the precision of the result may be doubtful even if `(Y — X) * Z` is a normalized floating-point number. Flushing to zero would, of course, always produce zero for this expression when `(Y — X)` underflows.

The process of rounding to nearest, hereafter just called *rounding,* is straightforward after multiplication. However, it is not so apparent what to do after addition, subtraction, or division. We consider addition/subtraction. Suppose, then, that we add the two numbers

```
     20   * 1.0000 . . . 0
+    2-23 * 1.1110 . . . 0
```

After unnormalizing the second number, we have

```
     20 * 1.0000 . . . 00
+    20 * 0.0000 . . . 01(111)
     20 * 1.0000 . . . 01(111)
```

(The enclosed bits are the bits beyond the 23 fractional bits of the significand.) The result, when rounded, yields 2^0 * 1.0 . . . 010. By examining a number of cases, one can see that only three bits need to be kept in the unnormalization process, namely,

g	r	s

where g is the *guard bit*, r is the *round bit*, and s is the *sticky bit*. When a bit b is shifted out of the significand in the unnormalization process,

b->g
g-> r (15)
s OR r -> s

Notice that if s ever becomes equal to 1 in the unnormalization process, it stays equal to 1 thereafter or "sticks" to 1. With these three bits, rounding is accomplished by incrementing the result by 1 if

g = 1 AND (r OR s = 1) (16)
or g = 1 AND (r AND s = 0)
and the least significant bit of the significand is 1

If adding the significands or rounding causes an overflow in the significand bits (only one of these can occur), a renormalization step is required. For example,

```
     2^0   * 1.1111 . . . 1
+    2^-23 * 1.1110 . . . 0
```

becomes, after rounding, 2^0 * 10.0 . . . 0 Renormalization yields 2^1 * 1.0 . . . 0, which is the correct rounded result, and no further rounding is necessary.

Actually, it is just as easy to save one byte for rounding as it is to save three bits, so that one can use six rounding bits instead of one, as follows.

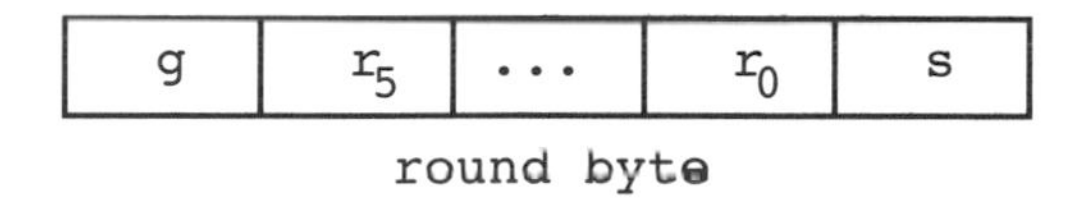

The appropriate generalization of (15) can be pictured as

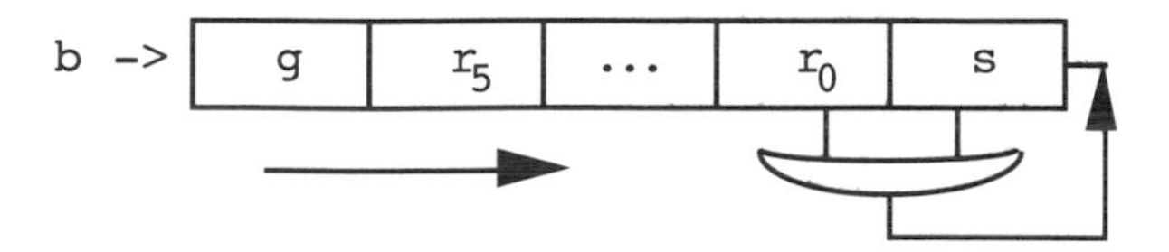

while (16) is exactly the same as before with r replaced by $r_5 \ldots r_0$

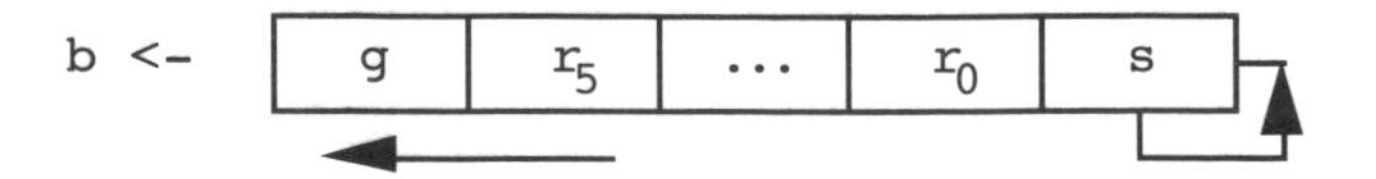

The rounding process for addition of numbers with opposite signs (e.g., subtraction) is exactly like that above except that the round byte must be included in the subtraction, and renormalization may be necessary after the significands are subtracted. In this renormalization step, several shifts left of the significand may be required where each shift requires a bit b for the least significant bit of the significand. It may be obtained from the round byte as shown below. (The sticky bit may also be replaced by zero in the process pictured without altering the final result. However, at least one round bit is required.) After renormalization, the rounding process is identical to (16). As an example,

$$\begin{array}{lll} & 2^{0} & * \; 1.1111 \; . \; . \; . \; 1 \\ - & 2^{-23} & * \; 1.1110 \; . \; . \; . \; 0 \\ \hline \end{array}$$

becomes

$$\begin{array}{lll} & 2^{0} & * \; 1.0000 \; . \; . \; . \; 00 \\ - & 2^{0} & * \; 0.0000 \; . \; . \; . \; 01(11100000) \\ \hline & 2^{0} & * \; 0.1111 \; . \; . \; . \; 10(00100000) \end{array}$$

which, after renormalization and rounding, becomes 2^{-1} * 1.1...100. Subroutines for floating-point addition and multiplication are given in Hiware's C and C++ libraries. To illustrate the principles without an undue amount of detail, the subroutines are given only for normalized floating-point numbers. Underflow is handled by flushing the result to zero and setting an underflow flag, and overflow is handled by setting an overflow flag and returning the largest possible magnitude with the correct sign. These subroutines conform to the IEEE standard but illustrate the basic algorithms, including rounding. The procedure for addition is summarized in Figure 7.20, where one should note that the significands are added as signed-magnitude numbers.

One other issue with floating-point numbers is conversion. For example, how does one convert the *decimal floating-point number* $3.45786*10^4$ into a binary floating-point number with the IEEE format? One possibility is to have a table of binary floating-point numbers, one for each power of ten in the range of interest. One can then compute the expression

$$3 * 10^4 + 4 * 10^3 + \; . \; . \; . \; + 6 * 10^{-1}$$

using the floating-point add and floating-point multiply subroutines. One difficulty with this approach is that accuracy is lost because of the number of floating point multiplies and adds that are used. For example, for eight decimal digits in the decimal significand, there are eight floating-point multiplies and seven floating-point adds used in the conversion process. To get around this, one could write $3.45786 * 10^4$ as $.345786 * 10^5$ and multiply the binary floating-point equivalent of 10^5 (obtained again from a table) by the binary floating-point equivalent of .345786. This, of course, would take only one floating-point multiply and a conversion of the decimal fraction to a binary floating-point number.

1. Attach a zero round byte to each significand and unnormalize the number with the smaller exponent.
2. Add significands of operands (including the round byte).
3. If an overflow occurs in the significand bits, shift the bits for the magnitude and round byte right one bit and increment the exponent.
4. If all bits of the unrounded result are zero, put the sign of the result equal to + and the exponent of the result to the most negative value; otherwise, renormalize the result, if necessary, by shifting the bits of the magnitude and round byte left and decrementing the exponent for each shift.
5. If underflow occurs, flush the result to zero, and set the underflow flag; otherwise, round the result.
6. If overflow occurs, put the magnitude equal to the maximum value, and set the overflow flag.

Figure 7.20. Procedure for Floating-Point Addition

Converting the decimal fraction into a binary floating-point number can be carried out in two steps.

1. Convert the decimal fraction to a binary fraction.
2. Convert the binary fraction to a binary floating-point number.

Step 2 is straightforward, so we concentrate our discussion on Step 1, converting a decimal fraction to a binary fraction.

Converting fractions between different bases presents a difficulty not found when integers are converted between different bases. For example, if

$$f = a_1 * r^{-1} + \ldots + a_m * r^m \quad (17)$$

is a base-r fraction, then it can happen that when f is converted to a base-s fraction,

$$f = b_1 * s^{-1} + b_2 * s^{-2} + \ldots \quad (18)$$

that is, b_i is not equal to 0 for infinitely many values of i. (As an example of this, expand the decimal fraction 0.1 into a binary one.) Rather than trying to draw analogies to the conversion of integer representations, it is simpler to notice that multiplying the right-hand side of (18) by s yields b_1 as the integer part of the result, multiplying the resulting fractional part by s yields b_2 as the integer part, and so forth. We illustrate the technique with an example.

Suppose that we want to convert the decimal fraction .345786 into a binary fraction so that

$$.345786 = b_1 * 2^{-1} + b_2 * 2^{-2} + \ldots$$

Then b_1 is the integer part of 2 * (.345786), b_2 is the integer part of 2 times the fractional part of the first multiplication, and so on for the remaining binary digits. More often than not, this conversion process from a decimal fraction to a binary one does not terminate after a fixed number of bits, so that some type of rounding must be done. Furthermore, assuming that the leading digit of the decimal fraction is nonzero, as many as three leading bits in the binary fraction may be zero. Thus, if one is using this step to convert to a binary floating-point number, probably 24 bits after the leading zeros should be generated with the 24th bit rounded appropriately. Notice that the multiplication by 2 in this conversion process is carried out in decimal so that BCD arithmetic with the DAA instruction is appropriate here much like the CVBTD subroutine of Figure 7.9.

The conversion of a binary floating-point number to a decimal floating-point number is a straightforward variation of the process above and is left as an exercise.

This section covered the essentials of floating-point representations, arithmetic operations on floating-point numbers, including rounding, overflow, and underflow, and the conversion between decimal floating-point numbers and binary floating-point numbers. HiWare's C and C++ libraries illustrate subroutines for adding, subtracting, and multiplying single-precision floating-point numbers. This section and these libraries should make it easy for you to use floating-point numbers in your assembly language programs whenever you need their power.

7.6 Fuzzy Logic

This section, taken from the Motorola CPU12 reference manual (Rev. 1), section 9, gives a general introduction to fuzzy logic concepts and illustrates an implementation of fuzzy logic programming. There are a number of fuzzy logic programming strategies; this discussion concentrates on the methods that use 6812 fuzzy logic instructions.

In general, fuzzy logic provides for set definitions that have fuzzy boundaries rather than the crisp boundaries of Boolean logic. A Boolean variable is either true or false, while in fuzzy logic, a *linguistic variable* has a *value* that is a degree of confidence between 0 and 1. A value can be "0.2 (or 20%) confident." For a specific input value, one or more linguistic variables may be confident to some degree at the same time, and their sum need not be 1. As an input varies, one linguistic variable may become progressively less confident while another becomes progressively more confident.

Fuzzy logic *membership functions* better emulate human concepts like "I got a B on the last quiz" than Boolean functions that are either absolutely true or absolutely false; that is, conditions are perceived to have gradual or fuzzy boundaries. Despite the term "fuzzy," a specific set of input conditions always deterministically produces the same result, just as in conventional control systems. Fuzzy sets provide a means of using linguistic expressions like "I got a B" in rules that can then be evaluated with numerical precision and repeatability. We will see that fuzzy membership functions help solve certain types of complex problems that have eluded traditional methods, as we study how fuzzy logic could compute a student's course grade from his or her quiz scores.

An application expert, without any microcontroller programming experience, can generate a knowledge base. In it, membership functions express an understanding of the system's linguistic terms. And in it, ordinary language statement *rules* describe how a human expert would solve the problem. These are reduced to relatively simple data structures (the knowledge base) that reside in the microcontroller memory.

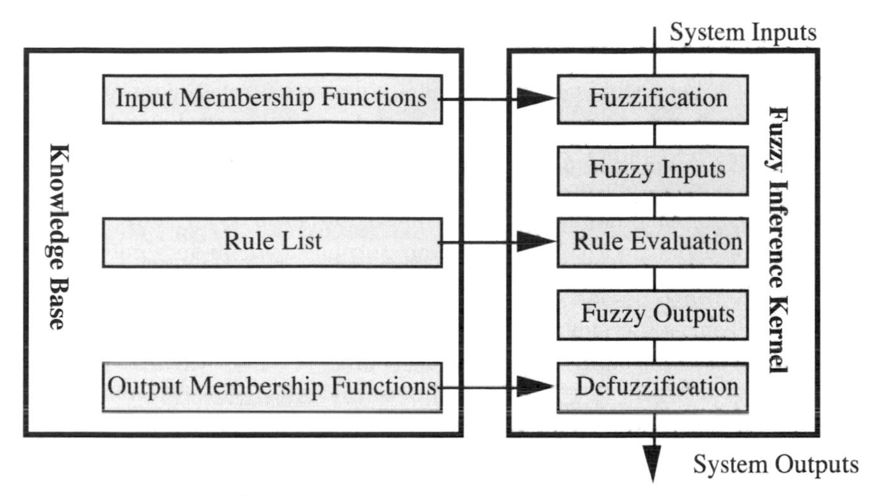

Figure 7.21. Fuzzy Logic Control System

A microcontroller-based fuzzy logic control system has a *fuzzy inference kernel* and a *knowledge base*. See Figure 7.21. The fuzzy inference kernel is executed periodically to determine system outputs based on current system inputs. The knowledge base contains membership functions and rules.

A programmer who does not know how the application system works can write a fuzzy inference kernel. One execution pass through the fuzzy inference kernel generates system output signals in response to current system input conditions. As in a conventional control system, the kernel is executed as often as needed to maintain control. If the kernel is executed too often, processor bandwidth and power are wasted;

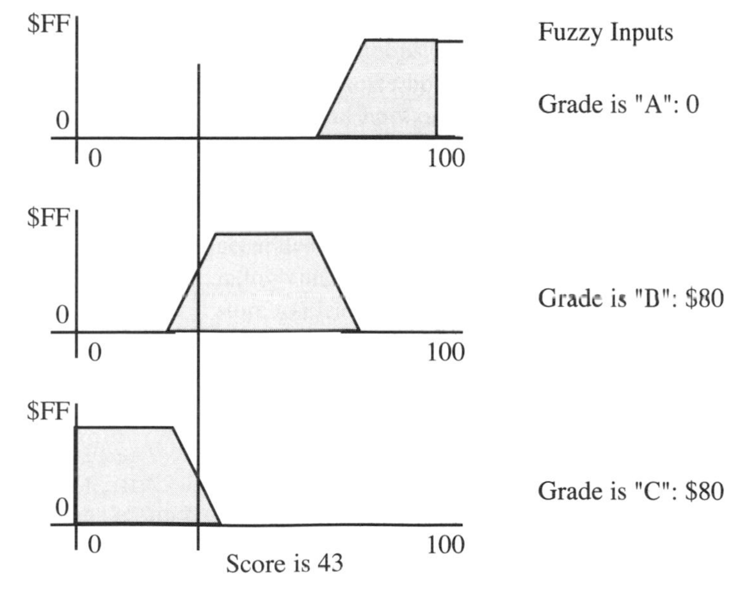

Figure 7.22. Membership Functions

but if too infrequently, the system gets too far out of control. The steps of this kernel are: fuzzification, rule evaluation, and defuzzification.

During fuzzification, the current system input values are compared against stored input membership functions, usually in a program loop structure, to determine the degree to which each linguistic variable of each system input is true. Three membership functions are indicated in Figure 7.22; these convert a student's quiz test score to a letter grade. This is accomplished by finding the y-value for the current input value on a trapezoidal membership function for each label of each system input. If a student's test score is 43, his or her membership in linguistic variable "Grade is A" is zero, in linguistic variable "Grade is B" is $80 (50%), and in linguistic variable "Grade is C" is $80 (50%). Fuzzy logic avoids the agony of missing a "B" by a point or two.

In our example, system inputs are each quiz's and homework assignment's numeric grades. Linguistic variable values are the degree of confidence that the student has a particular grade associated with each letter grade. If there are two quizzes and one homework assignment, and each has three grades, then there are nine linguistic variables.

The end result of the fuzzification step is a collection of fuzzy linguistic variables reflecting the system. This is passed to the rule evaluation phase, which processes a list of rules from the knowledge base using current fuzzy input values to produce a list of fuzzy output linguistic variables. These fuzzy outputs are considered raw suggestions for what the system output should be in response to the current input conditions. The following is an example of a typical rule:

If you get an A on quiz 1, an A on quiz 2, and an A on homework assignments, then you should get an A+ for the course.

The left portion of the rule is a statement of input conditions, *antecedents* connected by a *fuzzy AND* operator, and the right portion of the rule is a statement of output actions called *consequents*.

In an automotive antiskid braking system, about 600 such rules are used to compute the brake pressure to be applied. Analogous to the Boolean sum-of-products, rule evaluation employs a fuzzy AND operator, used to connect antecedents within a rule, and a *fuzzy OR* operator, which is implied among all rules affecting a given consequent. Each rule is evaluated sequentially, but the rules as a group are treated as if they were all evaluated simultaneously. The AND operator corresponds to the mathematical minimum operation, and the fuzzy OR operation corresponds to the maximum operation. Before evaluating any rules, all fuzzy outputs are set to zero (meaning none are true). As each rule is evaluated, the minimum antecedent is taken to be the overall confidence of the rule result. If two rules affect the same fuzzy output, the rule that is most true governs the value in the fuzzy output, because the rules are connected by an implied fuzzy OR. There is also a *fuzzy negate* operator, which complements each bit.

Each antecedent expression consists of the name of a system input, followed by "is," followed by a label name defined by a membership function in the knowledge base. Because "and" is the only operator allowed to connect antecedent expressions, there is no need to include these in the encoded rule. Each consequent expression consists of the name of a system output, followed by "is," followed by a label name; for example:

If quiz 1 is A, quiz 2 is A, homework assignment is A, then course is A+.

Rules can be weighted. The confidence value for a rule is determined as usual by finding the smallest rule antecedent. Before applying this value to the consequents for the rule, the value is multiplied by a fraction from zero (rule disabled) to $FF (rule fully enabled). The resulting modified confidence value is then applied to the fuzzy outputs. Equation (19) illustrates that the output for a set of rules is the output S_i for each rule times a weight F_i, divided by the sum of the weights.

The rules can reflect nonlinear, but deterministic, relationships. For instance, if a student gets a B on the first quiz and an A on the second quiz, the instructor may decide that the course grade is A–, but if a student gets an A on the first quiz and a B on the second quiz, the instructor may decide that the course grade is B+. But if the student also did poorly in lab work, then the course grade might be B–, to give the student some incentive to get to work. That is, if quiz 1 is A, quiz 2 is B, and lab is not C, then course is B+, but if quiz 1 is A, quiz 2 is B, and lab is C, then course is B–. In the limit, if there are n input variables and each input variable is associated with m linguistic

```
*
* Fuzzification step:
*
FUZZIFY:     LDX   #INPUT_MFS     ;Point at member function definitions
             LDY   #FUZ_INS       ;Point at fuzzy input vector
             LDAA  CURRENT_INS    ;Get first input value
             LDAB  #7             ;7 fuzzy values per input
GRAD_LOOP:   MEM                  ;Evaluate a member function
             DBNE  B,GRAD_LOOP    ;For 7 labels of 1 input
             LDAA  CURRENT_INS+1  ;Get second input value
             LDAB  #7             ;7 fuzzy values per input
GRAD_LOOP1:  MEM                  ;Evaluate a member function
             DBNE  B,GRAD_LOOP1   ;For 7 fuzzy values of 1 input
* Rule Evaluation step:
             LDAB  #7             ;Loop count
RULE_EVAL:   CLR   1,Y+           ;Clr a fuzzy out & inc ptr
             DBNE  B,RULE_EVAL    ;Loop to clr all fuzzy Outs
             LDX   #RULE_START    ;Point at first rule element
             LDY   #FUZ_INS       ;Point at fuzzy ins and outs
             LDAA  #$FF           ;Init A (and clears V-bit)
             REV                  ;Process rule list
* Defuzzification step:
DEFUZ:       LDY   #FUZ_OUT       ;Point at fuzzy outputs
             LDX   #SGLTN_POS     ;Point at singleton positions
             LDAB  #7             ;7 fuzzy outs per COG output
             WAV                  ;Calculate sums for weighted av
             EDIV                 ;Final divide for weighted av
             TFR   Y,D            ;Move result to accumulator D
             STAB  COG_OUT        ;Store system output
```

Figure 7.23. A Fuzzy Inference Kernel

variables, then there can be as many as n^m rules to produce each output linguistic variable, one to cover each case. That would be unattractive, but heuristics are used to substantially reduce this number of rules using common sense.

But before the results can be applied, fuzzy outputs must be further processed, or defuzzified, to produce a single output value that represents the combined effect of all of the fuzzy outputs. In our example, a single output would be a numerical grade for the course. The end result of the rule evaluation step is a collection of suggested or "raw" fuzzy outputs. These values were obtained by plugging current conditions (fuzzy input values) into the system rules in the knowledge base. The raw results cannot be supplied directly to the system outputs because they may be ambiguous. For instance, one raw output can indicate that a grade should be B+ with a degree of confidence of 50% while, at the same time, another indicates that the grade should be C– with a degree of confidence of 25%. A simple "max defuzzification" technique, which outputs the maximum of all the degrees of confidence, ignores all other degrees of confidence, and gives inferior results. A better defuzzification step resolves multiple-degree ambiguities by combining the raw fuzzy outputs into a composite numerical output using *singletons*. The singleton for a linguistic variable is a single value assigned to the output variable if the rules produce this linguistic variable with a degree of confidence of 1, and all other linguistic variables have a degree of confidence of zero. For instance, the singleton for A+ might be 100, the singleton for A might be 95, that for A– might be 90, and so on. Singletons S_i are weighted by fuzzy output degrees of confidence F_i, for n such outputs, normalized to a degree of confidence of one, in the expression (19)

$$\text{output} = \Sigma^n_{i=1} S_i F_i \; / \; \Sigma^n_{i=1} F_i \qquad (19)$$

Figure 7.23 shows a 6812 fuzzy inference kernel. When the fuzzification step begins, a MEM instruction fuzzifies the inputs. The current value of the system input is in accumulator A, index register X points to the first membership function definition, and a index register Y points to the first fuzzy input. As each fuzzy input is calculated by executing a MEM instruction, the result is stored to the fuzzy input, and both X and Y are updated automatically to point to the locations associated with the next fuzzy input. For each system input, a DBNE instruction executes as many MEM instructions as a system input has fuzzy input linguistic variables. Thus MEM and DBNE handle one system input's fuzzification. Repeated such program segments handle each system input.

Each trapezoidal membership function is defined by four 8-bit parameters, X1, X2, S1, and S2. X1 is where the trapezoid's left slope intercepts Y = 0, and X2 is where the trapezoid's right slope intercepts Y = 0. S1 is the trapezoid's left slope ($\Delta Y/\Delta X$), and S2 is the trapezoid's right slope ($-\Delta Y/\Delta X$), but a slope of 0 is defined as a vertical line.

More complicated membership functions can be evaluated by the TBL instruction, which interpolates arbitrary table functions. A 16-bit version of TBL, ETBL, permits handling 16-bit input variables and 16-bit degrees of confidence.

The rule evaluation step is likewise almost completely executed by the REV instruction. Before it is executed, fuzzy outputs are cleared, index register X points to a vector of the rules, Y points to the base address of fuzzy inputs and outputs, and accumulator A is set for maximum ($FF). Each antecedent expression is represented as an 8-bit relative offset from index register Y, to read an 8-bit fuzzy input. Antecedents are separated from consequents with reserved "offset" value $FE. Each consequent expression is represented as an 8-bit relative offset from index register Y, to write an 8-bit fuzzy output. The consequents end with a reserved value $FE, if more rules must be evaluated, or $FF, after the last rule is evaluated. The condition code V signifies whether antecedents are being processed (V = 0) or consequents are being processed (V = 1).

Besides REV, the more complex REVW instruction allows each rule to have a separate weighting factor, and EMIND and EMAXD can be used to implement 16-bit rule evaluation.

The WAV instruction calculates the numerator and denominator sums for weighted average of the fuzzy outputs. Before executing WAV, accumulator B must be loaded with the number of iterations, index register Y must be pointed at the list of singleton positions in the knowledge base, and index register X must be pointed at the list of fuzzy outputs in RAM. If the system has more than one system output, the WAV instruction is executed once for each system output. The final divide is performed with a separate EDIV instruction placed immediately after the WAV instruction.

The EMACS instruction can be used to evaluate 16-bit linguistic variables. A separate but simple program segment must calculate the sum of linguistic variables, which is automatically calculated by the WAV instruction.

The 6812 is currently the only microcontroller that has machine instructions that can be used to implement the complete fuzzy inference kernel. These machine instructions speed up execution of the kernel by a factor of about 10 over software evaluation that uses ordinary instructions. This feature makes the 6812 the microcontroller of choice for time-critical applications that use fuzzy logic.

In conclusion, fuzzy logic provides fuzzy rather than crisp boundaries. Linguistic variables indicate a degree of confidence. Combinations of variables are presumed to be true to the worst-case (minimum) degree of confidence, and alternatives are presumed to be true to the best-case (maximum) degree of confidence. Final values are weighted sums of typical values, using degrees of confidence as weights. Such fuzzy logic systems are currently being applied to automotive control and other rather complex control systems.

7.7 Summary

This chapter covered the techniques you need to handle integer and floating point arithmetic in a microcontroller. We discussed the conversion of integers between any two bases and then discussed signed and unsigned multiple precision arithmetic operations that had not been discussed in earlier examples. Floating-point representations of numbers with a fractional part, and the algorithms used to add and multiply floating-point numbers, were discussed together with the problems of rounding and conversion. The IEEE standard floating-point format was used throughout these discussions. We ended with the use of the stack for holding the arguments for arithmetic subroutines and showed how you can write a sequence of subroutine calls to evaluate any formula.

You should now be able to write subroutines for signed and unsigned arithmetic operations, and you should be able to write and use such subroutines that save results on the stack. You should be able to convert integers from one representation to another and write subroutines to do this conversion. You should be able to write numbers in the IEEE floating-point representation, and you should understand how these numbers are added or multiplied and how errors can accumulate.

Do You Know These Terms?

See the end of chapter 1 for instructions.

long data
float data
state
biasing
m-digit base-b representation
Polish notation
parsing tree
fixed-point representation
significand
exponential part
floating-point
hidden bit
single precision floating-point
floating-point numbers
bias
normalized number
denormalized number
unnormalization
renormalized
overflow
underflow
rounding toward plus infinity
rounding toward minus infinity
truncation
rounding toward zero
rounding to nearest
flushing to zero
rounding
guard bit
round bit
sticky bit
decimal floating-point number
linguistic variable
value
membership function
rule
fuzzy inference kernel
knowledge base
antecedent
fuzzy AND
consequent
fuzzy OR
fuzzy negate
singleton

PROBLEMS

1. How would you rewrite the subroutines of Figures 7.1 and 7.2 if you did not want any registers changed by the subroutines except D, the output parameter?

2. Write a shortest subroutine DIVS that divides the signed contents of B by the signed contents of A, putting the quotient in B and the remainder in A.

3. Rewrite the subroutine of Figure 7.4 so that, using INCH, the digits can be input from the terminal as a string. A carriage return should terminate the input string so that zero through five digits can be put in. (The empty sequence should be treated as zero.) Your subroutine should do the same thing as the one in Figure 7.4 as long as the number of digits put in is five or less. INCH inputs a character from the keyboard, returning it in A.

4. Rewrite the subroutine of Figure 7.7 so that OUTCH can be used to output the decimal digits to the terminal. OUTCH prints the character input in A.

5. Write a shortest subroutine INBCD using INCH that will input six ASCII decimal digits and place the equivalent 6-digit BCD number at the address passed in X. INCH inputs a character from the keyboard, returning it in A.

6. Write a shortest subroutine OUTBCD that will output the 6-digit BCD number pointed to by X to the terminal. The subroutine, using OUTCH, should display the equivalent 6-digit decimal with leading zeros suppressed. OUTCH prints the character in A.

7. Using the subroutine OUTCH, write another shortest subroutine OUTBCDA that puts out the BCD contents of accumulator A to the terminal. OUTCH prints the character in A.

8. Write a shortest subroutine that will replace the DAA instruction when two base-13 digits are stored per byte.

9. Give a sequence of subroutine calls for the following formula.

```
z = sqrt((17 + (x/y)) * (w - (2 + w/y)))
```

Provide a graphical parsing tree for this formula, and show assembly language statements (dc.b directives, etc.) for the storage of the parsing tree for z so that Problem 7.10 can be done.

10. Write a flow chart that will read the data structure of Problem 7.9 (or any similar formula tree stored in a linked list structure), and then write an assembly language source program (in ASCII) for the subroutine calls needed to evaluate the formula.

11. Give the graphical parsing trees needed to efficiently evaluate the formula

$$z = (4x^2 + y^2) + (y/2x)^2 \ln(\sqrt{x} + (4x + y)^2 + 11y)$$

Assume that you have subroutines to evaluate a square root and a natural logarithm. To do an efficient calculation. you should evaluate the common subexpression (e.g., x^2) first.

12. Show how the formulas

```
delta = delta + c                    s = s + (delta * delta)
```

can be more efficiently evaluated by passing arguments through registers instead of on the stack as was done in the text.

13. Modify the subroutine of Figure 7.14 so that the Z flag is returned correctly for a 4-byte negate.

14. Write a shortest subroutine that multiplies the 24-bit unsigned number in A (high byte) and X with the 24-bit unsigned number in B (high byte) and Y, returning the product in X (most-significant 16 bits), Y (middle-significant 16 bits), and D (least-significant 16 bits) .

15. Write a shortest subroutine that multiplies the 24-bit signed number in A (high byte) and X with the 24-bit signed number in B (high byte) and Y, returning the product in X (most-significant 16 bits), Y (middle-significant 16 bits), and D (least-significant 16 bits) .

16. Write a shortest subroutine DIVS that divides the signed contents of Y and D by the signed contents of the next word on the stack, putting the quotient in Y and D.

17. Write a shortest subroutine DIVS that divides the signed contents of Y and D by the signed contents of the next word on the stack, putting the remainder in Y and D.

18. Because the half-carry is not set correctly by the instructions SUBA and SBCA, how do you subtract multiple byte BCD numbers? Explain.

19. Write a subroutine to convert a decimal fraction .xyz . . . input from the terminal using INCH, to the closest 16-bit binary fraction returned in D. The decimal fraction can have up to five decimal digits terminated by a carriage return.

20. Write a subroutine to convert a 16-bit binary fraction in D to a decimal fraction .xyz . . . and output it from the terminal using OUTCH. The decimal fraction can have up to five decimal digits terminated by a carriage return.

21. Write a subroutine FPOUT to convert a single-precision binary floating-point number, which is a positive integer less than 100,000, popped from the top of the stack to a decimal floating-point number that is displayed on a terminal screen. Assume you have FPADD and FPMUL, and follow the steps in Figure 7.6.

22. Write a subroutine FPIN to convert a decimal floating-point number input on the keyboard in "scientific notation," which is a positive integer less than 100,000, to a single-precision binary floating-point number pushed on the top of the stack. Assume you have a subroutine FPADD and FPMUL. Follow the steps in Figure 7.5.

23. Give an example that shows that the round bit r can not be eliminated, that is, give an example that shows that it really is not superfluous.

24. Give an example of the addition of two floating-point numbers, each with a magnitude less than or equal to 1, that results in the biggest possible rounding error. Repeat the problem for multiplication.

25. Write a subroutine FPDIV to divide a single precision floating-point number on the top of the stack into the second number on the stack, popping both numbers and pushing the remainder on the top of the stack.

26. Write a subroutine FPADD to add a single precision floating-point number on the top of the stack into the second number on the stack, popping both numbers and pushing the result on the top of the stack.

27. Write a program to make your microcomputer a simple four-function calculator, using subroutines from Hiware's library, subroutine FPDIV of problem 7.24, and input and output subroutines FPIN and FPOUT of problems 20 and 21. Your calculator should use Polish notation and should only evaluate +, –, ~, and /, inputting data in the form of up to eight decimal digits and decimal point (e.g., 123.45, 1234567.8, 12., and so forth).

28. Write, in DC.B's, the parameters pointed to by X when the MEM instruction is executed for each of the linguistic variables in Figure 7.20, for the data in Figure 7.23. The highest C is 50, the lowest B is 40, the highest B is 70, and the lowest A is 60.

29. Assume a function f(x) is graphed as the grade B graph in Figure 7.22. Write a shortest assembly language subroutine F and a table T stored using a DC.B directive, to evaluate f(x) where $40 \le x \le 70$ is input in accumulator A, which leaves the result in accumulator A. [Hint: see Figure 9.12 in the CPU12 manual (CPU12RG/D)].

The Motorola M68HC12A4EVB board can implement all the experiments including those of Chapter 10. The wire-wrap area shows a shift-register that implements the device diagrammed in Figure 10.6.

8

Programming in C and C++

This chapter gives background material for Chapter 9, which shows how C or C++ statements are encoded in assembly language. Together, they illustrate what a programmer is doing when he or she writes high-level language programs. However, if you have already covered this material, it can be skipped.

The first section provides terminology and understanding of where to use high-level language compilers and interpreters. We then begin with a description of C, illustrating first operators and statements and then conditional and loop expressions. We give an example of a program that uses many of the features we need in the next chapter, and then discuss C++ and object-oriented programming.

8.1 Compilers and Interpreters

We first discuss the difference between an assembler and a compiler. A *compiler* is a program that converts a sequence of (ASCII) characters that are written in a *high-level language* into machine code or into assembly language that can be converted into machine code. A high-level language is different from an assembly language in two ways. First, a line of a high-level language statement will often generate five to a few tens of machine instructions, whereas an assembly-language statement will usually generate (at most) one machine instruction. Second, a high-level language is designed to be oriented to the specification of the problem that is to be solved by the program and to the human thought process, while a program in an assembly language is oriented to the computer instruction set and to the hardware used to execute the program. Consider the dot product subroutine used in the previous chapter, written in C below. Each line of the program generates many machine instructions or lines of assembly-language code. Each high-level language statement is designed to express an idea used in the statement of the problem and is oriented to the user rather than the machine. The compiler could generate the assembly-language program or the machine code produced by this program.

```
int dotprod(char v[], char w[]) { int i, dprd = 0;
      for(i = 0; i < 2; i++) dprd += v[i] * w[i];
      return dprd;
}
```

Compilers are used for different purposes than are assemblers. Studies have shown that a typical programmer can generate about ten lines of documented, debugged code per day, regardless of whether the program is written in a high-level language or an assembly language. Because a high-level language generates about an order of magnitude more machine instructions per line, a high-level language program should be an order of magnitude shorter (in the number of lines) and an order of magnitude cheaper to write than an assembly language program that does the same job.

However, a high-level language compiler usually produces inefficient code. For example, an instruction `STAA LOC1` might be immediately followed by the instruction `LDAA LOC1` in the compiler output. As the compiler generates code from each line of the program, line by line, the last operation of one line can generate the STAA instruction, and the first machine code generated by the next line might be the LDAA instruction, for the same variable. The compiler is usually unable to detect such an occurrence and to simplify the code produced by it. Such inefficient code is quite acceptable in a large computer where the slow execution and large memory space needed to store the program are traded against the cost of writing the program. Hardware is cheap and programmers are expensive, so this is a good thing. In a very small computer, which might be put in a refrigerator to control the cooling cycle or keep the time, memory space is limited because the whole computer is on just one chip. Inefficient code is unacceptable here because there is not much room for code and the cost of writing the program is comparatively small. The company that uses high-level languages for small microcomputers will not be able to offer all the features that are crammed into a competitor's product that is programmed in efficient assembly language; or, if it offers the same features, its product will cost more because more memory is needed.

Some compilers are called *optimizing*. They use rules to detect and eliminate the unnecessary operations such as the `STAA` and `LDAA` pair described above. They can be used to generate more efficient code than that generated by nonoptimizing compilers. But even these optimizing compilers produce some inefficient code. You should examine the output of an optimizing compiler to see just how inefficient it is, and you should ignore the claims as to how optimal the code is. Compilers are more powerful, and using them is like driving a car with an automatic transmission, whereas using assemblers is like driving a car with a standard transmission. An automatic transmission is easy to drive and appeals to a wider market. A standard transmission is more controlled and enables you to get the full capabilities out of the machine.

We now consider the differences between the compiler and the interpreter. An *interpreter* is rather like a compiler, being written to convert a high-level language into machine code. However, it converts a line of code one line at a time and executes the resulting code right after it converts it from the high-level language program. A pure interpreter stores the high-level language program in memory, rather than the machine code for the program, and reads a line at a time, interprets it, and executes it. A popular high-level language for interpreters is JAVA, and a JAVA program appears below, doing the same job as the previous programs in C. By design, it has the same syntax as C.

An interpreter reads and executes the source code expression `dprd += v[i] * w[i]` twice. A compiler interprets each source code expression just once, reading it and generating its machine code. Later the machine executes the machine code twice. Interpreters are slow. However, it is easy to change the program in memory and execute it again in an interpreter without having to go through the lengthy process of compiling the code. On the internet, programs can be sent to different servers to be executed. Servers on the internet can immediately interpret a JAVA program regardless of which server gets the code or from where it was sent. This interpretive language has proven to be a very powerful tool for the internet.

Some interpreters are almost compilers. The high-level language is stored in memory almost as written, but some words are replaced by *tokens*. For example, in the preceding program, the word `for` could be replaced by a token $81. All the bytes in the program would be ASCII characters, whose value would be below $7F. The original high-level language can be regenerated from the information in memory, because the tokens can be replaced by their (ASCII) character string equivalents. But as the program is executed, the interpreter can essentially use the token $81 as a command, in this case to set up a loop. It does not have to puzzle over what the (ASCII) characters `f`, `o`, and `r` are before it can decide what the line means. These interpreters have the convenience of a pure interpreter, with respect to the ease of changing the program, but they have speeds approaching those of compilers. They are really partly compiler, to get the tokens, and partly interpreter, to interpret the tokens and the remaining characters and have some of the better features of both.

The state-of-the-art 6812 clearly illustrates the need for programming microcontrollers in a high-level language and in object-oriented languages. Further, the 32K-byte flash memory of the MC68HC912B32 or the 4K-byte EEPROM memory of the MC68HC812A4 is large enough to support high-level language programs. Also, object-oriented features like modularity, information hiding, and inheritance will further simplify the task of controlling 6812-based systems.

This chapter illustrates C and C++ programming techniques. C programming is introduced first. The use of classes in C++ will be introduced at the end of this chapter. While this introduction is very elementary and rather incomplete, it is adequate for the discussion of how high-level languages generate machine code in the next chapter.

8.2 Operators and Assignment Statements

We first explain the basic form of a C *procedure,* the simple and the special numeric operators, conditional expression operators, conditional statements, loop statements, and functions. However, we do not intend to give all the rules of C you need to write good programs. A C program has one or more procedures, of which the first to be executed is called `main`, and the others are "subroutines" or "functions" if they return a value.

All the procedures, including `main`, are written as follows. Carriage returns and spaces (except in names and numbers) are not significant in C programs and can be used to improve readability. The periods (.) in the example below do not appear in C programs but are meant here to denote that one or more declaration or statement may appear. Each *declaration of a parameter or a variable* and each statement ends in a semicolon (;), and more than one of these can be put on the same line. Parameters and variables used in the 6812 are usually 8-bit (`char`), 16-bit (`int`), or 32-bit (`long`) signed integer types. They can be declared unsigned by putting the word *unsigned* in front of `char`, `int`, or `long`. In this and the next chapter we will not discuss `long` data. More than one variable can be put in a declaration; the variables are separated by commas (,). A vector having n elements is denoted by the name and square brackets around the number of elements n, and the elements are numbered 0 to $n - 1$. For example, the declaration `int a,b[10];` shows two variables, a scalar variable `a` and a vector `b` with ten elements. Variables declared outside the procedure (e.g., before the line with `procedure_name`) are global, and those declared within a procedure (e.g., between the curly brackets { and } after `procedure_name`) are local. Parameters will be discussed in §8.5. A *cast* redefines a value's type. A cast is put in parentheses before the value. If `i` is an `int`, `(char)i` is a `char`.

```
declaration of global variable;
declaration of global variable;
...
procedure_name(parameter_1,Parameter_2,...)
...
{
        declaration of local variable;
        declaration of local variable;
        ...
        statement;
        statement;
        ...
}
```

Table 8.1. Conventional C Operators Used in Expressions

Operator	Meaning
=	make the left side equal to the expression on its right
+	add
-	subtract
*	multiply
/	divide
%	modulus (remainder after division)
&	logical bit-by-bit AND
|	logical bit-by-bit OR
~	logical bit-by-bit negation
<<	shift left
>>	shift right

Statements may be algebraic expressions that generate assembly-language instructions to execute the procedure's activities. A statement may be replaced by a sequence of statements within a pair of curly brackets ({ and }). This will be useful in conditional and loop statements discussed soon. Operators used in statements include addition, subtraction, multiplication, and division, and a number of very useful operators that convert efficiently to assembly-language instructions or program segments. Table 8.1 shows the conventional C operators that we will use in this book. Although they are not all necessary, we use a lot of parentheses so we will not have to learn the precedence rules of C grammar. The following simple C procedure `fun` has (signed) 16-bit input parameter a and 32-bit local variable `b;` it puts 1 into `b` and then puts the `(a+b)` th element of the ten-element unsigned global 8-bit vector d into 8-bit unsigned global `c` and returns nothing (`void`) as is indicated by the data type to the left of the procedure name:

```
unsigned char c,d[10];
void fun(int a) { long b;
    b=1;
    c = d[a+b];
}
```

Some very powerful special operators are available in C. Table 8.2 shows the ones we use in this book. For each operator, an example is given together with its equivalent result using the simple operators of Table 8.2. The assignment operator = assigns the value on its right to the variable named on its left and returns the value it assigns so that value can be used in an expression to the left of the assignment operation: The example shows 0 is assigned to `c`, and that value (0) is assigned to `b`, and then that value is assigned to `a`. The increment operator `++` can be used without an assignment operator (e.g., `a++` just increments `a`). It can also be used in an expression in which it increments its operand after the former operand value is returned to be used in the expression. For example, `b = a[i++]` will use the old value of `i` as an index to put `a[i]` into `b`, then it will increment *i*. Similarly, the decrement operator -- can be used in expressions. If the ++ or -- appear in front of the variable, then the value returned by the expression is the updated value; `a[++i]` will first increment `i`, then use the incremented value as an index into *a*. The next row shows the use of the + and = operators used together to represent adding to a variable. The following rows show - | and & appended in front of = to represent subtracting from, ORing to, or ANDing to a variable. Shift << and >> can be used in front of the = sign too. This form of a statement avoids the need to twice write the name of, and twice compute addresses for, the variable being added to or subtracted from. The last two rows of Table 8.2 show shift left and shift right operations and their equivalents in terms of elementary shift and assignment operators.

A statement can be conditional, or it can involve looping to execute a sequence of statements that are written within it many times. We will discuss these control flow statements by giving the flow charts for them. See Figure 8.1 for conditional statements, 8.2 for case statements, and 8.3 for loop statements. These simple standard forms appear throughout the rest of the book, and we will refer to them and their figures.

Table 8.2. Special C Operators

<table>
<tr><th>operator</th><th>example</th><th>equivalent to:</th></tr>
<tr><td>=</td><td>a=b=c=0;</td><td>a=0;b=0;c=0;</td></tr>
<tr><td>++</td><td>a ++;</td><td>a=a+1;</td></tr>
<tr><td>- -</td><td>a --;</td><td>a=a-1;</td></tr>
<tr><td>+=</td><td>a += 2;</td><td>a=a+2;</td></tr>
<tr><td>- =</td><td>a -= 2;</td><td>a=a-2;</td></tr>
<tr><td>|=</td><td>a |= 2;</td><td>a=a|2;</td></tr>
<tr><td>&=</td><td>a &= 2;</td><td>a=a&2;</td></tr>
<tr><td><<=</td><td>a <<= 3</td><td>a=a<<3;</td></tr>
<tr><td>>>=</td><td>a >>= 3</td><td>a=a>>3;</td></tr>
</table>

Table 8.3.
Conditional Expression Operators

<table>
<tr><td>&&</td><td>AND</td></tr>
<tr><td>||</td><td>OR</td></tr>
<tr><td>!</td><td>NOT</td></tr>
<tr><td>></td><td>Greater Than</td></tr>
<tr><td><</td><td>Less Than</td></tr>
<tr><td>> =</td><td>Greater than or Equal</td></tr>
<tr><td>< =</td><td>Less Than or Equal</td></tr>
<tr><td>= =</td><td>Equal to</td></tr>
<tr><td>! =</td><td>Not Equal To</td></tr>
</table>

8.3 Conditional and Loop Statements

Simple conditional expressions of the form *if then* (shown in Figure 8.1a), full conditionals of the form *if then else* (shown in Figure 8.1b), and extended conditionals of the form *if then else if then else if then . . . else* (shown in Figure 8.1c), use conditional expression operators (shown in Table 8.3). In the last expression, the *else if* part can be repeated as many times as needed, and the last part can be an optional *else*. Variables are compared using *relational operators* (> and <), and these are combined using *logical operators* (`&&`). For example, `(a > 5) && (b < 7)` is true if `a > 5` and `b < 7`.

A useful alternative to the conditional statement is the *case statement*. (See Figure 8.2.) An expression giving a numerical value is compared to each of several possible comparison values; the matching comparison value determines which statement will be executed next. The case statement (Figure 8.2a) jumps into the statements just where the variable matches the comparison value and executes all the statements below it. The *break statement* can be used (as shown in Figure 8.2b) to exit the whole case statement after a statement in it is executed, rather than executing the remainder of it.

Loop statements can be used to repeat a statement until a condition is met. A statement within the loop statement will be executed repeatedly. The expressions in both the following loop statements are exactly like the expressions of the conditional statements, using operators as shown in Table 8.3.

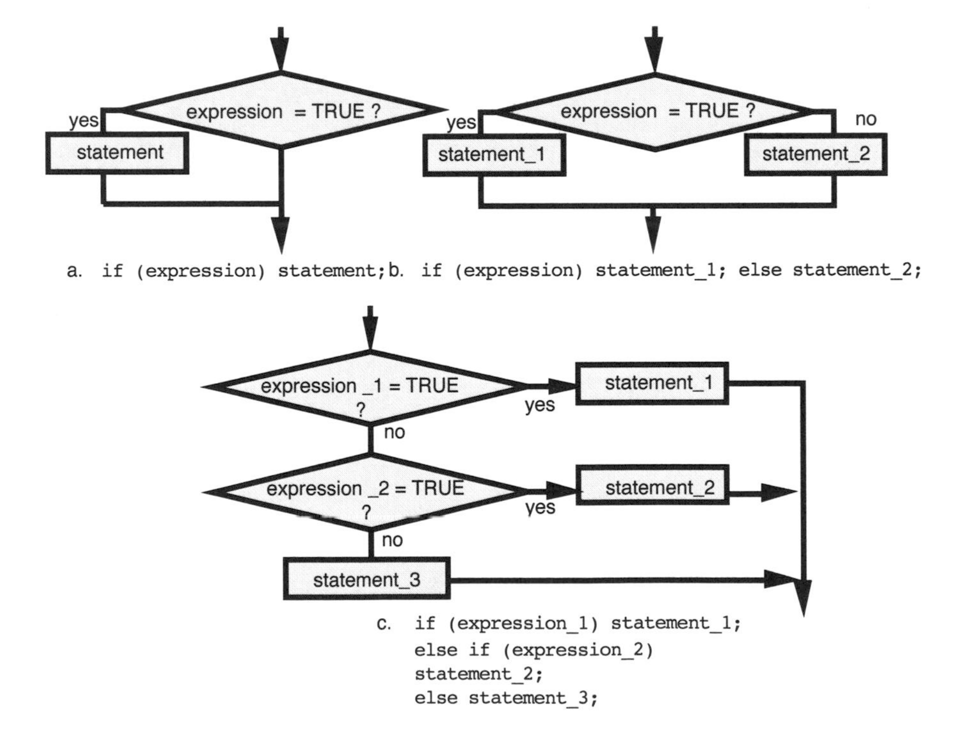

Figure 8.1. Conditional Statements

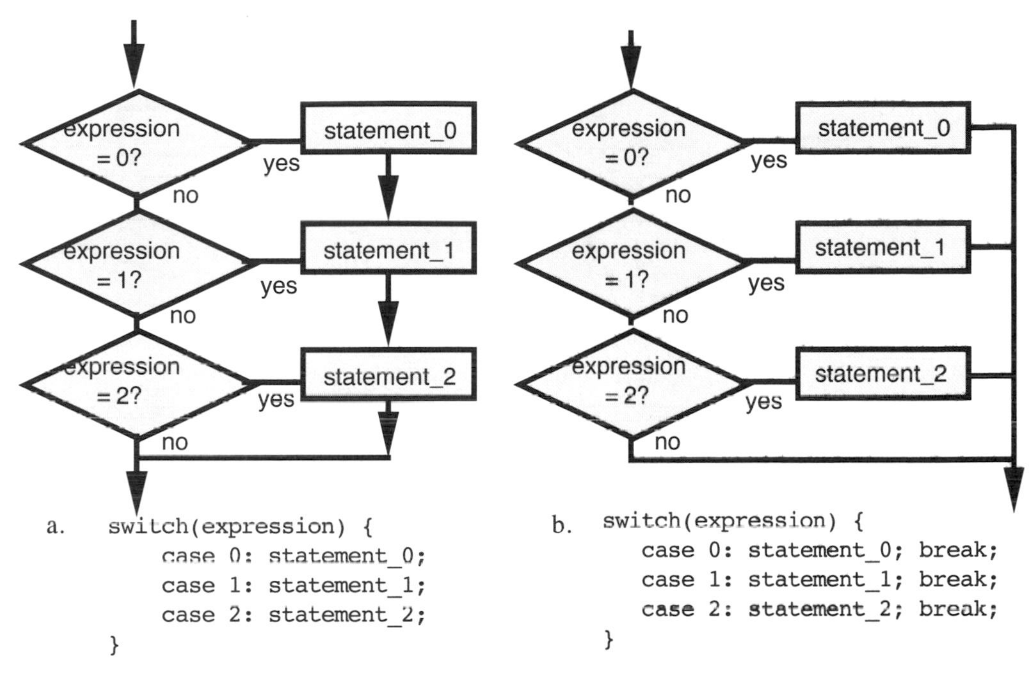
expression
= 0?
yes
no
statement_0
expression
= 1?
yes
no
statement_1
expression
= 2?
yes
no
statement_2
a.
switch(expression) {
case 0: statement_0;
case 1: statement_1;
case 2: statement_2;
}
b.
switch(expression) {
case 0: statement_0; break;
case 1: statement_1; break;
case 2: statement_2; break;
}

Figure 8.2. Case Statements

The *while statement* of Figure 8.3a tests the condition before the loop is executed and is useful if, for example, a loop may have to be done 0 times. The *do while statement* (shown in Figure 8.3b) tests the condition after the loop is executed at least once, but it tests the result of the loop's activities. It is very useful in I/O software. It can similarly clear `alpha[10]`. Though perhaps less clear, it usually leads to more efficient code. The `do while()` construct is generally more efficient than the `while()` construct because the latter has an extra branch instruction to jump to its end.

The more general *for statement* (shown in Figure 8.3c) has three expressions separated by semicolons (;). The first expression initializes variables used in the loop; the second tests for completion in the same style as the while statement; and the third updates the variables each time after the loop is executed. Any of the expressions in the `for` statement may be omitted. For example, `for(i = 0; i < 10; i++) alpha[i] = 0;` will clear the array `alpha` as the above loops did.

The break statement will cause the `for`, `while`, or `do while` loop to terminate just as in the case statement and may be used in a conditional statement. For instance, `for(;;) {i++; if(i == 30) break;}` executes the compound statement `{i++; if(i ==30) break; }` indefinitely, but the loop is terminated when `i` is 30.

8.4 Constants and Variables

An important feature of C, extensively used to access I/O devices, is its ability to describe variables and addresses of variables. If *a* is a variable, then `&a` is the address of `a`. If a is a variable that contains an address of another variable `b`, then `*a` is the

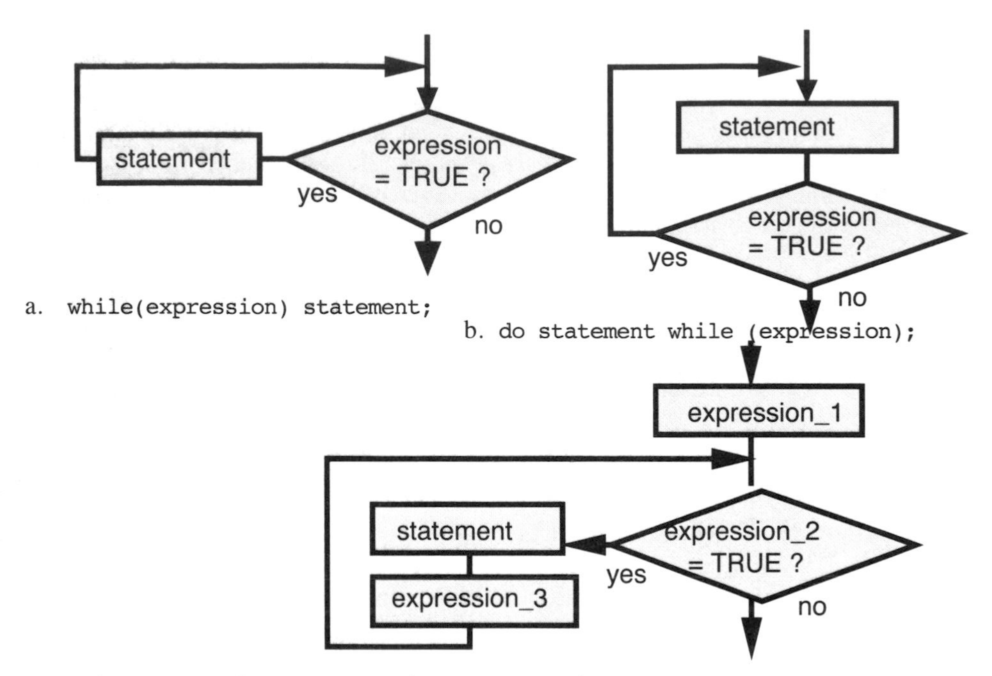

Figure 8.3. Loop Statements

contents of the word pointed to by *a*, which is the contents of `b`. (Note that `a*b` is `a` times `b` but `*b` is the contents of the word pointed to by `b`.) Whenever you see `&`, read it as "address of," and whenever you see `*`, read it as "contents of thing pointed to by." In a declaration statement, the statement `char *p;` means that the thing pointed to by `p` is a character, and `p` points to (contains the address of) a character. In an assignment statement, `*p = 1;` means that 1 is assigned to the value of the thing pointed to by *p*, whereas `p=1;` means that the pointer `p` is given the value 1. Similarly, `a=*p;` means that `a` is given the value of the thing pointed to by `p`, while `a=p;` means `a` gets the value of the pointer `p`. C compilers can give an error message when you assign an integer to a pointer. If that occurs, you have to use a cast. Write `p = (int *)0x4000;` to tell the compiler 0x4000 is really a pointer value to an integer and not an integer itself.

Constants can be defined by *define* or *enum statements*, put before any declarations or statements, to equate names to values. The `define` statement begins with the characters `#define` and does not end with a semicolon.

```
#define ALPHA 100
```

Thenceforth, we can use the label `ALPHA` throughout the program, and 100 will effectively be put in place of `ALPHA` just before the program is actually compiled. This permits the program to be better documented, using meaningful labels, and easier to maintain, so that if the value of a label is changed, it is changed everywhere it occurs.

A number of constants can be created using the enum statement. Unless reinitialized with an "=" sign, the first member has value 0, and each next member is one greater than the previous member. Hexadecimal values are prefixed with zero ex (0x):

```
enum { BETA, GAMMA, DELTA = 0x5};
```

defines `BETA` to have value 0, `GAMMA` to have value 1, and `DELTA` to be 5.

Any scalar variable can be declared and initialized by a "=" and a value; for instance, if we want global integers `i`, `j` and `k` to be initially 1, 2 and 3, we write a global declaration:

```
int i=1, j=2, k=3;
```

C procedures access global variables using direct addressing, and such global variables may be initialized in a procedure `_startup` that is executed just before `main` is started. Initialized local variables of a procedure should generate machine code to initialize them just after they are allocated each time the procedure is called. The procedure

```
void fun(){
        int i, j, k; /* allocate local variables */
        i = 1; j = 2; k = 3; /* initialize local variables */
}
```

is equivalent to the procedure

```
void fun(){
        int i = 1, j = 2, k = 3; /* allocate and init. local vars. */
}
```

A 16-bit element, three-element vector 31, 17, and 10, is generated by a declaration `int v[3]` and stored in memory as (hexadecimal):

001F
0011
000A

and we can refer to the first element as `v[0]`, which happens to be 31. However, the same sequence of values could be put in a vector of three 8-bit elements, generated by a declaration `char u[3]` and stored in memory as:

1F
11
0A

The declaration of a global vector variable can be initialized by use of an "=" and a list of values, in curly brackets. For instance, the three-element global integer vector `v` can be allocated and initialized by

```
int v[3] = {31, 17, 10};
```

The vector *u* can be similarly allocated and initialized by the declaration

```
char u[3] = {31, 17, 10};
```

The procedure `fun()` in §8.2 illustrated the accessing of elements of vectors in expressions. The expression `c = d[a + b],` accessed the `a + b` th element of the 8-bit 10-element vector `d.` When reading the assembly code generated by C, be wary of the implicit multiplication of the vector's precision (in bytes) when calculating offset addresses of elements of the vector. Because C does not check that indexes are within a vector, a C program must be able to implicitly or explicitly assure this to avoid nasty bugs, as when a vector's data is inadvertently stored outside memory allocated to a vector.

The C *structure* mechanism can store different-sized elements. The mechanism is implemented by a declaration that begins with the word *struct* and has a definition of the structure within angle brackets and a list of variables of that structure type after the brackets, as in

```
struct { char l1; int l2; char l3;} list;
```

A globally defined list can be initialized as we did with vectors, as in

```
struct { char l1; int l2; char l3;} list = {5,7,9};
```

The data in a list are identified by "dot" notation, where a dot (.) means "element." For instance, `list.l1` is the `l1` element of the list `list.` If *P* is a pointer to a `struct,` then arrow notation, such as `P->l1,` can access the element `l1` of the list. The `typedef` statement, though it can be used to create a new data type in terms of existing data types, is often used with `structs.` If `typedef a struct { char l1; int l2; char l3;} list;` is written, then `list` is a data type, like `int` or `char,` and can be used in declarations such as `list b;` that declare `b` to be an instance of type `list.` We will find the `typedef` statement to be quite useful when a `struct` has to be declared many times and pointers to it need to be declared as well. A structure can have `bit fields`, which are unsigned integer elements having less than 16 bits. Such a structure as

```
struct {unsigned a:1, b:2, c:3;}l;
```

has a one-bit field `l.a,` two-bit field `l.b`, and three-bit field `l.c.` A *linked list structure*, a list in which some elements are addresses of (the first word in) other lists, is flexible and powerful and is widely used in advanced software.

We normally think of an array as a two-dimensional pattern, as in

1	2	3
4	5	6
7	8	9
10	11	12

An *array* is considered a vector whose elements are themselves vectors, and C syntax reflects this philosophy. For instance, the global declaration

```
int ar1[4][3]={{1,2,3},{4,5,6},{7,8,9},{10,11,12}};
```

allocates and initializes a *row major* ordered array (rows occupy consecutive memory words) `ar1`, and `a = ar1[i][j];` puts the row-*i* column-*j* element of `ar1` into `a.`

A *table* is a vector of identically formatted structs. Tables often store characters, where either a single character or a collection of *n* consecutive characters is considered an element of the structs in the table. Index addressing is useful for accessing elements in a row of a table. If the address register points to the first word of any row, then the displacement can be used to access words in any desired column. Also, autoincrement addressing can be used to select consecutive words from a row of the table.

In C, a table `tbl` is considered a vector whose elements are structures. For instance, the declaration

```
struct {char l1;int l2;char l3;} tbl[3];
```

allocates a table whose rows are similar to the list `list` above. The dot notation with indexes can be used to access it, as in

```
a = tbl[2].l1;
```

In simple compilers, multidimensional arrays and `structs` are not implemented. They can be reasonably simulated using one-dimensional vectors. The user becomes responsible for generating vector index values to access row-column elements or `struct` elements.

Finally a comment is anything enclosed by `/*` and `*/` . These can be put anywhere in your program, except within quotation marks. Alternatively, everything after `//` on a source line is considered a comment. However, the latter syntax is not available on all C compilers.

8.5 Procedures and Their Arguments

A procedure in C may be called by another procedure in C as a procedure. The arguments may be the data themselves, which is call by value, or the address of the data, which is call by name. Call by reference is not used in C (it is often used in FORTRAN). Consider the following example: `RaisePower` computes `i` to the power `j`, returning the answer in `k` where `i`, `j`, and `k` are integers. The variable `i` is passed by value, while *j* and *k* are passed by name. The calling procedure would have `RaisePower(i,&j,&k);` and the called procedure would have

```
void RaisePower (int i, int *j, int *k) {   int n;
    for(*k = 1, n = 0; n < *j; n++) *k =* k * i;
}
```

Formal parameters are listed after the procedure name in parentheses, as in (i,j,k), and in the same order they are listed after the procedure name as they would be for a declaration of local variables. However, they are listed before the curly bracket ({).

Call by value, as `i` is passed, does not allow data to be output from a procedure, but any number of call by value input parameters can be used in a procedure. Actual parameters passed by name in the calling procedure have an ampersand (`&`) prefixed to them to designate that the address is put in the parameter. In the called procedure, the formal parameters generally have an asterisk (`*`) prefixed to them to designate that the data at the address are accessed. Observe that call by name formal parameters `j` or `k` used inside the called procedure all have a prefix asterisk. A call by name parameter can pass data into or out of a procedure, or both. Data can be input into a procedure using call by name, because the address of the result is passed into the procedure and the procedure can read data at the given address. A result can be returned from a procedure using call by name, because the address of the result is passed into the procedure and the procedure can write new data at the given address to pass data out of the procedure. Any number of call by name input/output parameters can be used in a procedure.

A procedure may be used as a function that returns exactly one value and can be used in the middle of algebraic expressions. The value returned by the function is put in a *return statement*. For instance, the function power can be written

```
int power(int i, int j) {  int k, n;
    for(n = 1, k = 0; k < j; k++) n = n * i;
    return n;
}
```

This function can be called within an algebraic expression by a statement *a=power(b,2)*. The output of the function named in the return statement is passed by call by result.

In C, the address of a character string can be passed into a procedure, which uses a pointer inside it to read the characters. For example, the string *s* is passed to a procedure `puts` that outputs a string by outputting to the user's display screen one character at a time using a procedure `putchar`. The procedure puts is written

```
void puts(s) char *s; {
    while(*s != 0) putchar(*(s++));
}
```

It can be called in either of three ways, as shown side by side:

```
void main(){                void main(){               void main(){
     char s[6]="ALPHA";         char s[6]="ALPHA";            puts("ALPHA");
     puts(&s[0]);               puts(s);               }
                            }
}
```

The first calling sequence, though permissible, is clumsy. The second is often used to pass different strings to the procedure, while the third is better when the same constant string is passed to the procedure in the statement of the calling program.

A *prototype* for a procedure can be used to tell the compiler how arguments are passed to and from it. At the beginning of a program we write all prototypes, such as

```
extern void puts(char *);
```

The word *extern* indicates that the procedure `puts()` is not actually here but is elsewhere. The procedure itself can be later in the same file or in another file. The argument `char *` indicates that the procedure uses only one argument and it will be a pointer to a character (i.e. the argument is called by name). In front of the procedure name a type indicates the procedure's result. The type *void* indicates that the procedure does not return a result. After the prototype has been declared, any calls to the procedure will be checked to see if the types match. For instance, a call `puts('A')` will cause an error message because we have to send the address of a character (string), not a value of a character to this procedure. The prototype for `power()` is:

```
extern int power(int, int);
```

to indicate that it requires two arguments and returns one result, all of which are call-by-value-and-result 16-bit signed numbers. The compiler will use the prototype to convert arguments of other types if possible. For instance, if x and y are 8-bit signed numbers (of type `char`) then a call `power(x,y)` will automatically extend these 8-bit to 16-bit signed numbers before passing them to the procedure. If a procedure has a `return n` statement that returns a result, then the type statement in front of the procedure name indicates the type of the result. If that type is declared to be `void` as in the `puts()` procedure, there may not be a *return n* statement that returns a result.

At the beginning of each file, prototypes for all procedures in that file should be declared. While writing a procedure name and its arguments twice, once in a prototype and later in the procedure itself, may appear clumsy, it lets the compiler check for improper arguments and, where possible, instructs it to convert types used in the calling routine to the types expected in the called routine. We recommend the use of prototypes.

The *macro* is similar to a procedure but is either evaluated at compile time or is inserted into the program wherever it is used, rather than being stored in one place and jumped to whenever it is called. The macro in C is implemented as a `#define` construct. As `#defines` were earlier used to define constants, macros are also "expanded" just before the program is compiled. The macro has a name and arguments rather like a procedure, and the rest of the line is the body of the macro. For instance

```
#define f( a, b, c) a = b * 2 + c
```

is a macro with name `f` and arguments `a, b,` and `c.` Wherever the name appears in the program, the macro is expanded and its arguments are substituted. For instance if `f( x, y, 3)` appeared, then `x = y * 2 + 3` is inserted into the program. Macros with constant arguments are evaluated at compile time, generating a constant used at run time.

8.6 An Example

A very nice coding scheme called the *Huffman code* can pack characters into a bit stream and achieve about a 75% reduction in storage space when compared to storing the characters directly in an ASCII character string. It can be used to store characters more compactly and can also be used to transmit them through a communications link more efficiently. As a bonus, the encoded characters are very hard to decode without a code description, so you get a more secure communications link using a Huffman code. Further, Huffman coding and decoding provide a rich set of examples of C techniques.

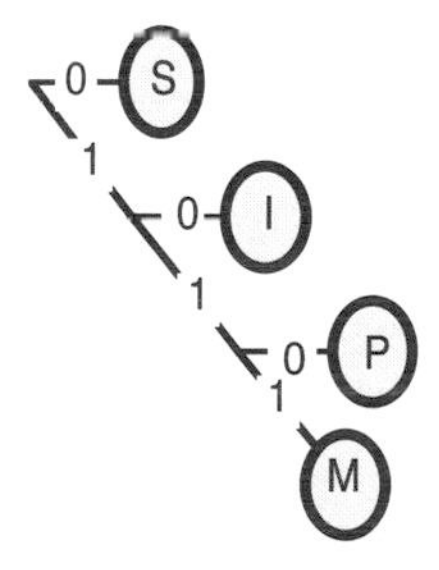

Figure 8.4. A Huffman Coding Tree

The code is rather like Morse code, in that frequently used characters are coded as short strings of bits, just as the often-used letter "e" is a single dot in Morse code. To insure that code words are unique and to suggest a decoding strategy, the code is defined by a tree having two branches at each branching point (*binary tree*), as shown in Figure 8.4. The letters at each end (leaf) are represented by the pattern of 1s and 0s along the branches from the left end (root) to the leaf. Thus, the character string MISSISSIPPI can be represented by the bit string 111100010001011011010. Note that the ASCII string would take 88 bits of memory while the Huffman string would take 21 bits. When you decode the bit string, start at the root, and use each successive bit of the bit string to guide you up (if 0) or down (if 1) the next branch until you get to a leaf. Then copy the letter, and start over at the root of the tree with the next bit of the bit string.

A C program for Huffman coding is shown below. The original ASCII character string is stored in the *char* vector *strng*. We will initialize it to the string *MISSISSIPPI* for convenience, although any string of M I S and P letters could be used. The procedure converts this string into a Huffman coded 48-bit bit string stored in the vector `code[3]`. It uses the procedure `shift()` to shift a bit into `code`. This procedure, shown after `main`, is also used by the decoding procedure shown after it.

```
    int code[3], bitlength; /* output code and its length */

    char strng[12] = "MISSISSIPPI"; /* input code, terminated in a NULL */
    struct table{ char letter; char charcode[4]; } codetable[4]
        = { 'S',"XX0", 'I',"X10", 'P',"110",'M',"111" };

void main(){ int row, i; char *point, letter;
    for (point=strng; *point ; point++ ){
        row = 0; do{
            if (((*point) & 0x7f) == codetable[row++].letter){
                i = 0; while(i < 3){
                    letter = codetable[row].charcode[i++];
                    if (letter != 'X')
                        { shift(); code[2]|=(letter&1); bitlength++;}
                }
            }
        } while(row < 4);
    }
    i= bitlength; while((i++)<48)shift(); /* shift out unchanged bits */
    }

    int shift() { int i;
        i = (0x8000 & code[0]) == 0x8000; code[0] <<= 1;
        if (code[1] & 0x8000) code[0]++; code[1] = code[1] <<1;
        if (code[2] & 0x8000) code[1]++; code[2] = code[2] <<1;
        return(i);
    }
```

Huffman decoding, using the same `shift()`, is done as follows:

```
int code[3] = {0xf116, 0xd000, 0}, bitlength = 21; /* input string */
char ary[3][2] = {{'S',1},{'I',2},{'P','M'}};
char strng[20]; /* buffer for output characters */

void main(){   int row,entry; char *point;
     point=strng; row =0;
     while((bitlength--)>=0){
         if((entry = ary[row][shift()]) < 0x20)  row = entry;
         else {row =0;  *(point++) = entry &0x7f; }
     }
     *point = '\0'; /* terminate C string with NULL character */
}
```

We suggest that you compile these procedures and step through them using a high-level debugger. We will discuss each of the features of this program below.

While a `for` loop can be used for the encoder's outermost, middle, and innermost loops, we have shown the loops using the different constructs to provide different examples. Each execution of the encoder's outermost `for` loop takes a character from *strng*. The next inner loop, a `do while` loop, looks up the character in `table`, a vector of `structs`. When it finds a matching character, the innermost loop, a `while` loop, copies the encoded bits into the 48-bit bit vector `code`.

Bits are stored in the global vector `code` by the subroutine `shift()`. The leftmost 16 bits are stored in `code[0]`, the next 16 bits are stored in `code[1]`, and the rightmost 16 bits are stored in `code[2]`. `shift()` shifts the bit vector left, outputting the leftmost bit. `shift()` first makes local variable *i* 1 if the bit vector's most significant bit is 1, otherwise *i* is 0 (*i* is the return value). `code[0]` is shifted left, clearing its least significant bit, then `code[0]` is incremented if `code[1]`'s most significant bit is 1. Note that incrementing shifts the most significant bit of `code[1]` into `code[0]`. Then `code[1]` is shifted, and then `code[2]` is shifted, in like manner. Observe that the least significant bit of `code[2]` is cleared by the last shift. The encoder procedure, after it calls `shift()`, ORs a 1 into this bit if it reads an ASCII character 1 from the `struct table`.

The decoder program shifts a bit out of *code*, for `bitlength` bits, using `shift()`. This bit is used as an index in the two-dimensional array `ary`. This array stores the next node of the tree shown in Figure 8.4. A row of the array corresponds to a node of the figure, and a column corresponds to an input code bit. The element stored in this array is an ASCII character to be put in the output buffer `string` if its value is above the ASCII code for space (0x20). If it is a character, decoding proceeds next at the root of the tree, or row zero of the array. Otherwise the value is the node number of the tree, or row number of the array, where decoding proceeds next.

Now that we have shown how nice the Huffman code is, we must admit a few problems with it. To efficiently store some text, the text must be statistically analyzed to determine which letters are most frequent, so as to assign these the shortest codes. Note that S is most common, so we gave it a short code word. There is a procedure for generating the best Huffman code, which is presented in many information theory books, but you have to get the statistics of each letter's occurrences to get that code.

Nevertheless, though less than perfect, one can use a fixed code that is based on other statistics if the statistics are reasonably similar. Finally, although the code is almost unbreakable without the decoding tree, if any bit in the bit string is erroneous, your decoding routine can get completely lost. This may be a risk you decide to avoid because the code has to be sent through a communications link that must be as error free as possible.

8.7 Object-Oriented Programming in C++

The concept of object-oriented programming was developed to program symbolic processes, database storage and retrieval systems, and user-friendly graphic interfaces. However, it provides a programming and design methodology that simplifies the programming of microcontrollers and systems that center on them.

Object-oriented programming began with the language SMALLTALK. Programmers using C wanted to use object-oriented techniques. Standard C cannot be used, but a derivative of C, called C++, has been developed to utilize objects with a syntax similar to that of C. Although a 6812 C++ compiler was not available to the author when this book was written, the Metrowerks C++ compiler was used to generate code for 68332 and 68340-based microcontrollers to check out the ideas described below.

C++ has a few differences from C. C++ permits declarations inside expressions, as in `for (int i = 0; i < 10; i++)`. Parameters can be passed by name using a PASCAL-like convention; & in front of a formal parameter is like VAR. See the actual parameter `a` and corresponding formal parameter `b` below:

```
void main(){ char a;            void f(char &b) {
   f(a);                            b = '1';
}                               }
```

An object's data are *data members*, and its procedures are *function members;* data and function members are *encapsulated* together in an *object*. Combining them is a good idea because the programmer becomes aware of both together and logically separates them from other objects. As you get the data, you automatically get the function members used on them. In the class for a character stack shown below, observe that data members `error, Bottom, Top,` and `Ptr` are declared much as in a C `struct,` and function members *push, pull,* and *error* are declared like prototypes are declared in C. Protection terms, `protected, public,` and `virtual,` will be soon explained.

```
class Cstack {                      // definition of a class
    protected:                      // members below are not available outside
        char Error;                 // a data member to record errors
        int *Bottom, *Top, *Ptr;    // data members to point to the stack
    public:                         // members below are available outside
        Cstack(char);               // constructor, used to initialize data members
        virtual void push(int);     // function member to push onto stack
        virtual int pull(void);     // function member to pull from stack
        virtual char error(void);   // function member to check on errors
};
```

A class's function members are written rather like C procedures with the return type and class name in front of two colons and the function member name.

```
void Cstack::push(int i){if(Ptr==Top){Error=1; return;} *(++Ptr)=i; }

int Cstack::pull(){if(Ptr==Bottom){ Error=1; return 0;} return *(Ptr--);}

char Cstack::error(){ char i; i = Error; Error = 0; return i; }
```

Any data member, such as `Top`, may be accessed inside any function member of class `Cstack`, such as `push()`. Inside a function member, when a name appears in an expression, the variable's name is first searched against local variables and function formal parameters. If the name matches, the variable is local or an argument. Then the variable is matched against the object data members and finally against the global variables. In a sense, object data members are global among the function members, because each of them can get to these same variables. However, it is possible that a data member and a local variable or argument have the same name such as `Error`. The data member can be identified as `this->Error`, using key word `this` to point to the object that called the function member, while the local variable or argument is just `Error`.

C++ uses constructors, allocators, destructors, and deallocators. An *allocator* allocates data member storage. A *constructor* initializes these variables; it has the same function name as the class name. Declaring or blessing an object automatically calls the allocator and constructor, as we will see shortly. A *destructor* terminates the use of an object. A destructor has the same function name as the class name but has a tilde (~) in front of the function member name. A *deallocator* recovers storage for data members for later allocation. We do not use a deallocator in our experiments; it is easier to reset the 6812 to deallocate storage. Here's *Cstack*'s constructor:

```
Cstack::Cstack(int i){Top=(Ptr=Bottom=(char*)allocate(i))+i;Qlen=
Error=0;}
```

Throughout this section, a conventional C procedure *allocate* provides buffer storage for an object's data members and for an object's additional storage such as its stacks. The contents of global variable `free` are initialized to the address just above the last global; storage between `free` and the stack pointer is subdivided into buffers for each object by the `allocate` routine. The stack used for return addresses and local variables builds from one end and the allocator builds from the other end of a common RAM buffer area. `allocate`'s return type `void *` means a pointer to anything.

```
char *free=0xb80;
void *allocate(int i) { void *p=free; free += i; return p; }
```

A global object of a class is declared and then used as shown below:

```
Cstack S(10);
void main() { int i;
    S.push(1); i = S.pull();
}
```

The object's data members, `Error, Bottom, Top,` and `Ptr,` are stored in memory just the way a global `struct` is stored. Suppose `S` is a stack object as described above, and `Sptr` is a pointer to a stack object. If a data member could be accessed in *main,* as in `i = S.Error` or `i = Sptr->Error` (we see later that it can't be accessed from *main*), the data member is accessed by using a predetermined offset from the base of the object exactly as a member of a C `struct` is accessed. Function members can be called using notation similar to that used to access data in a `struct; S.push(1)` calls the `push` function member of `S` to push 1 onto `S`'s stack. The "`S.`" in front of the function member is rather like a first actual parameter, as in `push(S,1),` but can be used to select the function member to be run, as we will see later, so it appears before the function.

The class's constructor is executed before the main procedure is executed, to initialize the values of data members of the object. This declaration `S(10)` passes actual parameter `10` to the constructor, which uses it, as formal parameter `i,` to allocate 10 bytes for the stack. The stack is stored in a buffer assigned by the allocate routine.

Similarly a local object of a class can be declared and then used as shown below:

```
void main() { int i; Cstack S(10);
   S.push(1); i = S.pull();
}
```

The data members `Error, Bottom, Top,` and `Ptr,` are stored on the hardware stack, and the constructor is called just after `main` is entered to initialize these data members; it then calls `allocate` to find room for the stack. The function members are called the same way as in the first example when the object was declared globally.

Alternatively, a pointer `Sptr` to an object can be declared globally or locally; then an object is set up and then used as shown below.

```
void main() { Cstack * Sptr; int i;
        Sptr = new Cstack (20);
        Sptr ->push(1); i = Sptr ->pull();
}
```

In the first line, `Sptr,` a pointer to an object of class `stack,` is declared here as a local variable. (Alternatively it could have been declared as a global variable pointer.) The expression `Sptr = new Cstack (20);` is put anywhere before the object is used. This is called *blessing* the object. The allocator and then the constructor are both called by the operator *new*. The allocator *allocate* automatically provides room for the data members `Error, Bottom, Top,` and `Ptr.` The constructor explicitly calls up the allocate procedure to obtain room for the object's stack itself, and then initializes all the object's data members. After it is thus blessed, the object can be used in the program. An alternative way to use a pointer to an object is with a `#define` statement to insert the asterisk as follows:

```
#define S (*Sptr)
void main() { int i;
    Cstack *Sptr = new Cstack(20);
    S.push(1); i = S.pull();
}
```

Wherever the symbolic name `S` appears, the compiler substitutes `(*Sptr)` in its place. Note that `*ptr.member` is the same as `ptr->member.` So this makes the syntax of the use of pointers to objects match the syntax of the use of objects most of the time. However, the blessing of the object explicitly uses the pointer name.

A hierarchy of derived and base classes, inheritance, overriding, and factoring are all related ideas. These are described below, in that order.

A class can be a *derived class* (also called *subclass*) of another class, and a hierarchy of classes can be built up. We create derived classes to use some of the data or function members of the base class, but we can add members to, or replace some of the members of, the base class in the derived class. For instance the aforementioned class `Cstack` can have a derived class `Istack` for `int` variables; it declares a potentially modifiable constructor and different function members `pull` and `push` for its stack. When defining the class `Istack` the *base class* (also called *superclass*) of `Istack` is written after its name and a colon as *: public Cstack*. A class such as `Cstack`, with no base class, is called a *root class;* it has no colon and base class shown in its declaration.

```
class Istack:public Cstack
 {public:Istack(char);virtual void push(int);virtual int pull(void);};

Istack::Istack(char i) : Cstack(i & ~1) {}

void Istack:: push (int i)
 {if(Ptr==Top) {Error=1; return;} *(++Ptr) = i>>8;  *(++Ptr)=i; }

int Istack:: pull ()
 {int i;if(Ptr==Top){Error=1;return 0;} return *(Ptr--)|(*(Ptr-)<<8);}
```

The notion of *inheritance* is that an object will have data and function members defined in the base class(es) of its class as well as those defined in its own class. The derived class inherits the data members or function members of the parent that are not redefined in the derived class. If we execute `Istack::error` then the function member `Cstack::error` is executed, because `Istack` does not declare a different `error` function member. If a function member cannot be found in the class that the object was declared or blessed for, then its base class is examined to find the function member to be executed. In a hierarchy of derived classes, if the search fails in the class's base class, the base class's base class is searched, and so on, up to the root class. *Overriding* is the opposite of inheritance. If we execute `Sptr->push(1);`, function member `Istack::push` is executed rather than `Cstack::push`, because the class defines an overriding function member. Although we did not need additional variables in the derived class, the same rules of inheritance and overriding would apply to data members as to function members.

Most programmers face the frustration of several times rewriting a procedure, such as one that outputs characters to a terminal, wishing they had saved a copy of it and used the earlier copy in later programs. Commonly reused procedures can be kept in a library. However, when we collect such common routines, we will notice some common parts in different routines. Common parts of these library procedures can be put in one place by *factoring*. Factoring is common to many disciplines—or instance, to algebra. If you have *ab* + *ac* you can factor out the common term a and write a (b + c), which has fewer multiplications. Similarly, if a large number of classes use the same function member, such a function member could be reproduced for each. Declaring such a function member in one place in a base class would be more statically efficient, where all derived classes would inherit it. Also, if an error were discovered and corrected in a base class's function member, it is automatically corrected for use in all the derived classes that use the common function member. `Istack's` constructor, using the notation `:Cstack(i & ~1)` just after the constructor's name `Istack::Istack(char i)` before the constructor's body in *{}*, calls the base class's constructor before its own constructor is executed. In fact, `Istack`'s constructor does nothing else, as is denoted by the empty procedure *{}*. All derived classes need to declare their constructor, even if that constructor does nothing but call its base class's constructor. Other function members can call their base's function members by the key word *inherited* as in `inherited::push(i);` or by explicitly naming the class, in front of the function call, as in `Cstack::push(i);`

Consider the hypothetical situation where a program can declare classes `Cstack` and `Istack`. Inside `main`, are statements `Sptr->push(1);` and `i = Sptr->pull();`. At compile time, either of the objects can be declared for either `Cstack` or `Istack`, using conditional compilation; for instance, the program on the left:

```
void main(){ int i;
#ifdef mode
    Cstack S(10);
#else
    Istack S(10);
#endif
    S.push(1); i = S.pull();
}
```

```
void main(){ int i; Cstack *Sptr;
#ifdef mode
    Sptr = new Cstack(10);
#else
    Sptr = new Istack (10);
#endif
    Sptr->push(1); i = Sptr->pull();
}
```

declares *S* a class `Cstack` object if `mode` is `#declared`, otherwise it is a class `Istack` object. Then the remainder of the program is written unchanged. Alternatively, at compile time, a pointer to objects can be blessed for either the `Cstack` or the `Istack` class. The program above right shows this technique.

Moreover, a pointer can be blessed to be objects of different classes at run time. At the very beginning of *main*, assume a variable called *range* denotes the actual maximum data size saved in the stack:

```
void main(){ int i, range; Cstack *Sptr;
   if(range >= 128) Sptr = new Istack(10); else Sptr = new Cstack(10);
   Sptr->push(1); i = Sptr->pull();
}
```

`Sptr->push(1);` and `i = Sptr->pull();` will use the stack of 8-bit members if the range is small enough to save space, otherwise they will use a stack that has enough room for each element to hold the larger data, as will be explained shortly.

Polymorphism means that any two classes can declare the same function member name and argument, especially a class and its inherited classes. It means that simple intuitive names like `push` can be used for interchangeable function members of different classes. Polymorphism will be used later when we substitute one object for another object; the function member names and arguments do not have to be changed. You don't have to generate obscure names for functions to keep them separate from each other. Moreover, in C++, the number and types of operands, called the function's signature, are part of the name when determining if two functions have the same name. For instance, `push(char a)` is a different function than `push(int a)`.

A function member that is declared *virtual* will be called indirectly instead of by a BSR or equivalent instruction. The program jumps indirectly through a list of function member entry addresses, called a *vtable*. The blessing of the object links the calls to the object's function members through this vtable. If an object is blessed for a different class, another class's vtable is linked to the object, so that calling the object's function members will go to the class's function members for which it is blessed.

When object pointers are blessed at run time and have virtual function members, if a virtual function member appears for a class and is overridden by function members with the same name in its derived classes, the sizes and types of all the arguments should be the same, because the compiler does not know how an object will be blessed at run time. If they were not, the compiler would not know how to pass arguments to the function members. For this reason, we defined the arguments of `Cstack's push` and `pull` function members to be `int` rather than `char`, so that the same function member name can be used for the `int` version or a `char` version. This run-time selection of which class to assign to an object isn't needed with declarations of objects, but only with blessing of object pointers, because the run-time program can't select at compile time which of several declarations might be used. Also the pointer to the object must be declared an object of a common base class if it is to be used for several class.

Information hiding limits access to data or function members. A member can be declared *public*, making it available everywhere, *protected*, making it available only to function members of the same class or a derived class of it, or *private*, making it available only to the same class's function members and hiding it from other functions. These words appearing in a class declaration apply to all members listed after them until another such word appears; the default if no such words appear is private. The data member `Error` in the class `Cstack` cannot be accessed by a pointer in `main` as in `i =S.Error` or `i =Sptr->Error` because it is not `public`, but only through the public function member `error()`. This way, the procedure `main` can read (and automatically clear) the `Error` variable but cannot accidentally or maliciously set `Error`, nor can it read it, forgetting to clear it. You should protect your data members to make your program much more bug-proof. Declare all data and function members as `private` if they are to be used only by the class's own function members, declare them `protected` if they might be used by derived classes, and declare them `public` if they are used outside the class and its derived classes.

Templates generalize object-oriented programming. A *template class* is a class that is defined for an arbitrary data type, which is selected when the object is blessed or declared. We will define a templated class `Stack`. The class declaration and the function members have a prefix like `template <class T>` to allow the user to bless or declare the object for a specific class having a particular data type, as in `S = new Stack<char>(10)`. The generalized class definition is given below; you can substitute the word `char` for the letter *T* everywhere in declarations or class function members.

The following class also exhibits another feature of C++, which is the ability to write the function member inside the declaration of the class. The function is written in place of the prototype for the function. This is especially useful when templates are used with int function members, because otherwise the notation `template <class T>` and the class name `Stack::` would have to be repeated before each function member.

```
template <class T> class Stack{ private:T *Bottom,*Top,*Ptr;char Error;
  public :Stack(int i)
    { Top = ( Bottom = Ptr  = (T*)allocate( i ) ) + i ); Error = 0; }
  virtual void  push (T i){if(Ptr==Top){Error=1; return;} *(++Ptr)=i;}
  virtual T  pull(void){if(Ptr==Bottom){Error=1; return 0;}return *Ptr--;}
  virtual char error () { char i; i = Error; Error = 0; return i; }
};
```

If you declare `Stack<char> S(10);` or bless `Sptr = new Stack<char>(10);` then a stack is implemented that stores 8-bit data, but if you declare `Stack<int> S(10)` or bless `Sptr = new Stack<int>(10);` then a stack is implemented that stores 16-bit data. Templates permit us to define one generalized class that can be declared or blessed to handle 8-bit, 16-bit, or 32-bit signed or unsigned data when the program is compiled. This selection must be made at compile time, because it generates different calls.

Operator overloading means that the same operator symbol generates different effects depending on the type of the data it operates on. The C compiler already effectively loads its operators. The + operator generates an ADDB instruction when adding data of type `char`, and an ADDD instruction when adding data of type *int*. What C++ does but C cannot do is to overload operators so they do different things when an operand is an object, which depends on the object's definition. In effect, the programmer can provide a new part of the compiler that generates the code for symbols, depending on the types of data used with the symbols. For instance, the << operator used for shift can be used for input or output if an operand is an I/O device. The expression `S << a` can be defined to output the character `a` to the object `S`, and `S >> a` can be defined to input a character from the object `S` and put it into `a`. This type of operator overloading is used in I/O streams for inputting or outputting formatted character strings. Without this feature, we simply have to write our function calls as `a=S.Input()` and `S.Output(a)` rather than `S<<a` or `S>>a`. However, with overloading we write a simpler program; for instance we can write an I/O stream `S << a << " is the value of " << b;` Overloading can also be used to create arithmetic-looking expressions that use function members to evaluate them. Besides operators like + and –, C++ considers the cast to be an operator, as well as the assignment =. In the following example, we overload the cast operator as shown by `operator T ();` and the assignment operator as shown by `T operator = (T);`. `T` will be a cast, like `char`, so `operator T ();` will become `operator char ();` whenever the compiler has an explicit cast like `(char)i`, where `i` is an object, or an implicit cast where object `i` appears in an expression needing a `char`, the compiler calls the user-defined overloaded operator to perform the cast function. Similarly, wherever the compiler has calculated an expression that has a `char` value but the assignment statement has an object `i` on its left, the compiler calls up the overloaded = operator the user specifies with `T operator = (T);`

```
template <class T> class Ostack : public Stack<T> {  char Index;
    public: Ostack(int i):Stack(i) { }/* constructor, calls base constructor */
    operator T ()                              /* overloaded cast operator */
        {if(index>(Ptr-Bottom)){Error=1;return 0;} return Ptr[-index];}
    T operator = (T data)                      /* overloaded assignment operator */
       {if(index>(Ptr-Bottom)){Error=1;return 0;} return Ptr[-index]=data;}
    T operator [](char data){index=data;return *this; }/* index operator */
};
```

The overloaded index operator `[]` illustrates another C++ feature. This overloaded operator is called whenever the compiler sees an index `[]` to the right of an object, as in `S[0]`, whether the object and index are on the left or right of an assignment statement `=`. It executes the overloaded operator `[]` before it executes the overloaded cast or overloaded assignment operator. This overloaded operator simply stores what is inside the square brackets, *0* in our example, in an object data member *index*. Then the following overloaded cast or assignment operator can use this saved value to supply the offset to the stack. Then `S[i]` would read or write the *i*th element from the top of the stack.

Now, whenever the compiler sees an object on the left side of an equal sign when it has evaluated a number for the expression on the right side and it would otherwise be unable to do anything correctly, the compiler looks at your declaration of the overloaded assignment operator, to determine that the number will be pushed onto the stack. The expression `S = 1;` will do the same thing as `S.push(1);`, and `*Sptr = 1;` will do the same thing as `Sptr->push(1);`. Similarly, whenever the compiler sees an object anywhere on the right side of an equal sign when it is trying to get a number and it would otherwise be unable to do anything correctly, the compiler looks at your declaration of the overloaded cast operator to determine that the number will be pulled from the stack. The expression `i = S;` will do the same thing as `i = S.pull();`, and `i = *Sptr;` will do the same thing as `i = Sptr->pull();`. Now if a stack *S* returns a temperature in degrees centigrade, you can write an expression like `degreeF = (S * 9) / 5 + 32;` or `degreeF = (*Sptr * 9) / 5 + 32;`, and the compiler will pull an item from the stack each time it runs into the `S` symbolic name. While overloading of operators isn't necessary, it provides a mechanism for simplifying expressions to look like common algebraic formulas.

A derived class usually defines an overloaded assignment operator even if its base class has defined an overloaded assignment operator in exactly the same way, because the (Metrowerks) C++ compiler can get confused with the "=" sign. If `S1` and `S2` are objects of class `Cstack<char>`, then `S1 = S2;` won't pop an item from `S2` and push it onto `S1`, as we would wish when we use overloaded assignment and cast operators, but "clones" the object, copying `device2`'s contents into `device1` as if the object were a `struct`. That is, if `S1`'s class's base class overrides "=" but `S1`'s class itself does not override "=", `S1 = S2;` causes `S2` to be copied into `S1`. However, if "=" is overridden in `S1`'s class definition, "=" is an overridden assignment operator, and `S1 = S2;` pops an item from `S2` and pushes it onto `S1`. The derived class has to override "=" to push data. The "=" operator, though useful, needs to be carefully handled. All our derived classes explicitly define `operator =` if "=" is to be overridden.

C++ object-oriented programming offers many useful features. Encapsulation associates variables with procedures that use them in classes, inheritance permits factoring out of procedures that are common to several classes, overriding permits the redefinition of procedures, polymorphism allows common names to be used for procedures, virtual functions permit different procedures to be selected at run time, information hiding protects data, template classes generalize classes to use different data types, and operator overloading permits a program to be written in the format of algebraic expressions. If the programmer doesn't have C++ but has a minimal C compiler, many of the features of object-oriented programming can be simulated by adhering to a set of conventions. For instance, in place of a C++ call `Cstack.push()`, one can write instead `StackPush()`. Information hiding can be enforced by only accessing variables like `QptrError` in procedures like `StackPush()`. C++ gives us a good model for useful C conventions.

Object-oriented programming has very useful features for designing state-of-the-art microcomputer's I/O device software, as proposed by Grady Booch in his tutorial *Object-Oriented Computing*. Encapsulation is extended to include not only instance variables and methods, but also the I/O device, digital, analog, and mechanical systems used for this I/O. An object is these parts considered as a single unit. For instance, suppose you are designing an automobile controller. An object (call it `PLUGS`) might be the spark plugs, their control hardware, and procedures. Having defined `PLUGS`, you call function members (for instance, `SetRate(10)` to `PLUGS`), rather like connecting wires between the hardware parts of these objects. The system takes shape in a clear intuitive way as the function members are defined. In top-down design, you can specify the arguments and the semantics of the methods that will be executed before you write them. In bottom-up design, the object `PLUGS` can be tested by a driver as a unit before it is connected to other objects.

An object can be replaced by another object, if the function calls are written the same way (polymorphism). If you replace your spark plug firing system with another, the whole old `PLUGS` object can be removed and a whole new `PLUGS1` object inserted. You can maintain a library of classes to construct new products by building on large pretested modules. Having several objects with different costs and performances, you can insert a customer-specified one in each unit.

In this context, protection has clear advantages. If interchangeable objects avoid mismatch problems, then all public function and data members have to be defined exactly the same. Private and protected function and data members, by contrast, do not need to be defined exactly the same in each of the objects because they cannot be accessed outside the object. Make a function or data member public only if you will maintain the member's appearance and meaning in all interchangeable classes. Because there is no need to make private and protected members the same in all classes, you have more flexibility, so you make function and data members public only if you have to.

Virtual functions have further advantages, if an object can be blessed to be one of a number of polymorphic classes at run time. For instance, output from a microcontroller can be sent to a liquid crystal display or a serial printer. The user can cause the output to be sent to either device without reloading, or in any way modifying, the microcontroller's program.

Classes can be used in a different way to simplify programming rather complex 6812 I/O systems. Some basic routines, available in a library of classes, will be needed to initialize the device, to exchange data with the device, or to terminate a device's use. These routines can be put into operating systems as device drivers. Alternatively, they can be implemented as classes. Then as larger systems are implemented, such as `PLUGS`, that use the device, new classes can be defined as derived classes of these existing classes, to avoid rewriting the methods inherited from the classes in the library.

8.8 Summary

This chapter gives some background in C and C++, so that the next chapter can illustrate how each C or C++ expression or mechanism can be assembled into 6812 assembly language, which we dwell on in the next chapter. We will also include C or C++ programs to illustrate data structures, arithmetic operations, and I/O operations in the remaining chapters. This chapter has served to give you enough background to be able to fully appreciate the points to be made in the remaining chapters.

Do You Know These Terms?

See the end of chapter 1 for instructions.

compiler
high-level language
optimizing compiler
interpreter
tokens
procedure
declaration of a parameter or a variable
cast
statement
relational operators
logical operators
case statement
break statement
while statement
do while statement
for statement
define statement
enum statement
structure
struct
linked list structure
array
row major
table
call by value
return statement
prototype
macro
Huffman code
binary tree
data member
function members
encapsulate
object
allocator
constructor
destructor
deallocator
allocate
blessing
new
derived class
subclass
base class
superclass
root class
inheritance
overriding
factoring
polymorphism
virtual
vtable
information hiding
public
protected
private
template class
operator overloading

Problems

Problems in this chapter and many in later chapters are C and C++ language programming problems. We recommend the following guidelines for problems answered in C: In main() or "self-initializing procedures" each statement must be limited to C operators and statements described in this chapter, should include all initialization operations, and should have comments as noted at the end of §8.4, and subroutines should follow the style for C procedures recommended in §8.5. Unless otherwise noted, you should write programs with the greatest static efficiency.

1. Write a shortest C procedure `void main()` that will find `x` and `y` if `ax + by = c` and `dx + ey = f`. Assume that *a*, `b, c, d, e, f` are global integers that somehow are initialized with the correct parameters, and answers, `x` and `y`, are stored in local variables in `main()`. (You might verify your program with a source-level debugger.)

2. Write a shortest C procedure `void main()` that sorts five numbers in global integer vector `a[5]` using an algorithm that executes four passes, where each pass compares each `a[i], i` running from 0 to 3. During each pass, `a[i]` is compared to each `a[j], j` running from `i+1` to 4, putting the smaller in `a[i]` and larger in `a[j]`.

3. Write a C procedure `void main()` to generate the first five Fibonacci numbers **F**(i), (**F**(0) = **F**(1) = 1 and for i>1, **F**(i) = **F**(i–1) + **F**(i–2)) in global integers `a0,a1,a2,a3,a4` so that `ai` is **F**(i). Compute **F**(2), **F**(3), and **F**(4).

4. A two-dimensional array can be simulated using one-dimensional vectors. Write a shortest C procedure `void main()` to multiply two 3 x 3 integer matrices, **A** and **B**, putting the result in **C**, all stored as one-dimensional vectors in row major order. Show the storage declarations/directives of the matrices, so that **A** and **B** are initialized as

$$\mathbf{A} = \begin{matrix} 1 & 2 & 3 \\ 4 & 5 & 6 \\ 7 & 8 & 9 \end{matrix} \qquad \mathbf{B} = \begin{matrix} 10 & 13 & 16 \\ 11 & 14 & 17 \\ 12 & 15 & 18 \end{matrix}$$

5. A *long* can be simulated using one-dimensional `char` vectors. Suppose `A` is a zero-origin 5-by-7 array of 32-bit numbers, each number stored in consecutive bytes most significant byte first, and the matrix stored in row major order, in a 140-byte `char` vector. Write a C procedure `int get(char *a, unsigned char i, unsigned char j, char *v)`, where *a* is the storage array, `i` and `j` are row and column, and *v* is the vector result. If `0 ≤ i < 5` and `0 ≤ j < 7`, this procedure puts the `i`th row, `j`th column 32-bit value into locations `v, v+1, v+2`, and `v+3`, most significant byte first, and returns 1; otherwise it returns a 0 and does not write into `v`.

6. A `struct` can be simulated using one-dimensional `char` vectors. The `struct{long v1; unsigned int v2:4, v3:8, v4:2, v5:1};` has, tightly packed, a 32-bit element `v1`, a 4-bit element `v2`, an 8-bit element *v3*, a 2-bit element *v4*, a 1-bit element *v5*, and an unused bit to fill out a 16-bit `unsigned int`. Write shortest C procedures `void getV1(char *s, char *v), void getV2( char *s, char *v),` void getV3(char *s, char *v), void getV4(char *s, char *v), void getV5(char *s, *v), void putV1(char *s, char *v), void putV2(char *s, char *v), void putV3(char *s, char *v), void putV4(char *s, char *v), void putV5(char *s, *v), in which get... will copy the element from the struct to the vector and put... will copy the vector into the struct; e.g., getV2(s, v) copies element V2 into *v*, and putV5(s, v) copies v into element V5.

7. Write a shortest C procedure void main() and procedures it calls, without any assembly language, which will first input up to 32 characters from the keyboard to the 6812 (using getchar()) and will then jump to one of the procedures, given below, whose name is typed in (the names can be entered in either upper or lower case, or a combination of both, but a space is represented as an underbar). The procedures: void start(), void step_up(), void step_down(), void recalibrate(), and void shut_down(), just type out a message; for instance, start() will type out "Start Entered" on the host monitor. The main() procedure should generate the least number of bytes of object code possible and should run on the host. Although you do not have to use a compiler and target machine to answer this problem, you can use it without penalty, and it may help you get error-free results faster.

8. Suppose a string such as "SEE THE MEAT," "MEET A MAN," or "THESE NEAT TEAS MEET MATES" is stored in char string[40];. Using one dimensional vector rather than linked list data structures to store the coding/decoding information:

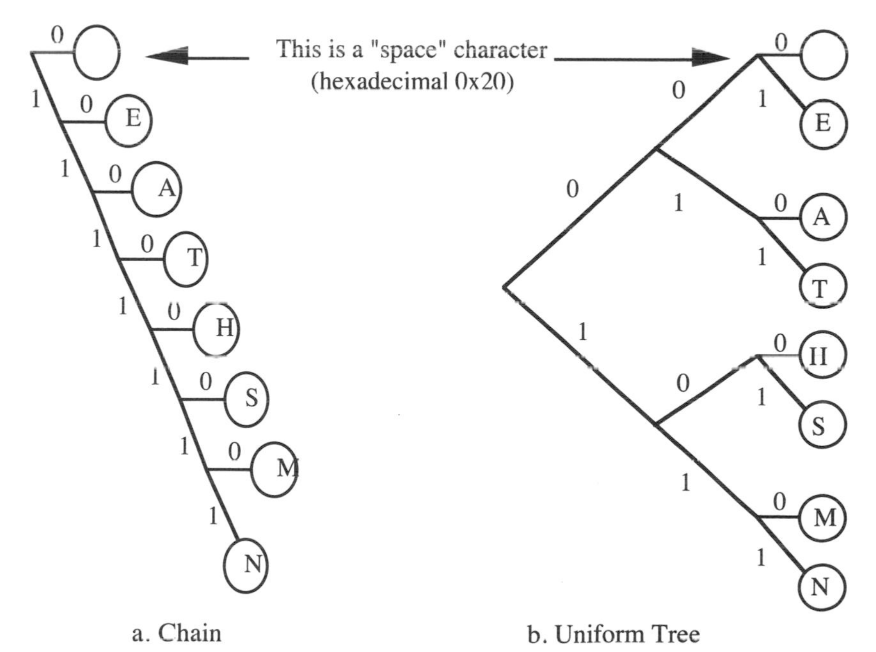

Figure 8.5. Other Huffman Codes

a. Write a C procedure `encode()` to convert the ASCII `string` to Huffman code, as defined by the coding tree in Figure 8.5a, storing the code as a bit string, first bit as most significant bit of first element of `int code[16];`.

b. Write a C procedure `decode()` that decodes such a code in `int code[16]`, using the coding tree in Figure 8.5a, putting the ASCII string back as it was in `char string[40]`.

9. Repeat Problem 8 for the Huffman coding tree in Figure 8.5b.

10. Write an initialization and four shortest C procedures `void pstop(int)` push to top, `int pltop()` pull from top, `psbot(int)` push to bottom, `int plbot()` pull from bottom, of a ten-element 16-bit word deque. The deque's buffer is `int deque[10]`. Use global `int` pointers, `top` and `bottom`. Use global `char` varaibles for the size of the deque, `size`, and error flag `errors` which is to remain cleared if there are no errors and to be 1 if there are underflow or overflow errors. Note that C always initializes global variables to zero if not otherwise initialized. The procedures should manage the deque correctly as long as *errors* is zero. Procedures `pstop()` and `psbot()` pass by value, and procedures `pltop()` and `plbot()` pass by result.

11. Write a C procedure `get(char *a, int i)`, whose body consists entirely of embedded assembly language, which moves `i` bytes following address `a` into a `char` global vector `v`, assuming `v` has a dimension larger than or equal to `i`. To achieve speed, use the MOVB and DBNE instructions. The call to this procedure, `get(s, n)`, is implemented:

```
ldx  s
pshx
ldx  n
jsr get
leas 4,sp
```

12. Write a shortest C procedure *hexString(unsigned int n, char *s)* that runs in a target machine to convert an unsigned integer *n* into printable characters in *s* that represent it in hexadecimal so that *s[0]* is the ASCII code for the 1000's hex digit, *s[1]* is the code for the 100's hex digit, and so on. Suppress leading 0s by replacing them with blanks.

13. Write the shortest procedure `int inhex()` in C to input a four-digit hexadecimal number from the keyboard (the letters A through F may be upper or lower case; typing any character other than 0...9, a...f, A...F, or entering more than four hexadecimal digits terminates the input and starts the conversion) and convert it to a binary number, returning the converted binary number as an unsigned int. Although you do not have to use a compiler and target machine to answer this problem, you can use it without penalty, and it may help you get error-free results faster.

14. Write a shortest C program `int check(int base, int size, int range)` to write a checkerboard pattern in a vector of `size` $s = 2^n$ elements beginning at `base`, and then check to see that it is still there after it is completely written. It returns 1 if the vector is written and read back correctly; otherwise it returns 0. A checkerboard pattern is *range* $r = 2^k$ elements of 0s, followed by 2^k element of $FF, followed by 2^k elements of 0s, . . . for k < n, repeated throughout the vector. (This pattern is used to check dynamic memories for pattern sensitivity errors.)

15. Write a class `BitQueue` that is fully equivalent to the class `Cstack` in §8.3, but pushed, stores, and pulls 1-bit values, and all sizes are in bits rather than 16-bit words. The bits are stored in 16-bit *int* vector allocated by the `allocate()` procedure.

16. Write a class `ShiftInt` that is fully equivalent to the class `Cstack` in §8.3, but the constructor has an argument *n,* and function member `j = obj.shift(i);` shifts an `int` value `i` into a shift register of `n ints` and shifts out an `int` value to `j`.

17. Write a class `ShiftChar` that is a derived class of the class `ShiftInt` in Problem 18, where function member `j = shift(i);` shifts a `char` value `i` into a shift register of *n* `chars` and shifts out a `char` value to `j`. `ShiftChar` uses `ShiftInt`'s constructor.

18. Write a class `ShiftBit` that is fully equivalent to the class `ShiftInt` in Problem 19, but shifts 1-bit values, and all sizes are in bits rather than 16-bit words. The bits are stored in 16-bit `int` vector allocated by the `allocate()` procedure.

19. Write a templated class `Deque` that is a derived class of templated class `Cstack`, and that implements a deque that can be pushed into and pulled from either end. The member functions `pstop()` push to top, `pltop()` pull from top, `psbot()` push to bottom, and `plbot()` pull from bottom. Use inherited data and function members wherever possible.

20. Write a templated class `IndexStack` which is a derived class of templated class *Cstack* (§8.7), that implements an indexable stack, in which the *i*th member from the top can be read. The member functions `push()` pushes, `pull()` pulls, `read(i)` reads the `i`th element from the top of the stack. Function member `read(i)` does not move the stack pointer. Use inherited data and function members wherever possible.

21. Write a templated class `IndexDeque` that is a derived class of templated class *Queue,* that implements an indexable deque that can be pushed into and pulled from either end, and in which the `i`th member from the top or bottom can be read. The member functions `pstop()` push to top, `pltop()` pull from top, `psbot()` push to bottom, `plbot()` pull from bottom, `rdtop(i)` reads the `i`th element from the top, and rdbot(i) reads the ith element from the bottom of the deque. Function members rdtop(i) and rdbot(i) do not move the pointers. Use inherited data and function members wherever possible.

22. Write a templated class Matrix that implements matrix addition and multiplication for square matrixes (number of rows = number of columns). Overloaded operator + adds two intervals resulting in an interval, overloaded operator * multiplies two matrixes resulting in a matrixes, and overloaded operators = and cast with overloaded operator *[]* writes or reads elements; for instance, if M is an object of class Matrix, then M[i][j] = 5; will write 5 into row i, column j, of matrix M, and k = M[i][j]; will read row i, column j, of matrix M into k. Matrix's constructor has an argument size that is stored as a data member size and allocates enough memory to hold a size by size matrix of elements of the template's data width, using a call to the procedure allocate.

23. Intervals can be used to calculate worst-case possiblilities, for instance in determining if an I/O device's setup and hold times are satisfied. An interval <a,b>, a ≤ b, is a range of real numbers between a and b. If <a,b> and <c,d> are intervals A and B, then the sum of A and B is the interval <a+c, b+d>, and the negative of A is <–b, –a>. Interval A contains interval B if every point in A is also in B. Write a templated class *Interval* having public overloaded operators + for adding two intervals resulting in an interval, – for negating an interval resulting in an interval, and an overloaded operator > returning a *char* value 1 if the left interval contains the right interval, otherwise it returns 0. If A, B, and C are of class Interval, the expression A = B + C; will add intervals A and B and put the result in C, A = - B; will put the negative of A into B, and the expression if(A > B) i = 0; will clear i if A contains B. The template allows for the values such as a or b to be char, int, or long. The class has a public variable error that is initially cleared and set if an operation cannot be done or results in an overflow.

24. Write a templated class Interval, having the operators of Problem 25 and additional public overloaded operators * for multiplying two intervals to get an interval, and / for dividing two intervals to get an interval, and a procedure sqrt(Interval), which is a friend of Interval, for taking the square root. Use the naive rule for multiplication, where all four terms are multiplied, and the lowest and highest of these terms are returned as the product interval, and assume there is already a procedure long sqrt(long) that you can use for obtaining the square root (do not write this procedure). If A, B, and C are of class Interval, the expression A = B * C; will multiply intervals *B* and *C* and put the result in A; A = B/C will divide B by C, putting the result in A, and A = sqrt(B); will put the square root of *B* into *A*. Note that a/b is a * (1/b), so the multiply operator can be used to implement the divide operator, a - b is a + (-b), so the add and negate operators can be used to implement the subtract operator; and 4 * a is a + a + a + a, so scalar multiplication can be done by addition. Also Interval has a public data member error that can be set if we invert an interval containing 0 or get the square root of an interval containing a negtive value. Finally, write a *main()* procedure that will initialize intervals `a` to <1,2>, `b` to <3,4>, and `c` to <5,6>, and then evaluate the result of the expression `( -b + sqrt( b * b - 4 * a * c ) ) / (a + a)`.

A memory expansion card, **Adapt812 MX1**, plugs onto the rear of Adapt 812, offering the user up to 512K of Flash and 512K of SRAM. A real-time clock/calendar and battery back-up for the SRAM is included, as well as a prototyping area for the user's own application circuitry. A versatile dual-slot backplane/adapter couples the memory card to the micro-controller card so that the entire assembly can be plugged into a solderless breadboard.

9

Implementation of C Procedures

This chapter is perhaps the most important chapter in this book. We show how the 6812 assembly language implements C expressions and statements. We will use the HiWare C++ compiler in our examples. While different compilers will generate different code for the same C statement, studying one such implementation prepares you well to understand other implementations.

We first discuss how C allocates and accesses global and local variables. Then we consider how variables of different types are correctly coded in expressions and assignment statements. Next we discuss the implementation of conditional statements. We then describe how arrays and structs are accessed and then how loops are executed. Finally we discuss procedure calls and arguments, and we present our conclusions.

After you study this chapter, you will be able to read the assembly-language output of a C compiler with ease. One of the incidental benefits of this chapter is that you will see how to implement many operations in assembly language, by reading a "definition" of the problem to be solved in a C expression or statement and seeing the "solution" to the problem in assembly language. You will also be able to write C code that produces more efficient assembly-language code. As a further benefit, you will be able to fine-tune a C procedure by replacing parts of it with assembly-language code that can be embedded in the C procedure. Also, you will learn that you can write a C procedure that you can debug on a personal computer and hand-compile it into an assembly-language program. The C source program statements can be written in assembly-language comments to document your assembly-language program. This is one way to quickly write complex assembly-language programs.

This is therefore a very interesting chapter to complete the earlier chapters. You will really understand how hardware, which we showed in Chapters 1 to 3 implemented the 6812 instruction set, becomes a powerful machine that executes C and C++ procedures, in which you can express complex algorithms. You should be comfortable writing in a high-level language like C or C++, knowing what really happens, right down to the machine level, whenever you write an expression in your program.

We point out that the examples in this chapter are generated by a specific version (5.0.8) of the HiWare C++ compiler, with selected optimization options. You can expect to get slightly different code using different compilers, versions, or optimization options.

9.1 Global and Local Variables

This section shows how a variable is allocated and accessed in C. The first part of this section shows how to write a constant into a variable, which highlights the allocation mechanism, and the addressing mode and instruction data width used with each variable. The shorter second part describes what is done before the main procedure is executed, and a short third part discusses access to I/O ports, which are treated as global variables in C.

In the first part of this section, we will use a short procedure `main` as an example that we briefly show first. The names of the variables are abbreviations of their characteristics; `guc` represents a global `unsigned char`, and `gsi` represents a global signed `int`. The C program in Figure 9.1a, which writes some constants into these variables, is compiled into the assembly-language program in Figure 9.1b. Global variables, which are declared outside a procedure, can be allocated by the compiler by the equivalent of assembly-language `ORG` statement and `DS` statements. For instance, a possible global declaration for Figure 9.1b is shown in Figure 9.1c.

```
unsigned char guc; int gsi;
main(){ char lsc; unsigned int lui;
    guc = 0; gsi = 5; lsc = 0; lui = 7;
}
```

a. A C Program

```
     4: main(){ char lsc; unsigned int lui;
0000095B 1B9D              LEAS   -3,SP
     5:     guc = 0;
0000095D 790800            CLR    $0800
     6:     gsi = 5;
00000960 C605              LDAB   #5
00000962 87                CLRA
00000963 7C0801            STD    $0801
     7:     lsc = 0;
00000966 6A82              STAA   2,SP
     8:     lui = 7;
00000968 C607              LDAB   #7
0000096A 6C80              STD    0,SP
    10: }
0000096C 1B83              LEAS   3,SP
0000096E 3D                RTS
```

b. Assembly Language Developed From Part (a)

```
      ORG 2048 ; put global data at the beginning of RAM ($800)
guc:  DS.B 1   ; allocate a byte for scalar char variable guc
gsi:  DS.W 1   ; allocate two bytes for scalar int variable gsi
```

c. Declarations for Part (b)

Figure 9.1. A Simple Program

In Figure 9.1c the ORG statement's operand is set to a RAM location, such as $800. char or unsigned char variables are allocated one byte by DS.B directives, and int or unsigned int variables are allocated two bytes by DS.W directives.

Global variables, which in the 6812 are located in SRAM at $800 to $bff, generate 16-bit direct addressed instructions. The global unsigned char variable guc can be written into by a STAB, MOVB, or CLR instruction. The statement guc = 0; is implemented in assembly-language code as

```
CLR $800      ; clear global variable guc
```

Similarly, a char variable can be cleared, because signed and unsigned variables are coded as all zeros when intialized as zero. The global int variable gsi can be written into by a STD, MOVW, or a pair of CLR instructions. The statement gsi = 5; is implemented in assembly-language code as

```
LDAB  #5
CLRA
STD   $0801
```

Local variables, which are declared within, and generally at the beginning of, a procedure, are generally allocated at run time by means of a LEAS statement. For instance, the local declaration char lsc, unsigned int lui; requires three bytes on the stack, so it is allocated by the instruction

```
LEAS -3,SP          ; allocate local variables
```

immediately upon entry into the procedure, and deallocated by the instruction

```
LEAS 3,SP           ; deallocate local variables
```

at the end of the procedure, just before RTS is executed. Local variables generate index addressed instructions. The local char variable lsc can be written into by a STAB, MOVB, or CLR instruction. The variable lsc is at 2,SP. Because the compiler knows that accumulator A must be clear, as a result of the previous operation, the statement lsc = 0; is implemented in assembly-language code as

```
STAA  2,SP          ; clear local variable lsc
```

The global unsigned int variable gui can be written into by a STD, MOVD, or a pair of CLR instructions. The variable lui is at 0,SP. Because the compiler knows that accumulator A must be clear, the statement lui = 7; is efficiently implemented in assembly-language code as

```
LDAB  #7
STD   0,SP   ; write 7 into local variable lui
```

A C program can have global variables, and it can have global constants. Global constants are initialized by writing them into RAM when the program is downloaded or in ROM in a stand-alone microcontroller. In the HiWare compiler, a `#pragma into_rom` preceeds the storage of such constant variables declared as `const char,` and the like.

Global variables (as opposed to global constants) must be initially cleared, unless they are indictated as being initialized to some other value, before `main()` is executed. The following program segment can be used to clear 39 bytes beginning at $800:

```
    LDX  #$800       ; initialize pointer to beginning of global storage
    LDD  #39         ; initialize counter to number of bytes in global storage
L:  CLR  1,X+        ; clear a byte
    DBNE D,L         ; loop until all bytes cleared
```

If a global variable is declared and initialized, as in the statement

```
unsigned char guc = 4; int gsi = 4;
```

then the following assembly-language program segment should be executed after the above described program segment that clears all global variables, before `main` is called.

```
    LDD  #4          ; generate the constant 4 for both assignments
    STAB guc         ; store low-order 8 bits into char variable
    STD  gsi         ; store all 16 bits into int variable
```

Finally, we discuss I/O access as a variant of global addressing. The 6812's main I/O ports are on page zero. They can be accessed as if they are global variables:

```
volatile unsigned char PORTA, PORTB, DIRA, DIRB
```

which are linked to a segment at location 0. This is equivalent to an assembler sequence:

```
        org 0     ; put this "global data" at the beginning of I/O (0)
PORTA   ds.B 1    ; port A is at location 0
PORTB   ds.B 1    ; port B is at location 1
DIRA    ds.B 1    ; port A direction register is at location 2
DIRB    ds.B 1    ; port B direction register is at location 3
```

An output statement to output 5 to port A is written `PORTA = 5;` which is implemented

```
LDAB #5      ; generate constant 5
STAB PORTA   ; write it to the output port
```

An input statement `guc = PORTA;` to input port A to `guc` is implemented

```
LDAB PORTA   ; read input data
STAB guc     ; write it into global variable guc
```

The input and output statements use page-zero addressing, which provides improved static and dynamic efficiency over direct adddressing. Note that the MOVB instruction is not useful for accessing these I/O ports, because there is no page-zero address option in MOVB. The LDAB and STAB instructions above are more efficient than a MOVB instruction.

The assignment of I/O ports to global variable names should be written and executed before true global variables are assigned, because the origin will be set to the beginning of RAM (at $800) to assign true global variables. The declaration of globally defined I/O ports is often put in an #include file, which is inserted in a program before globals are defined in the program.

9.2 Expressions and Assignment Statements

In this section, we illustrate how operators are used in expressions. We will look at addition and subtraction statements that use same-width and different-width operands in a discussion of upcasting and downcasting. We will then study statements that use logical and arithmetic operators. We will carefully consider the increment and decrement operators and then look at expressions that save temporary results on the hardware stack.

The program in Figure 9.2a has several local and global variables, some of which are signed and others of which are unsigned, and some of which are 8-bit and others of which are 16-bit. Figure 9.2b shows assembly language developed from this program. Observe that each variable's name is an abbreviation of its characteristics; gsi is a global signed integer.

Many C statements are easily and efficiently translated into assembly language. This is especially true when all the variables in a statement are 8-bit char or unsigned char variables or when all the variables in a statement are 16-bit int or unsigned int variables. Assume the following statements are written in Figure 9.2a's main. Figure 9.2's statement gsi = lui + 12; is easily encoded as

```
LDX    0,SP     ; get 16-bit local variable lui
LEAX   12,X     ; add 12 (note that this is shorter than addd #12)
STX    $0801    ; put into 16-bit global variable gsi
```

and similarly the statement guc = lsc - 33; is simply encoded as

```
LDAB   2,SP     ; get 8-bit local variable lsc
SUBB   #33      ; subtract 33
STAB   $0800    ; put into 8-bit global variable guc
```

If a statement gets an int variable and writes a char variable, the source is truncated when it is read. Figure 9.2's statement guc = lui + 9; is encoded as

```
LDAB   1,SP     ; get low byte of 16-bit local variable lui
ADDB   #9       ; add 9
STAB   $0800    ; put into 8-bit global variable guc
```

An optimizing compiler can change the instruction ADDD #9 to ADDB #9 because the result will not be altered (reducing the precision is called *downcasting*).

However, without optimization, the intermediate values are generally computed using the largest precision of the variables in the expression (extending the precision called *upcasting*). Although many C compilers do all arithmetic in the largest precision, 16 bits in our case, HiWare efficiently operates upon numbers in the smallest possible precision.

```
unsigned char guc; int gsi;
main(){ char lsc; unsigned int lui;
   gsi = lui + 12;    guc = lsc - 33;
   guc = lui + 9;     gsi = gsi + lsc;    lui = guc - 17;
}
```

a. C Program

```
  4: main(){ char lsc; unsigned int lui;
main:
0000095B 1B9D                LEAS  -3,SP
  5:      gsi = lui + 12;
0000095D EE80                LDX   0,SP
0000095F 1A0C                LEAX  12,X
00000961 7E0801              STX   $0801
  6:      guc = lsc - 33;
00000964 A682                LDAA  2,SP
00000966 8021                SUBA  #33
00000968 7A0800              STAA  $0800
  7:      guc = lui + 9;
0000096B E681                LDAB  1,SP
0000096D CB09                ADDB  #9
0000096F 7B0800               STAB  $0800
  8:      gsi = gsi + lsc;
00000972 A682      LDAA  2,SP
00000974 B704      SEX   A,D
00000976 F30801    ADDD  $0801
00000979 7C0801    STD   $0801
  9:      lui = guc - 17;
0000097C F60800    LDAB  $0800
0000097E 87                  CLRA
0000097F 830011    SUBD  #17
00000983 6C80      STD   0,SP
 10: }
00000984 1B83      LEAS  3,SP
00000986 3D        RTS
```

b. Assembly Language Generated by Part (a)

Figure 9.2. A C Program with Local and Global Variables

If a statement gets a char variable and writes an int or unsigned int variable, the result is sign extended when it is read. Figure 9.2's statement gsi = gsi + lsc; or equivalently gsi += lsc; is simply encoded as

```
LDAA  2,SP    ; get 8-bit global variable lsc
SEX   A,D     ; upcast from char to int or unsigned int
ADDD  $0801   ; add in 16-bit global variable gsi
STD   $0801   ; put into 16-bit global variable gsi
```

But if a statement gets an unsigned char variable and writes an int or unsigned int variable, the high byte is cleared before the unsigned char variable is read. Figure 9.2's statement lui = guc - 17; is encoded as

```
LDAB  $0800   ; get 8-bit global variable guc saved earlier
CLRA          ; upcast from unsigned char to int or unsigned int
SUBD  #17     ; subtract 17
STD   0,SP    ; put into 16-bit global variable lui
```

You should observe that the declaration char or int affects the instruction data length, and char and unsigned char determine whether, on upcasting, the 8-bit data is sign extended with an SEX instruction or filled with zeros using a CLRA instruction.

The previous examples should indicate to the C programmer how to decide how a variable is to be type cast. If its range of values is 0 to 127, declare it to be an unsigned char, because upcasting is done with a short CLRA instruction rather than a longer SEX instruction. If its range is 0 to 256, declare it to be an unsigned char, but if the range is –128 to 127, declare it a char, to save space and time. Otherwise declare it to be int.

To discuss how common operators are handled, we use the following main as an example; it merely ANDs, ORs, multiplies, and divides some variables. Figure 9.3a's program is compiled into the assembly-language program in Figure 9.3b.

Logical bit-by-bit ANDing is illustrated in Figure 9.3 by the expression lsc = lsc& guc; or equivalently by lsc &= guc;, which is realized by

```
LDAA  2,SP    ; get local variable lsc
ANDA  $0800   ; AND with global variable guc
STAA  2,SP    ; put into local variable lsc
```

However, if one of the operands is constant, the BCLR instruction can be used. The expression in Figure 9.3, lsc = lsc& 0x12;,or equivalently lsc &= 0x12; is realized by

```
BCLR  2,SP,#237  ; AND local variable lsc with inverted constant 0x12
```

Note that the complement of the constant is used in the operand of BCLR. Logical bit-by-bit ORing is illustrated in Figure 9.3 by the expression gsi = gsi | lui; or equivalently by gsi |= lui;, which is realized by

```
LDD   $0801   ; get global variable gsi
ORAA  1,SP    ; OR with high byte of local variable lui
ORAB  0,SP    ; OR with low byte of local variable lui
STD   $0801   ; put into  global variable gsi
```

```
unsigned char guc; int gsi;
main(){ char lsc; unsigned int lui;
    lsc = lsc & guc;  lsc &= 0x12; gsi = gsi | lui;
    gsi = gsi & 0x1234;  lui = lsc * guc;  lui = lui / guc;
}
```

a. A C program

```
   4: main(){ char lsc; unsigned int lui;
0000095B 1B9D           LEAS  -3,SP
   5:    lsc = lsc & guc; lsc &= 0x12;
0000095D A682           LDAA  2,SP
0000095F B40800         ANDA  $0800
00000962 6A82           STAA  2,SP
00000964 0D82ED         BCLR  2,SP,#237
   6:    gsi = gsi | lui; gsi = gsi | 0x1234;
00000967 FC0801         LDD   $0801
0000096A EA81           ORAB  1,SP
0000096C AA80           ORAA  0,SP
0000096E 7C0801         STD   $0801
00000971 1C080112       BSET  $0801,#18
00000975 1C080234       BSET  $0802,#52
   7:    lui = lsc * guc;
00000979 A682           LDAA  2,SP
0000097B B706           SEX   A,Y
0000097D F60800         LDAB  $0800
00000980 87             CLRA
00000981 13             EMUL
00000982 6C80           STD   0,SP
   7:    lui = lui / guc;
00000984 F60800         LDAB  $0800
00000987 87             CLRA
00000988 B745           TFR   D,X
0000098A ECB1           LDD   0,SP
0000098C 1810           IDIV
0000098E 6E80           STX   0,SP
   8: }
00000990 1B83           LEAS  3,SP
00000992 3D             RTS
```

b. Assembly Language Generated by Part (a)

Figure 9.3. A C Program with Some Operators

However, if one of the operands is constant, `BSET` can be used. The expression `gsi = gsi | 0x1234;` or equivalently `gsi |= 0x1234;` is realized by

```
BSET  $0801,#18  ; OR high byte of global variable gsi with 0x12
BSET  $0802,#52  ; OR global variable low byte gsi with 0x34
```

One can complement accumulator D by complementing accumulator A and then complementing accumulator B. We might want to implement NEGD in the same way with a pair of instructions NEGA NEGB, but 1 is always added to A whether a carry is generated by adding 1 to B or not. Because NEGB sets C if the contents of B is nonzero, we can cancel the addition of 1 to A by NEGA, except when the contents of B is 0, by

```
NEGA
NEGB
SBCA #0
```

Multiplication uses a multiply instruction EMUL or EMULS, depending on whether the operations is signed; for instance the expression in Figure 9.3, lui = lsc * guc, is

```
LDAA    2,SP    ; get global variable lsc
SEX     A,Y     ; upcast from char to int, copy to register Y
LDAB    $0800   ; get guc
CLRA            ; upcast from unsigned char to int
EMUL            ; multiply unsigned
STD     0,SP    ; put 16-bit result into local lui
```

Similarly, division can be implemented using a divide instruction IDIV, EDIV, or EDIVS, depending on whether the operation is signed; for instance in Figure 9.3 the expression lui = lui / guc; (or equivalently, lui /= guc;) is implemented

```
STD     2,-SP   ; save numerator from previous statement
LDAB    $800    ; get denominator guc
CLRA            ; upcast from unsigned char to unsigned int
TFR     D,X     ; put in X for IDIV
LDD     2,SP+   ; get global variable lui
IDIV            ; divide X into D, putting quotient in X
STX     0,SP    ; put into local variable lui
```

To discuss how increment and decrement operators are handled, we use main in Figure 9.4 as an example; it merely increments and decrements some variables.

The simple increment and decrement instructions can be used directly with char or unsigned char variables.The statement in Figure 9.4, guc++;, is implemented

```
INC     $0800   ; increment global variable guc
```

and the statement lsc--; in Figure 9.4 is implemented

```
DEC     0,SP    ; decrement local variable lsc
```

However, the increment and decrement instructions on int or unsigned int variables use a longer sequence. The first problem with this operator is that there is no 16-bit memory operand version of it, and the second problem is that the carry bit is not affected by the 8-bit version of this operator. The statement gsi++; in Figure 9.4 is coded as

```
LDX   $0801   ; get gsi
INX           ; add 1
STX   $0801   ; put it back
```

and the statement `lui--;` in Figure 9.4 is implemented

```
LDX   1,SP    ; get lui
DEX           ; subtract 1
STX   1,SP    ; put it back
```

When the increment or decrement operator appears in a larger expression, the initial or final value of the variable used in the outer expression depends on whether the pair of "+" or "–" signs appears before or after the variable. `lsc = ++ guc;` is implemented

```
unsigned char guc; int gsi;
main(){ char lsc; unsigned int lui;
    guc++; lsc--;  gsi++;  lui--;  lsc = ++guc;  lsc = guc++;
}
```

a. A C Program

```
   4: main(){ char lsc; unsigned int lui;
0000095B 1B9D           LEAS  -3,SP
   5:      guc++;
0000095D 720800         INC   $0800
   6:      lsc--;
00000960 6380           DEC   0,SP
   7:      gsi++;
00000962 FE0801         LDX   $0801
00000965 08             INX
00000966 7E0801         STX   $0801
   8:      lui--;
00000969 EE81           LDX   1,SP
0000096B 09             DEX
0000096C 6E81           STX   1,SP
   9:      lsc = ++guc;
0000096E 720800         INC   $0800
00000971 B60800         LDAA  $0800
00000974 6A80           STAA  0,SP
  10:      lsc = guc++;
00000976 A680           LDAA  0,SP
00000978 720800         INC   $0800
0000097B 6A80           STAA  0,SP
0000097D 1B83           LEAS  3,SP
0000097F 3D             RTS
```

b. Assembly Language Generated by Part (a)

Figure 9.4. A C Program with Incrementing and Decrementing

```
INC     $0800  ; increment variable guc
LDAA    $0800  ; get value
STAA    0,SP   ; store variable lsc
```

while *lsc = guc++;* in Figure 9.4 is implemented

```
LDAA    $0800  ; get value
INC     $0800  ; increment variable guc
STAA    0,SP   ; store variable lsc
```

To discuss how the stack is used to store temporary results, we use `main` in Figure 9.5 as an example; it merely ORs, ANDs, and shifts some variables. This C program `main` is compiled into the assembly-language program in Figure 9.5b.

A statement involving several operations saves intermediate values on the stack. If an operand of an instruction like ADD or SUB has to be zero-filled or sign extended, then the instruction's other operand, in accumulator B or D, may have to be temporarily moved somewhere. It can be conveniently pushed on the stack and then pulled and operated on. The statement `lui = lui - lsc;` in Figure 9.5, or equivalently, `lui -= lsc;` is implemented by sign extending `lsc` and pushing the result on the stack. Then the variable `lui` is recalled into accumulator D, and the extended value of `lsc` is subtracted.

```
LDAA  2,SP    ; get local variable lsc
SEX   A,Y     ; sign extend
LDD    0,SP   ; get local variable lui
PSHY          ; save on stack
SUBD 2,SP+    ; pull from stack, subtract from accumulator D
STD   0,SP    ; put into global variable lui
```

From Figure 9.5, we next offer an example that inserts three bits into a 16-bit local `int`. Note that, due to pushing the two-byte temporary variable on the stack, the stack address to recall *lui* is `4,SP`. The statement `lui = (lui & 0xfc7f) | ((lsc << 7) & 0x380);` is compiled

```
PSHD            ; save on stack (previous statement just computed lui)
LDAB   #128     ; we will multiply by 2**7 to shift left
CLRA            ; seven bits, so get this constant ready
PSHY            ; save lsc from previous statement
EMUL            ; shift it
ANDA   #3       ; mask off low two bits of accumulator A
ANDB   #128     ; mask off high bit of accumulator B
PSHD            ; save temporary result
LDD    4,SP     ; get lui (notice offset adjustment)
ANDB   #127     ; mask all low-order bits
ANDA   #252     ; mask all high-order bits
ORAA   1,SP+    ; combine new and old values
ORAB   1,SP+    ; in both bytes
STD    2,SP     ; write out new value of lui
```

```
unsigned char guc; int gsi;
main(){ char lsc; unsigned int lui;
    lui -= lsc; lui = (lui & 0xfc7f) | ((lsc << 7) & 0x380);
    lui = (lui << 3) + (lui << 1) + lsc -'0';
}
```

a. C Program

```
    4: main(){ char lsc; unsigned int lui; lui -= lsc;
0000095B 1B9D          LEAS  -3,SP
0000095D A682          LDAA  2,SP
0000095F B706          SEX   A,Y
00000961 EC80          LDD   0,SP
00000963 35            PSHY
00000964 A3B1          SUBD  2,SP+
00000966 6C80          STD   0,SP
    6:     lui = (lui & 0xfc7f)| ((lsc << 7) & 0x380);
00000968 3B            PSHD
00000969 C680          LDAB  #128
0000096B 87            CLRA
0000096C 35            PSHY
0000096D 13            EMUL
0000096E C480          ANDB  #128
00000970 8403          ANDA  #3
00000972 3B            PSHD
00000973 EC84          LDD   4,SP
00000975 C47F          ANDB  #127
00000977 84FC          ANDA  #252
00000979 AA81          ORAA  1,SP+
0000097B EA81          ORAB  0,SP+
0000097D 6C82          STD   2,SP
    7:     lui = (lui << 3) + (lui << 1) + lsc -'0';
0000097F 59            ASLD
00000980 59            ASLD
00000981 59            ASLD
00000982 B745          TFR   D,X
00000984 EC82          LDD   2,SP
00000986 59            ASLD
00000987 1AE6          LEAX  D,X
00000989 B754          TFR   X,D
0000098B E380          ADDD  0,SP
0000098D 830030        SUBD  #48
00000990 6C84          STD   4,SP
00000992 1B87          LEAS  7,SP
00000994 3D            RTS
```

b. Assembly Language Generated by Part (a)

Figure 9.5. A C Program with ORing, ANDing and Shifting

The statement `lui = (lui << 3) + (lui << 1) + lsc -'0';` in Figure 9.5 which can be used to build a decimal number from ASCII characters, is compiled into

```
STD   0,SP     ; save lui which was left in D
LSLD           ; shift left three places
LSLD           ; in order to
LSLD           ; multiply by eight
TFR    D,X     ; save this intermediate result in a register
PULD           ; get lui again.
LSLD           ; shift left to double it
LEAX   D,X     ; add both parts
TFR    X,D     ; move to D to complete the addition
ADDD 0,SP      ; add lsc, which is left on the stack
SUBD #48       ; subtract the constant for ASCII '0'
STD   4,SP     ; save result in lui: note the offset
```

Temporary results can be saved in registers, as we saw in the `TFR D,X` instruction. The stack provides another place to temporarily save the data in accumulator D and can save essentially any number of such values. Note again that the stack pointer offset changes as temporary results are saved on the stack. At the subroutine's end, the stack pointer is adjusted, not only to deallocate local variables but also to deallocate temporary variables, using the instruction `LEAS 7,SP`. Deallocating at the end saves instructions that should deallocate temporary variables when they are no longer needed, to improve static efficiency, at the expense of using up more of the stack than would be needed if temporary variables were promptly deallocated when they were no longer needed.

9.3 Conditional Statements

A statement can be conditional, or it can control looping to execute a sequence of statements that are written within it many times. We first present Boolean operators that generate a 1 (true) or 0 (false) variable. We then give assembly-language program segments for an example of several of C's control statements.

To illustrate Boolean operators, the expression `main` in Figure 9.6 compares some variables. Many branch instructions such as `BEQ *+5` are used to indicate a branch that is five bytes ahead of the (beginning of the) `BEQ` instruction. This current location counter is used to avoid generating a lot of labels for local branching.

In Figure 9.6, the C procedure's first expression `guc = lsc > -3;` results in

```
LDAA   3,-SP  ; allocate 3 bytes for local variables and get variable lsc
CMPA   #253   ; if greater than -3 as a signed number
BGT    *+4    ; then proceed to "true" program segment
CLRA          ; if false, clear guc. If true,
CPS    #34305 ; then skip over operand  jump to operand which is LDAA #1
STAA   $0800  ; store the result
```

```
unsigned char guc; int gsi;
main(){ char lsc; unsigned int lui;
    guc = lsc > -3; lsc = lui > 5; lui = gsi >= 0;
    gsi  = lui >= 0;  gsi = lsc == 0; lsc = (gsi & 4) == 0;
}
```

a. A C Program

```
   5:      guc = lsc > -3; lsc = lui > 5;
0000095B A6AD          LDAA  3,-SP
0000095D 81FD          CMPA  #253
0000095F 2E02          BGT   *+4    ;abs = 0963
00000961 87            CLRA
00000962 8F8601        CPS   #34305
00000965 7A0800        STAA  $0800
00000968 EC81          LDD   1,SP
0000096A 8C0005        CPD   #5
0000096D 2202          BHI   *+4    ;abs = 0971
0000096F 87            CLRA
00000970 8F8601        CPS   #34305
00000973 6A80          STAA  0,SP
    7:     lui = gsi >= 0;
00000975 FC0801        LDD   $0801
00000978 2A02          BPL   *+4    ;abs = 097C
0000097A C7            CLRB
0000097B 8FC601        CPS   #50689
0000097E 87            CLRA
0000097F 6C81          STD   1,SP
    8:     gsi  = lui >= 0;
00000981 C601          LDAB  #1
00000983 87            CLRA
00000984 7C0801        STD   $0801
    9:     gsi = lsc == 0;
00000987 A680          LDAA  0,SP
00000989 2702          BEQ   *+4    ;abs = 098D
0000098B C7            CLRB
0000098C 8FC601        CPS   #50689
0000098F 87            CLRA
00000990 7C0801        STD   $0801
   10:     lsc = (gsi & 4) == 0;
00000993 1F08020402    BRCLR $0802,#4,*+7    ;abs = 099A
00000998 87            CLRA
00000999 8F8601        CPS   #34305
0000099C 6AB2          STAA  3,SP+
```

b. Assembly Language Generated by the body of Part (a)

Figure 9.6. A Program with Boolean Operators

Note that signed numbers use branches like `BGT`. Unsigned number comparisons use branches like `BHI`. One of the more difficult problems of accurately translating C into assembly language is that of choosing the correct kind of conditional branch instruction to take care of signed or unsigned comparisons.

In one of the more peculiar operations, `CPS #34305` is used to skip a two-byte operand, `LDAA #1`. Suppose the instruction is at location $962. If the entire instruction is executed, the only effect is that the condition codes are set, but they are not tested in subsequent instructions, so the `CPS #34305` is a no-op. However, if a branch to $963 is made, the constant `34305` is executed as an opcode. This instruction is `LDAA #1`, and the result is to load 1 into accumulator A. The HiWare C compiler uses this technique to make accumulator A either a 1 (T) or a 0 (F). We see several examples in the program in Figure 9.6. It is also used in a `case` statement in which several of the cases load different values into the same variable. Such a `case` statement appears in Figure 9.9.

The expression `lsc = lui > 5;` in Figure 9.6 results in the following code:

```
LDD     1,SP    ; get 16-bit variable lui
CPD     #5      ; if less than 5 as an unsigned number
BHI     *+4     ; branch to the operand of the CPS instruction
CLRA            ; otherwise clear the value
CPS     #34305  ; skip, or else set result to 1.
STAA    0,SP    ; store the result
```

Note that the constant 1 (T) or a 0 (F) is generated in accumulator A to reflect the Boolean value of the test `lui > 5`. Signed number comparisons that test just the sign use branches like `BPL`. The expression `lui = gsi >= 0;` in Figure 9.6 results in the following code:

```
LDD     $0801   ; preclear result
BPL     *+4     ; if nonnegative then
CLRB            ; clear and skip
CPS     #50689  ; skip or set result to 1.
CLRA            ; high result is always 0
STD     1,SP    ; put result in lui
```

Note that there is no test for the expression `gsi = lui >= 0;`, because unsigned numbers are always nonnegative. This is an error made by many programmers. Be careful when you determine the data type of a variable and when you test that variable, so that you avoid the situation where you test a variable declared to be an unsigned number for a value less than zero or a value greater or equal to zero.

Comparisons for equality or inequality can often use `TST` and branches like `BEQ`. The expression `gsi = lsc == 0;` in Figure 9.6 results in the following code:

```
LDAA    0,SP    ; test 8-bit variable lsc
BEQ     *+4     ; if nonzero then
CLRB            ; clear and skip
CPS     #50689  ; skip or set result to 1.
CLRA            ; high result is always 0
STD    $0801    ; put result in gsi
```

Certain bit tests can often use `BRSET` or `BRCLR` branches. In Figure 9.6, the expression `lsc = (gsi & 4) == 0;` results in the following code:

```
BRCLR  $0802,#4,*+2  ; if bit 4 is not zero then
CLRA                 ; clear result
CPS    #34305        ; skip, or set result to 1
STAA   3,SP+         ; put result in lui and deallocate local variables
```

The Boolean result of the test, a value of 1 (T) or a 0 (F), is actually not usually generated but may be used to branch to a different location. For instance `if(lui > 5)` results in the following code:

```
LDD    1,SP     ; get 16-bit variable lui
CPD    #5       ; if less than 5 as an unsigned number
BLS    L        ; branch around the expression if lui is higher than 5
```

Simple conditional expressions of the form *if then,* full conditionals of the form *if then else,* and extended conditionals of the form *if then else if then else if then . . . else,* use conditional expression operators. In the last expression, the *else if* part can be repeated as many times as needed, and the last part can be an optional *else*. Variables are compared using *relational operators* (> and <), and these are combined using *logical operators* (`&&`). We give examples of common simple conditionals first.

The C program in Figure 9.7a is compiled into the assembly language program shown in Figure 9.7b. A statement `if(! lsc) guc = 0;` or equivalently `if(lsc == 0) guc = 0;` is encoded as

```
LDAA   3,-SP    ; allocate and set condition codes for variable lsc
BNE    *+5      ; if nonzero, skip over next instruction
CLR    $0800    ; otherwise, if zero,  clear variable guc
```

Where the condition applies to a complex expression of many statements, the branch instructions can be converted to long branch instructions. For instance,

```
if(gsi < 5) { ... /* many instructions */ }
```

can be implemented

```
        LDD    $0801    ; get variable gsi
        CPD    #5       ; if greater than or equal to 5 as an unsigned number
        LBGE   L1       ; then skip over next several instructions

        ...             ; many instructions generated between { } appear here
L1:     EQU    *        ; located after the latter } matching the if statement's {
```

A simple C compiler can always implement the conditional operation using the long branch instructions like `LBHS,` but an optimizing C compiler will get the size of the branch offset. It uses a long branch instruction when the label cannot be reached by the corresponding shorter branch instruction.

```
unsigned char guc; int gsi;
main(){ char lsc; unsigned int lui;
     if(lsc == 0) guc = 0;
     if(gsi < 5) { ... /* many instructions */   }
     if(lsc + guc) gsi = 0;
     if(guc < 5) lsc = 0; else lsc = 9;
}
```

a. A C Program

```
   4: main(){ char lsc; unsigned int lui; if(lsc == 0) guc = 0;
0000095B A6AD          LDAA  3,-SP
0000095D 2603          BNE   *+5    ;abs = 0962
0000095F 790800        CLR   $0800
    6:     if(gsi < 5) { /* many instructions */ lui = 0;  }
00000962 FC0801        LDD   $0801
00000965 8C0005        CPD   #5
00000968 2C04          BGE   *+6    ;abs = 096E
0000096A C7            CLRB
0000096B 87            CLRA
0000096C 6C81          STD   1,SP
    7:     if(lsc + guc) gsi = 0;
0000096E A680          LDAA  0,SP
00000970 B704          SEX   A,D
00000972 B745          TFR   D,X
00000974 F60800        LDAB  $0800
00000977 87            CLRA
00000978 1AE6          LEAX  D,X
0000097A 044504        TBEQ  X,*+7    ;abs = 0981
0000097D C7            CLRB
0000097E 7C0801        STD   $0801
    8:     if(guc < 5) lsc = 0; else lsc = 9;
00000981 B60800        LDAA  $0800
00000984 8105          CMPA  #5
00000986 2404          BCC   *+6    ;abs = 098C
00000988 6980          CLR   0,SP
0000098A 2004          BRA   *+6    ;abs = 0990
0000098C C609          LDAB  #9
0000098E 6B80          STAB  0,SP
    9: }
00000990 1B83          LEAS  3,SP
00000992 3D            RTS
```

b. Assembly Language Generated by Part (a)

Figure 9.7. A Program with If-Then Expressions

```
unsigned char alpha, beta , gamma, delta, epsilon, zeta;
main(){
  if((alpha < 5)&&(beta == 0)) gamma = 0;
  if((alpha < 5)||(beta == 0)) gamma = 0;
  if(alpha != 0) beta = 10; else if(gamma == 0) delta++;
  else if((epsilon != 0)&&(zeta==1)) beta=beta << 3; else beta=0;
}
```

a. A C Program

```
   6:     if((alpha < 5)&&(beta == 0)) gamma = 0;
0000095B B60800        LDAA  $0800
0000095E 8105          CMPA  #5
00000960 2C08          BGE   *+10   ;abs = 096A
00000962 B60801        LDAA  $0801
00000965 2603          BNE   *+5   ;abs = 096A
00000967 790802        CLR   $0802
   7:     if((alpha < 5)||(beta == 0)) gamma = 0;
0000096A B60800        LDAA  $0800
0000096D 8105          CMPA  #5
0000096F 2D05          BLT   *+7   ;abs = 0976
00000971 B60801        LDAA  $0801
00000974 2603          BNE   *+5   ;abs = 0979
00000976 790802        CLR   $0802
   8:     if(alpha != 0) beta = 10;
00000979 B60800        LDAA  $0800
0000097C 2707          BEQ   *+9   ;abs = 0985
0000097E C60A          LDAB  #10
00000980 7B0801        STAB  $0801
00000983 201C          BRA   *+30   ;abs = 09A1
   9:     else if(gamma == 0) delta++;
00000985 B60802        LDAA  $0802
00000988 2605          BNE   *+7   ;abs = 098F
0000098A 720803        INC   $0803
0000098D 2012          BRA   *+20   ;abs = 09A1
  10:     else if((epsilon!=0)&&(zeta==1)) beta=beta<<3; else beta=0;
0000098F B60804        LDAA  $0804
00000992 270A          BEQ   *+12   ;abs = 099E
00000994 B60805        LDAA  $0805
00000997 042004        DBNE  A,*+7   ;abs = 099E
0000099A 0764          BSR   *+102   ;abs = 0A00
0000099C 2003          BRA   *+5   ;abs = 09A1
0000099E 790801        CLR   $0801
```

b. Assembly Language Generated by Part (a)

Figure 9.8. Assembly Language for a Decision Tree

An operation result may be in accumulator B or D. It can be tested by the TBEQ or TBNE instructions. In Figure 9.7, the statement `if(lsc + guc) gsi = 0;` similarly encodes as

```
LDAA  0,SP    ; get value of lsc
SEX   A,D     ; upcast to 16 bits
TFR   D,X     ; use X as accumulator
LDAB  $0800   ; get value of guc
CLRA          ; upcast to 16 bits
LEAX  D,X     ; add values
TBEQ X,*+7    ; check value of sum. If zero
CLRB          ; generate a 16-bit zero (A is already clear)
STD   $0801   ; store to clear variable gsi
```

In Figure 9.7, the `else` part of a conditional expression is easily implemented by a BRA instruction. The statement `if(guc < 5) lsc = 0; else lsc = 9;` encodes as

```
LDAA  $0800   ; get variable guc
CMPA  #5      ; if greater than or equal to 5 as an unsigned number
BHS   *+6     ; then skip over next instruction (this is BCC)
CLR   0,SP    ; otherwise, if zero,  clear variable lsc
BRA   *+6     ; now skip over next two instructions
LDAB  #9      ; write 9 into variable lsc
STAB  0,SP
```

A conditional expression can be a logical OR or a logical AND of tests described above. The logical OR test will check each case, from left to right, for a true result, and will execute the statement when it finds the first true result. The logical AND checks each case, from left to right, for a false, and bypasses the statement the first time it finds a false test. If `alpha`, `beta`, and `gamma` are signed global `char` variables, the statement in Figure 9.8 `if((alpha < 5) && (beta == 0)) gamma = 0;` encodes as

```
LDAA  $0800   ; get variable alpha
CMPA  #5      ; if less than 5 a signed number
BGE   *+10    ; then skip to CLR instruction
LDAA   $0801  ; if beta is nonzero
BNE   *+5     ; then skip over next instruction
CLR   $0802   ; if you get here,  clear variable gamma
```

and `if((alpha < 5) || (beta == 0)) gamma = 0;` is encoded as

```
LDAA  $0800   ; get variable alpha
CMPA  #5      ; if less than 5 as a signed number
BLT   *+7     ; then skip to CLR instruction
LDAA  $0801   ; if beta is nonzero
BNE   *+5     ; then skip over next instruction
CLR   $0802   ; if you get here,  clear variable gamma.
```

As seen in the previous examples, the ANDing of conditions is affected by branching around the "then" code if either condition is false, and the ORing of conditions is affected by branching to the "then" code if either condition is true.

Many *else if* expressions can be inserted between an *if* expression and the final *else* expression. The branch instructions jump out of a statement that is executed to the statement beyond the final *else* statement. Moreover, the final *else* expression may be omitted. Obviously, one can have more than two OR or AND tests, and one can nest OR tests within AND tests, or one can nest AND tests within OR tests, and so on.

One of the common errors in C is to confuse bit-wise logical OR with the OR test discussed above. The expression `if((alpha < 5) | (beta == 0)) gamma = 0;` encodes as

```
        LDAA  alpha   ; get variable alpha
        CMPA  #5
        BLT   L0      ; if greater or equal to 5
        LDX   #0      ; generate zero
        BRA   L1      ; and skip
L0:     LDX   #1      ; otherwise generate one
L1:     LDAA  beta    ; test variable beta
        BEQ   *+4     ; if nonzero, branch to middle of CPS
        CLRB          ; otherwise clear B
        CPS   #50689  ; address mode is actually LDAB #1
        CLRA          ; high-order byte is always zero
        PSHX          ; OR X into D
        ORAB  1,SP    ; by pushing X
        ORAA  2,SP+   ; then pulling it and ORing it into D
        TBEQ  D,*+6   ; if the result is nonzero
        CLR   gamma   ; clear variable gamma
```

What a difference a single character makes! Although the same answer is obtained with the statement `if((alpha < 5) || (beta == 0)) gamma = 0;` as with `if((alpha < 5) | (beta == 0)) gamma = 0;`, the assembly language generated by the latter is significantly less efficient than that generated by the former statement.

Another of the common errors in C is to confuse assignment with equality test. The expression `if(beta == 0) gamma = 0;` encodes as

```
        TST   beta    ; test variable beta
        BNE   *+5     ; if the result is nonzero then
        CLR   gamma   ; clear variable gamma
```

The expression `if(beta = 0) gamma = 0;` encodes as

```
        CLR   beta    ; clear variable beta (note: this is an assignment statement)
        BNE   *+5     ; if the result is nonzero (it isn't) then
        CLR   gamma   ; clear variable gamma
```

From the rest of Figure 9.8, note how a string of *else if (. . .) . . . else . . . ;* statements cause the tests we have already discussed to be done, and when one is successful, so its following statement is executed, a branch is made to the end of the series of *else if (. . .) . . . else . . . ;* statements. Incidentally, the subroutine branched to by `BSR *+102` shifts the byte in beta left three places.

The case statement is a useful alternative to the conditional statement. Consider an expression like `switch(n){ case 1: i=1; break; case 3: i=2; break; case 6: i=3;break;}`. This is compiled into assembly language by calling a subroutine `switch` to evaluate the case and providing the cases as addresses below its call, as shown in Figure 9.9a. This technique is used in HiWare's compiler when the cases are, or are nearly, consecutive numbers. Another technique used in some compilers is to implement the test using a sequence of `CMPA` and `BNE` instructions. Assembly language in Figure 9.9b implements the same switch statement with branch instructions. This technique is used in HiWare's compiler when the cases are not consecutive numbers.

```
* switch(n){case 0:i=1; break; case 1:i=2; break; case 2:i=3; break;}
    00000867 EC81              LDD    1,SP
    00000869 072D              BSR    switch
    0000086B 03        L:      DC.B   L0-L
    0000086C 06                DC.B   L1-L
    0000086D 09                DC.B   L2-L
  5:                     case 1: i=1; break;
    0000086E C601   L0:        LDAB   #1
    00000870 8FC602            CPS    #50690  ; is SKIP2 L1:LDAB #2
  6:                     case 3: i=2; break;
    00000873 8FC603            CPS    #50691  ; is SKIP2 L2:LDAB #3
    00000876 6B80              STAB   0,SP
                         * . . .
    00000898 30    switch: PULX
    00000899 E6E6              LDAB D,X
    0000089B 05E5              JMP B,X
```

a) Using a Subroutine and Argument List

```
*     switch(n){case 1:i=1; break;case 3:i=2;break;case 6: i=3;break;}
          LDAA   $0800   ; get switch operand alpha
          CMPA #6        ; check for last case
          BHI    *+27    ; branch over the rest of the cases
          CMPA #1        ; check for first case
          BEQ *+12       ; if found, go to LDAB instruction
          CMPA   #3      ; check second case
          BEQ    *+11    ; if so go to middle of first CPS instruction
          CMPA   #6      ; check last case
          BEQ    *+10    ; if so go to middle of second CPS instruction
          BRA    *+13    ; if not matched, skip over cases
          LDAB   #1      ; this is for the case one
          CPS    #50690 ;  skip, or LDAB #2
          CPS    #50691 ;  skip, or LDAB #3
          STAB   $0801   ; store result in beta
```

b) Using a Sequence of `CMP` and conditional branch instruction Pairs

Figure 9.9. Alternative Assembly Language for a Case Statement

9.4 Loop Statements, Arrays, and Structs

In this section, we show how statements within a loop can be repeated until a condition is met, governed by an expression much like the expressions of the conditional statements. First, we consider an accumulation of variables incremented or decremented in the loop, in different loop constructs. Then we discuss array elements accessed using indexes. Struct elements are accessed using AND, OR, and shift operations. We will access a two-dimensional array using indexes in `for` loops and a `struct` in a `do while` loop.

The `for` and `while` statements test the condition before the loop is executed and are useful if, for example, a loop may have to be done 0 times. The `do while` statement performs the statement once before it tests the condition. See Figure 9.10.

```
int i;
main(){ int j, k;
            for(j = k = 0; j != i; j++) k += j;
            while(j != 0) k += --j;
            do k += j--; while (j != i);
}
```

a. A C Program

```
* 3:         for(j = k = 0; j != i; j++) k += j;
00000802 C7              CLRB
00000803 87              CLRA
00000804 B745            TFR   D,X
00000806 2005            BRA   *+7      ;abs = 080D
00000808 1AE6            LEAX  D,X
0000080A C30001          ADDD  #1
0000080D BC0800          CPD   $0800
00000810 26F6            BNE   *-8      ;abs = 0808
* 4:         while(j != 0) k += --j;
00000812 2005            BRA   *+7      ;abs = 0819
00000814 830001          SUBD  #1
00000817 1AE6            LEAX  D,X
00000819 0474F8          TBNE  D,*-5      ;abs = 0814
* 5:         do k += j--; while (j != i);
0000081C 1AE6            LEAX  D,X
0000081E 09              DEX
0000081F BC0800          CPD   $0800
00000822 26F8            BNE   *-6      ;abs = 081C
00000824 3D              RTS
```

b. Assembly Language developed from Part (a)

Figure 9.10. For, While, and Do While Loops

In Figure 9.10 a `for` loop has an initialization expression, shown first in the `for` list of expressions; a test expression, shown in the middle; and a loop termination expression, shown last:

```
* 3:        for(j = k = 0; j != i; j++) k += j;
       CLRB              ; j in D
       CLRA
       TFR    D,X        ; k in X
       BRA    L1         ; do the test before the loop is done once
L0:    LEAX   D,X        ; this is the statement that is executed in the loop
       ADDD   #1         ; this is the expression done after each loop is done
L1:    CPD    $0800      ; this is the loop test
       BNE    L0
```

A `while` loop has a test expression that is executed before the loop is executed once:

```
* 4:        while(j != 0) k += --j;
       BRA    L3         ; do the test first
L2:    SUBD   #1         ; these are the two statements
       LEAX   D,X        ; that are executed in the loop
L3:    TBNE   D,L2       ; this is the loop test
```

A `do while` loop has a test expression that is executed after the loop is executed:

```
* 5:        do k += j--; while (j != i);
L4:    LEAX   D,X        ; these are the two statements
       DEX               ; that are executed in the loop
       CPD    $0800      ; this is the loop test
       BNE    L4
```

Figure 9.11 illustrates nested `for` loops and the two-dimesional array index addressing mechanism. This example shows how loop statements can themselves be loops, in a nested loop construction, and how optimizing compilers make loops more efficient. The outer `for` loop, `for(i = sum = 0; i < 10; i++)` is encoded as an initialization:

```
CLRA              ; generate 0
CLRB              ; in high and low bytes
STD    $081E      ; store to clear sum
STAB   1,SP       ; store to clear i
```

and by the outer loop termination:

```
INC    1,SP       ; count up
LDAA   1,SP       ; get the variable to be tested
CMPA   #10        ; compare against 10
BCS    *-40       ; loop as long as i is less than 10
CMPA   #3         ; check if another iteration is to be done
BCS    *-30       ; if so, branch to the instruction following the initialization
```

```
unsigned char a[10][3]; int sum;
main() { unsigned char i, j;
    for(i = sum = 0; i < 10; i++)
        for(j=0; j < 3; j++)
            sum += a[i][j];
}
```

a. A C Program

```
   4: main() { unsigned char i, j;
0000095B 3B            PSHD
   5:      for(i = sum = 0; i < 10; i++)
0000095C C7            CLRB
0000095D 87            CLRA
0000095E 7C081E        STD   $081E
00000961 6B81          STAB  1,SP
   6:                for(j = 0; j < 3; j++)    sum += a[i][j];
00000963 6980          CLR   0,SP
00000965 E681          LDAB  1,SP
00000967 87            CLRA
00000968 CD0003        LDY   #3
0000096B 13            EMUL
0000096C B745          TFR   D,X
0000096E E680          LDAB  0,SP
00000970 87            CLRA
00000971 1AE6          LEAX  D,X
00000973 E6E20800      LDAB  2048,X
00000977 F3081E        ADDD  $081E
0000097A 7C081E        STD   $081E
0000097D 6280          INC   0,SP
0000097F A680          LDAA  0,SP
00000981 8103          CMPA  #3
00000983 25E0          BCS   *-30    ;abs = 0965
   5:      for(i = sum = 0; i < 10; i++)
00000985 6281          INC   1,SP
00000987 A681          LDAA  1,SP
00000989 810A          CMPA  #10
0000098B 25D6          BCS   *-40    ;abs = 0963
   7: }
0000098D 30            PULX
0000098E 3D            RTS
```

b. Assembly Language developed from Part (a)

Figure 9.11. Array Manipulation Program

The initialization should branch to the loop termination, but because the compiler determines that the first loop will satisfy the termination test, this step is skipped to improve efficiency. The inner loop is almost identically constructed. Generally, the `for` loop is fundamentally a `while` loop with a built-in initialization and a built-in stepping operation that is executed after the statement governed by the *for* loop is executed and before the `while` condition is tested. The general and natural *for* loop is the most widely used looping mechanism in C and C++.

The more general `for` statement is used with instructions that access a two-dimensional array using accumulator D index addressing. The program in Figure 9.11 adds all the elements of a global `unsigned char` two-dimensional array into a global `int` variable. The two-dimensional array is declared as `a[10][3]`, so it is a ten-row, three-column array. Elements in rows are stored in consecutively accessed locations, and whole rows are stored in consecutive groups of these memory words (this is called *row-major order*). The middle of the assembly-language code from Figure 9.12 is shown below:

```
LDAB  1,SP     ; get row index
CLRA
LDY   #3       ; number of bytes per row
EMUL           ; gets the relative location of row i in array a
TFR   D,X
LDAB  0,SP
CLRA
LEAX  D,X      ; add column number
LDAB  2048,X   ; get element
```

It reads out element `i, j`, into accumulator D. Observe that the calculation of the location of the `i, j` th element is obtained by multiplication and addition operations (a polynomial expression).

The `do while` statement can produce efficient assembly-language code using C code that may look somewhat awkward. Figure 9.12 illustrates an efficient way to clear a vector. Note that the loop counter is decremented, and tested after it has been decremented, in order to use the instruction `DBNE`. In order to use this loop counter as an index into a vector, one is subtracted from the counter to make it the vector index. This produces the tightest loop to clear a vector in the 6812.

```
char alpha[10];
void main() { char i = 5; do alpha[i - 1] = 0; while(--i); }
```

a. A C Program

```
00000959 8605        LDAA  #5
0000095B B705        SEX   A,X
0000095D 69E207FF    CLR   2047,X
00000961 0430F7      DBNE  A,*-6    ;abs = 095B
00000964 3D          RTS
```

b. Assembly Language developed from Part (a)

Figure 9.12. Clearing a Vector with a Do While Loop

```
struct spiDevice {
    unsigned int spie:1,spe:1,swom:1,mstr:1,cpol:1,cpha:1,ssoe:1,
        lsbf:1;
} *spiPtr = (struct spiDevice *)0xd0;
#define spi (*spiPtr)

main() { spi.spe = 1; do ; while(spi.spe); }
```

a. A C Program

```
0000095B 3B          PSHD
0000095C FE0800      LDX   spiPtr
0000095F 6E80        STX   0,SP
00000961 0C0040      BSET  0,X,#64
00000964 EE80        LDX   0,SP
00000966 0E0040FA    BRSET 0,X,#64,*-6   ;abs = 0964
0000096A 30          PULX
0000096B 3D          RTS
```

b. Assembly Language developed from Part (a)

Figure 9.13. A Program Setting and Testing a Bit

The `do while` statement in Figure 9.13a tests the condition after the loop is executed at least once, but it tests the result of the loop's activities. This is very useful in I/O software because it lets you get data within the statement and test the result in the conditional expression, which is not executed until the statement is executed at least once. See Figure 9.13. The program sets a bit of a `struct` and tests it repeatedly until it is cleared (by hardware). It is compiled into assembly language shown in Figure 9.13b. The `struct` definition shown below merely defines accesses using the pointer `spiPtr`. The elements can be accessed using "arrow" notation. For instance the bit `spe` can be set using `spiPtr->spe = 1;` However, by declaring `#define spi (*spiPtr)`, the expression `spi.spe = 1;` can be used instead. This is encoded into assembly language using the `BSET` instruction:

```
BSET 0,X,#64           ; set bit 6 of location pointed to by X ( the spi port )
```

The statement `do ; while(spi.spe);` is implemented with

```
BRSET 0,X,#64,*-6  ; wait while bit 6 of location 0xD0 is 1
```

Generally, if the bitfield is more than one bit, data to be inserted will have to be shifted to the correct bit position, and masked parts of it are ORed with masked parts of bits in other fields, to be written into the memory. This code looks like the code for statement `lui = (lui << 3) + (lui << 1) + lsc -'0';` that we studied at the end of Section 9.2. Data read from such a bitfield will have to be shifted and masked in like manner.

9.5 Procedure Calls and Arguments

A C procedure is generally called using a `JSR` or `BSR` instruction, with the input arguments pushed on the stack. The return value of a function is generally left in accumulator D. If the input argument is a vector, or an "&" sign appears before the name, then the address is passed on the stack, using call-by-name; otherwise the data itself is pushed on the stack, using call-by-value. However, the rightmost argument is passed into a function through a register. The function might push this register value inside it, as a local variable. Passing one argument this way improves efficiency, because even if it is pushed on the stack inside the function itself, its code therein appears just once in a program, rather than each time the subroutine is called. As an example, the procedure `power` can be called by `main()` in Figure 9.14a. Figure 9.14b shows the calling procedure's assembly language.

Figure 9.15 shows the stack within the procedure that was called; its assembly language is shown Figure 9.14c. The `while` loop requires the test at the end of the loop and a branch at the beginning of the `while` loop to that test program sequence. Observe from Figure 9.14c that call-by-value argument `j` is generally in accumulator A. The test requires checking the argument `j` before it is decremented, so the instruction `PSHA` saves `j`, the `DEC` instruction decrements `j`, and the `BNE` instruction tests the value obtained by the `LDAB 1,SP+` instruction.

The `EMUL` instruction multiplies the value in D by the value in Y. We passed the address of argument `i` to `power`, merely to show how call-by-name can be handled. It was pushed on the stack. Note from Figure 9.15 that this address is at `2,SP.` The `int` value at that location can be read into index register Y by `LDY 2,SP LDD 0,Y` and the other multiplier, a local variable, is read into Y by the `TFR X,Y.` The data are multiplied and the result stored in the local variable using `TFR D,X`. Note that the final returned value is passed in accumulator D.

It is also possible to pull the return address and deallocate the procedure's arguments at the end of the procedure before returning to the main program. This is similar to the passing of the rightmost argument in a register. In some sense these optimization techniques are just minor modifications. However, they can improve static efficiency. If a procedure is called from ten different places in the main program, then putting push and pull instructions within the called procedure removes these instructions from ten places in the calling sequence and puts only one copy in the called procedure. Moreover, the technique of putting the first input argument in accumulator D works especially well for small procedures with only one argument; we may not need to save the argument on the stack at all, merely use the value in accumulator D. However, the last technique of pulling the program counter and balancing the stack inside the called procedure has a significant limitation. It is not possible to have a procedure with an arbitrary number of arguments when the called procedure removes the same number of bytes from the stack whenever it is called. The C `printf` procedure allows an arbitrary number of arguments, so it would not be able to pull the program counter and balancing the stack inside it.

```
int a;
void main() { int b; b = power(&a, 2); }
int power(int *i, unsigned char j) {  int n = 1;
      while( j-- )   n = n * *i;  return n;
}
```

a. A C Procedure calling a Subroutine

```
00000976 CC0800        LDD   #2048
00000979 3B            PSHD
0000097A C602          LDAB  #2
0000097C 07DD          BSR   *-33    ;abs = 095B
0000097E 3A            PULD
0000097F 3D            RTS
```

b. Assembly Language for the Calling Procedure in Part (a)

```
0000095B CE0001        LDX   #1
    4:     while( j-- )   n = n * *i;
0000095E B710          TFR   B,A
00000960 200B          BRA   *+13    ;abs = 096D
00000962 ED82          LDY   2,SP
00000964 36            PSHA
00000965 EC40          LDD   0,Y
00000967 B756          TFR   X,Y
00000969 13            EMUL
0000096A B745          TFR   D,X
0000096C 32            PULA
0000096D 36            PSHA
0000096E 43            DECA
0000096F E6B0          LDAB  1,SP+
00000971 26EF          BNE   *-15    ;abs = 0962
    5:     return n;
00000973 B754          TFR   X,D
00000975 3D            RTS
```

c. Assembly Language for the Called Procedure in Part (a)

Figure 9.14. A Subroutine to Raise a Number to a Power

```
                 |                       |
                 |-----------------------|
            SP-> | high return address   |
inside the       | low return address    |
subroutine       | high address of i     |
                 | low address of i      |
                 |-----------------------|
                 |                       |
```

Figure 9.15. Stack for power Procedure

C++ generally has similar operators and implements them in assembly language in similar ways. However, C++ has a calling mechanism, where a member function is designated `virtual`, that permits run-time substitutions of one class and its function members for another class and its function members. If we do not insert the word `virtual` in front of a function member in the class declaration, then the function is directly called by a JSR or BSR instruction, like C procedures discussed above.

If a function member is declared `virtual`, then to call it, we look its address up in a *vtable* associated with the class, as is shown on the right side of Figure 9.16. This table is used because generally a lot of objects of the same class might be declared or blessed, and they might have many virtual function members. For instance there could be stacks for input and for output and stacks holding temporary results in the program. A single table holds the function member addresses for all of a class's objects in one place. Suppose Q is a pointer to an object, and the object stores data members in the block pointed to by Q. The hidden pointer, at location Q, points to the jump table. Then, data members are easily accessed by the pointer Q, and virtual function members are almost as easily accessed by means of a pointer to a pointer.

The operator `new` blesses a pointer to an object. A subroutine `allocate` will return a location where the data members and the hidden pointer can be stored in index register X. Then if `Ctbl` is the jump table for the class `Cstack`, the statement `Sptr = new Cstack;` is implemented as

```
JSR  allocate     ; return value in accumulator D is a pointer to object data
STX  Sptr         ; save address in pointer to the object
MOVW #Ctbl,0,X    ; put the class Cstack jump table address in the hidden pointer
```

If `new` blesses a pointer to an object to make it an object of a different class, `Istack`, then if `Itbl` is the jump table for the class `Istack`, the statement `Sptr = new Istack;` is implemented as

```
JSR  allocate     ; return value in accumulator D is a pointer to object data
STX  Sptr         ; save address in pointer to the object
MOVW #tbl1,0,X    ; put class CharQueue's jump table address in the hidden pointer
```

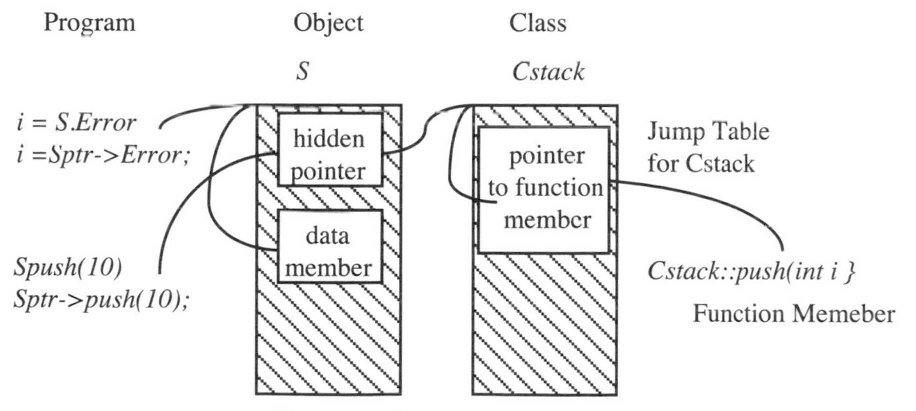

Figure 9.16. An Object and Its Pointers

If a function member is executed, as in `Sptr ->pull()`, the object's hidden pointer has the address of a table of function members; the specific function member is jumped to by using a predetermined offset from the hidden pointer. If we wish to call the pull member function, which might be at location 2 in this jump table, we can execute:

```
LDX [Sptr,PCR]; get the address of the jump table for the object pointed to by Sptr
JSR [2,X]      ; call the procedure at location 2 in the jump table (pull)
```

It will go to the jump table for the class for which Q was blessed by the new operator, to get the address of the function member to pull data from the queue.

Observe that different objects of the same class point to the same table, but each class has its own separate table. Note that data members of different objects of a class are different data items, but function members of different objects of a class are common to all the objects of the same class via this table.

A final technique used in C++ is the templated class such as Stack<T>. Such a templated class can potentially generate many classes such as Stack<char>, Stack<unsigned char>, Stack<int>, and so on. Rather than generate, and store in the microcontroller, all possible classes that can be obtained with different types for the template, a templated class generates real code only when it is declared or blessed.

9.6 Examples from Character String Procedures

In order to provide additional examples of C compiled into assembly language, we will compile some common C procedures that are used to handle character strings.

We first show `strlen` which is used to determine the number of characters in a null-terminated string. See Figure 9.17. Notice how the argument `str`, passed in accumulator D, is saved as a local variable right after the local variable `s` is pushed, and is then on top of the stack.

```
    1: int  strlen(char *str){ char *s = str;
0000088A 3B              PSHD
0000088B 3B              PSHD
0000088C 6C82            STD   2,SP
    2:   while(*str++);
0000088E EE80            LDX   0,SP
00000890 E630            LDAB  1,X+
00000892 6E80            STX   0,SP
00000894 0471F7          TBNE  B,*-6    ;abs = 088E
    3:   return (str - s - 1);
00000897 B754            TFR   X,D
00000899 A382            SUBD  2,SP
0000089B 830001          SUBD  #1
    4: }
0000089E 1B84            LEAS  4,SP
000008A0 3D              RTS
```

Figure 9.17. The Strlen Procedure

```
    3: char *strchr(char *str, int chr){
0000088A 3B             PSHD
    4:   while (*str) {
0000088B 200B           BRA    *+13    ;abs = 0898
    5:       if(*str == chr) return (str);
0000088D B715           SEX    B,X
0000088F AE80           CPX    0,SP
00000891 2711           BEQ    *+19    ;abs = 08A4
    6:       ++str;
00000893 EE84           LDX    4,SP
00000895 08             INX
00000896 6E84           STX    4,SP
    4:   while (*str) {
00000898 EE84           LDX    4,SP
0000089A E600           LDAB   0,X
0000089C 26EF           BNE    *-15    ;abs = 088D
    8:   if(*str == chr) return str;
0000089E B715           SEX    B,X
000008A0 AE80           CPX    0,SP
000008A2 2603           BNE    *+5    ;abs = 08A7
000008A4 EC84           LDD    4,SP
000008A6 8FC787         CPS    #51079
   10: }
000008A9 30             PULX
000008AA 3D             RTS
```

Figure 9.18. The Strchr Procedure

The procedure `strchr` searches for a character in a null-terminated string. See Figure 9.18. The first argument specifies the string. The second argument is a character. The procedure searches the string for a matching character; if it finds the character, it returns the address of the character in the string, otherwise it retuns a null (0).

We now show `strncpy` which is used to copy characters from and to a null-terminated string. See Figure 9.19. We show the calling routine for this example to illustrate the passing of more than three arguments. The `main` procedure calls the `strncpy` procedure with three arguments. Notice how arguments are pushed in order or their appearance from left to right, so the leftmost string, pushed first, is at 8,SP inside `strncpy`. You should step through the `while` loop to see how each C statement is compiled into assembly language. Note, however, that the pointers keep getting reloaded into X and Y registers from their local variable storage locations. You can do a lot better by writing the program in assembler language. But you can use this code, produced by the Hiware C++ compiler, as a starting point for a tightly coded assembler language program.

The procedure `strncmp` compares characters in two null-terminated strings, specified by the first two arguments, up to a number of characters specified in the third argument. See Figure 9.20. Observe the condition used to execute the `while` loop. If any of the three conditions are false, the subroutine terminates.

```
    char *strncpy(char *str_d,char *str_s,int count){char *sd = str_d;
0000088A 3B             PSHD
0000088B 3B             PSHD
0000088C EC88           LDD   8,SP
0000088E 6C82           STD   2,SP
    5:   while(count--) {
00000890 201A           BRA   *+28    ;abs = 08AC
    6:       if(*str_s) *str_d++ = *str_s++;
00000892 EE86           LDX   6,SP
00000894 E600           LDAB  0,X
00000896 270E           BEQ   *+16    ;abs = 08A6
00000898 EE88           LDX   8,SP
0000089A ED86           LDY   6,SP
0000089C E670           LDAB  1,Y+
0000089E 6B30           STAB  1,X+
000008A0 6E88           STX   8,SP
000008A2 6D86           STY   6,SP
000008A4 2006           BRA   *+8    ;abs = 08AC
    7:       else *str_d++ = '\0';
000008A6 EE88           LDX   8,SP
000008A8 6930           CLR   1,X+
000008AA 6E88           STX   8,SP
    5:   while(count--) {
000008AC EE80           LDX   0,SP
000008AE 191F           LEAY  -1,X
000008B0 6D80           STY   0,SP
000008B2 0475DD         TBNE  X,*-32    ;abs = 0892
    9:   return (sd);
000008B5 EC82           LDD   2,SP
   10: }
000008B7 1B84           LEAS  4,SP
000008B9 3D             RTS
   13: void main() { strncpy(s1, s2, 5);
000008BD CC080B         LDD   #2059    ; this is s1
000008C0 3B             PSHD
000008C1 CE0800         LDX   #2048    ; this is s2
000008C4 34             PSHX
000008C5 C605           LDAB  #5       ; this is the rightmost argument
000008C7 87             CLRA
000008C8 07C0           BSR   *-62    ;abs = 088A
000008CA 1B84           LEAS  4,SP
   15: }
000008D2 3D             RTS
```

Figure 9.19. The Strncpy Procedure

```
    4: int strncmp(char *str1, char *str2, int count) {
0000088A 6CAE           STD   2,-SP
    5:   if (!count) return 0;
0000088C 2618           BNE   *+26    ;abs = 08A6
0000088E C7             CLRB
0000088F 87             CLRA
00000890 203B           BRA   *+61    ;abs = 08CD
    7:       if (*str1 != *str2) break;
00000892 EE86           LDX   6,SP
00000894 E600           LDAB  0,X
00000896 EE84           LDX   4,SP
00000898 E100           CMPB  0,X
0000089A 261F           BNE   *+33    ;abs = 08BB
    8:       ++str1; ++str2;
0000089C EE86           LDX   6,SP
0000089E 08             INX
0000089F 6E86           STX   6,SP
000008A1 EE84           LDX   4,SP
000008A3 08             INX
000008A4 6E84           STX   4,SP
    6:   while(count-- && *str1 && *str2 ){
000008A6 EE80           LDX   0,SP
000008A8 191F           LEAY  -1,X
000008AA 6D80           STY   0,SP
000008AC 04450C         TBEQ  X,*+15    ;abs = 08BB
000008AF EE86           LDX   6,SP
000008B1 E600           LDAB  0,X
000008B3 2706           BEQ   *+8    ;abs = 08BB
000008B5 EE84           LDX   4,SP
000008B7 E600           LDAB  0,X
000008B9 26D7           BNE   *-39    ;abs = 0892
   10:   return (*str1 - *str2);
000008BB EE86           LDX   6,SP
000008BD E600           LDAB  0,X
000008BF B714           SEX   B,D
000008C1 EE84           LDX   4,SP
000008C3 3B             PSHD
000008C4 E600           LDAB  0,X
000008C6 B715           SEX   B,X
000008C8 34             PSHX
000008C9 EC82           LDD   2,SP
000008CB A3B3           SUBD  4,SP+
   11: }
000008CD 30             PULX
000008CE 3D             RTS
```

Figure 9.20. The Strncmp Procedure

9.7 Summary

In this chapter, we have shown how C constructs are encoded in assembly language. We showed how variables are allocated and accessed. We saw how simple expressions, and then more complex expressions, are implemented. Assembly-language implementations of conditional expressions were then shown. Implementation of indexed and sequential structures were covered along with implementation of looping statements. We then considered the implementation of procedures and the passing of arguments. Finally, the mechanism for handling a C++ virtual procedure call was considered.

This chapter provides the reader with a basic understanding of what is being done at the machine level when a high-level language statement is encoded. It should give the reader the understanding necessary to write efficient high-level language programs.

Do You Know These Terms?

See the end of chapter 1 for instructions.

downcasting	logical operators	virtual
upcasting	row-major order	vtable
relational operators		

PROBLEMS

In all the following problems, assume `lui`, `gsc`, etc., are declared as they are used throughout this chapter (see §9.1).

1. A global variable declared as `long alpha;` is loaded from global variables below. Show assembly-language program segments to load `alpha` from:

a. `unsigned int gui` b. `int gsi` c. `unsigned char guc` d. `char gsc`

2. A global variable declared as `long alpha;` is stored into global variables below. Show assembly-language program segments to store `alpha` to each variable, and indicate an assembly-language test that sets `char error` to 1 if an overflow occurs.

a. `unsigned int gui` b. `int gsi` c. `unsigned char guc` d. `char gsc`

3. A global variable declared as `unsigned long alpha;` is stored into global variables below. Show assembly-language program segments to store `alpha` to each variable, and indicate a test on the value of `alpha` that will result in an error.

a. `unsigned int gui` b. `int gsi` c. `unsigned char guc` d. `char gsc`

4. C local variables are not cleared. However, write a shortest program segment that clears all *N* local variables of a subroutine, where *N* is a constant.

5. Write a shortest program segment to execute each of the following C statements.

a. `gui=lsi+lsc;` b. `lsi=gsi+lsc;` c. `lsc = luc + gsc;` d. `gui += lsi;`

6. Write a shortest program segment to execute each of the following C statements.

a. `gui = lsi ^ lsc;` b. `lsi /= gsi;` c. `lsc = ~ luc;` d. `lui /= gsi;`

7. Global variables are declared as `long alpha, beta, v[10];`. Write a shortest program segment to execute each of the following C statements.

a. `beta = v[alpha];` b. `beta = v[++alpha];` c. `beta = v[--alpha];`

8. Global variables are declared as `struct { unsigned int alpha:3, beta:7, gamma:6 } s; int i;`. A `struct` with bit fields is packed from leftmost bit for the first field named on the left, through consecutive fields, toward the right. Write a shortest program segment to execute each of the following C statements.

a. `i = s.alpha;` b. `s. beta = i;` c. `s.alpha = s.gamma;`

9. Global variables are declared as `struct { unsigned int alpha:3, beta:7, gamma:6 } *p; int i;`. A `struct` with bit fields is packed from leftmost bit for the first field named on the left, through consecutive fields, toward the right. Write a shortest program segment to execute each of the following C statements.

a. `i = p->alpha;` b. `p->beta = i;` c. `p->alpha = p->gamma;`

10. Write a shortest program segment to execute each of the following C statements.

```
a. gui = (gui & 0xc7ff) + ((lsc << 11) & 0x3800);
b. lui = (lui & 0xffc7) | ((gsc << 3) & 0x38);
c. lui = (lui & 0xc7c7) + ((gsc<<3)&0x38) | ((lsc<<11)&0x3800);
```

11. Write a shortest program segment to execute each of the following C statements.

```
a. guc = gui >= lsc ;
b. luc = lui < gsc ;
c. lui = (gui >= lsc) || (lui < gsc) ;
```

12. Write a shortest program segment to execute each of the following C statements.

```
a. if( gui >= lsc ) lui++;
b. if( ! ( gui ^ lsc) ) lui *= 10;
c. if(( gui >= lsc ) && ( ! ( (gui ^ lsc) & gsc) ) ) lui ^= gui;
```

13. Write a shortest program segment to execute each of the following C statements.

```
a. if(( gui <= lsc )||( gui >=( lsc + 7))) lui++;
b. if( ( gui > lsc) && ( gui < (lsc + 3) ) ) lui *= 10;
c. if(( gui >= 0 ) && ( lsi < 0 ) && ( gui > lsi ) ) lui ^= gui;
```

14. Write the case statement below according to the conventions of Figure 9.9a.

```
switch(guc){case 2: gui = -1; break;case 4:lsc = -1;default:lsi = -1;}
```

15. Repeat Problem 14 according to the conventions of Figure 9.9b.

16. Rewrite the assembly-language program of Figure 9.11b for a main program, like Figure 9.11a, in which the declaration `int sum;` is replaced by `int k;`, and the statement `sum += a[i][j];` is replaced by `if (k > a[i][j]) k = a[i][j];`

17. Write the C program and the resulting assembly-language program that transposes a two-dimensional matrix of size 4 by 4, following the approach of Figure 9.11.

18. Write the C program and the resulting assembly-language program that stays in a do while loop as long as both bits 7 and 6 of the byte at location $d0 are zero, following the approach of Figure 9.13b.

19. Write the C program and the resulting assembly-language program that calls a procedure with prototype `unsigned int par(unsigned int R1, unsigned int R2);` to compute the resistance of two parallel resistors R1 and R2, following the approach of Figure 9.14, returning the result in accumulator D.

20. Write the C program and the resulting assembly-language program that calls a procedure with prototype `unsigned int inner(unsigned int *v, unsigned int *w);` to compute the inner product of two two-element vectors `v` and `w`, following the approach of Figure 9.14, returning the result in accumulator D.

21. Hand-compile the C procedure `strncat` below. Put the 6812 instructions under each C statement that generates them. `main` calls `strncat` which concatenates the second argument string on the end of the first argument string, but copies at most the number of characters given in the third argument. For full credit, store all parameters and local variables on the stack, even though they can be left in registers to make the program shorter, and do not optimize between statements, but provide the most statically efficient assembler language code for each C statement. Your solution should be reentrant, but need not be position independent. Assume that arguments which are put on the stack are pushed in the order that they appear from left to right.

```
char *strncat(char *str_d,char *str_s,int count){char *sd=str_d;
     while (*str_d++) ;
     str_d--;
     while (--count) { if (!(*str_d++ = *str_s++)) return sd; }
     *str_d = '\0'; return sd;
}
```

22. Hand-compile the C procedure `memchr` below. Put the 6812 instructions under each C statement that generates them. `main` calls `memchr` which searches the first argument string for the second argument character, but searches at most the number of characters given in the third argument. If it finds the second argument, it returns the address of that character in the string. Otherwise it returns a null (0). For full credit, store all parameters and local variables on the stack, even though they can be left in registers to make the program shorter, and do not optimize between statements, but provide the most statically efficient assembler language code for each C statement. Your solution should be reentrant, but need not be position independent. Assume that arguments which are put on the stack are pushed in the order that they appear from left to right.

```
char *memchr(char *buffer,char chr,int count){char *ptr=buffer;
     while(count--) { if( *ptr == chr ) return ptr; ++ptr; }
     return 0;
}
```

The Axiom PB68HC12A4 board is fitted with female harmonica plugs and a prototyping area for a laboratory developed for this book. Experiments can be quickly connected by pushing 22-gauge wire into the harmonica plugs and prototyping areas.

10

Elementary Data Structures

In all the earlier chapters, we have used data structures along with our examples. While you should therefore be somewhat familiar with them, they need to be systematically studied. There are endless alternatives to the ways that data are stored, and so there is a potential for disorder. Before you get into a crisis due to the general disarray of your data and then convince yourself of the need for data structures, we want you to have the tools needed to handle that crisis. In this chapter, we systematically cover the data structures that are most useful in microcomputer systems.

The first section discusses what a data structure is in more detail. Indexable structures, including the frequently used vector, are discussed in the second section. The third section discusses sequential structures, which include the string and the stack structures. The linked list is briefly discussed next, only to give you an idea of what it is, while the conclusions summarize the chapter with recommendations for further reading on data structures.

At the end of this chapter, you should be able to use simple data structures, such as vectors and strings, with ease. You should be able to handle deques and their derivatives, stacks and queues, and you should know a linked list structure when you see one. This chapter should provide you with the tools that you need to handle most of the problems that cause confusion when storing data in your microcomputer programs.

10.1 What a Data Structure Is

In previous chapters, we described a data structure as the way data are stored in memory. While this description was adequate for those earlier discussions, we now want to be more precise. A *data structure* is more or less the way data are stored and accessed. This section expands on this definition.

A data structure is an abstract idea that is used as a reference for storing data. It is like a template for a drawing. For example, a vector is a data structure that we have used since Chapter 3. Several sets of data can be stored in a vector in the same program and the same "template" is used to store each set. You may write or see a program that uses vectors that have five 1-byte elements. While writing another program, you may recognize the need for a vector that has five 1-byte elements and, by using the same template or data structure that you used earlier, you can quickly copy appropriate parts of the old program to handle the vector in your new program. Moreover, another program may need a similar structure that has ten 1-byte elements, or three 2-byte elements or even a vector whose elements are themselves vectors. Rather than having a different template around for each possible vector, you will, with some understanding, be able to modify a program that handles a vector with five 1-byte elements to handle these other cases, too. In a sense, data structures are elastic templates that can be stretched to accommodate different sizes.

We have used the analogy with a template to describe how data are stored in a data structure. The description of a data structure is completed when we describe how the data of the structure can be read or written, that is, accessed. A simple example will make this clear. A vector Z of N 1-byte elements can be stored in consecutive bytes of a buffer created with the DS directive:

```
Z: DS N
```

This buffer begins at location Z (see Figure 10.1). By pointing X to the buffer's first byte, we can easily access any byte of the vector. For instance, LDX #Z followed by LDAA 3,X, will read the fourth byte of the vector into accumulator A. We can also access it by an instruction LDAA Z+3, or if accumulator A has 3 and index register X has the address Z, we can access it using the instruction LDAA A,X. Suppose, however, that our N bytes were not stored in a buffer but were stored on a tape that, when read, moves forward one byte. The constraint here is that we can access only the "current" byte on the tape. This situation is exactly analogous to a person sitting at a terminal typing characters to be input to a computer. To remind us of this, the data structure of N consecutive bytes, which can be accessed only at some "current" position, is called a string. Of course, once a string is put into memory by storing it in consecutive bytes of a buffer, it can be accessed like a vector. This distinction becomes important in applications. Is one accessing a string in memory, or accessing a string from a terminal or some other hardware device? Thus the programmer should consider what data structure is appropriate for the application, which includes considering the constraints placed on accessing that structure.

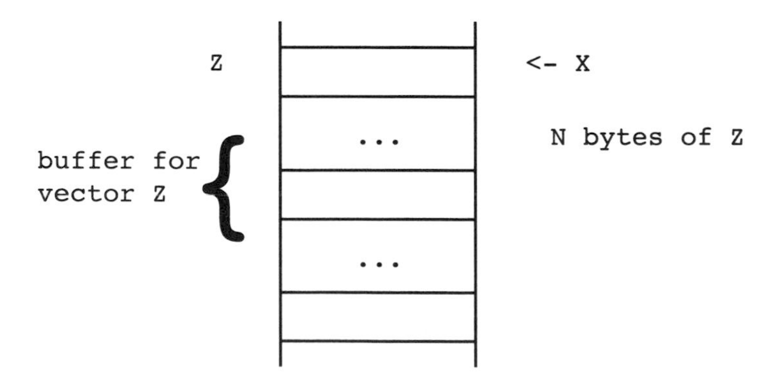

Figure 10.1. A Vector Z

The data structure, as we observed in Chapter 3, affects both static and dynamic efficiency. Besides the compactness and speed of the program itself, the compactness of the data may be affected by the structure used to store it. In a small microcomputer with 1K bytes of RAM, a program may not work at all if the data are stored in a structure that requires 2K bytes of RAM, but it may work if the correct data structure is used and that structure requires only 100 bytes of RAM. Using the right data structure can also improve clarity, because the techniques to access the structure for a particular program may be more transparent than with a less appropriate structure.

10.2 Indexable Data Structures

We have already used the vector data structure, which is the most common example of an indexable data structure. A *vector* is a sequence of *elements* (or *components*), each of which is labeled by a number called an *index*. The indexes of successive elements are successive integers, and each element is represented by the same number of bytes. The number of bytes representing each element of the vector is termed its *precision,* the number of elements in the vector is its *length,* and the index of the first element is its *origin* (or *base).* For example, a vector Z with N elements would usually have its element labeled with i denoted Z(i), where i runs between the origin and the (origin + N – 1). For an origin of 1, the elements are labeled Z(1), . . . , Z(N) while, for an origin of 0, the elements are labeled Z(0), . . . , Z(N – 1). If the origin is understood, we refer to the element Z(i) as the ith element.

Vectors stored in memory are stored in buffers, putting successive elements in successive memory locations. If the elements of the vector have more than one byte of precision and represent integers or addresses, we have adopted the Motorola convention that the elements are stored most significant byte first.

In C programs, a vector data structure of any length is obviously handled by C vector notation. Zero-origin 8-bit and 16-bit precision vectors are directly handled; for instance, a global vector Z of N 16-bit elements is declared as `int Z[N]`, and an element `i` is accessed as `Z[i]`. If the origin is changed, for instance to 1, then an element `i` is accessed as `Z[i - 1]`. If the precision is changed, unless memory space is critical, the next higher precision, 8-bit or 16-bit precision, would be used.

In assembly language, a buffer to hold vector Z is established with directive

```
Z:    DS    20                                  (1)
```

With this directive, we have a buffer that will hold a vector of up to twenty 1-byte elements, a vector of up to ten 2-byte elements, a vector of up to five 4-byte elements, and so on. Although the directive (1) establishes the buffer to hold the vector, it does not specify the origin, precision, or the length of the vector stored in the buffer. Any element of a vector can be accessed and, to access the ith element, the programmer must know the precision and origin of the vector. For example, if global vector Z has origin 1 and 1-byte precision, then if i is in accumulator B, Z(i) can be loaded into A with

```
LDX  #Z       ; Point X to Z                    (2)
DECB          ; i-1 into B
LDAA B,X      ; Z(i) into A
```

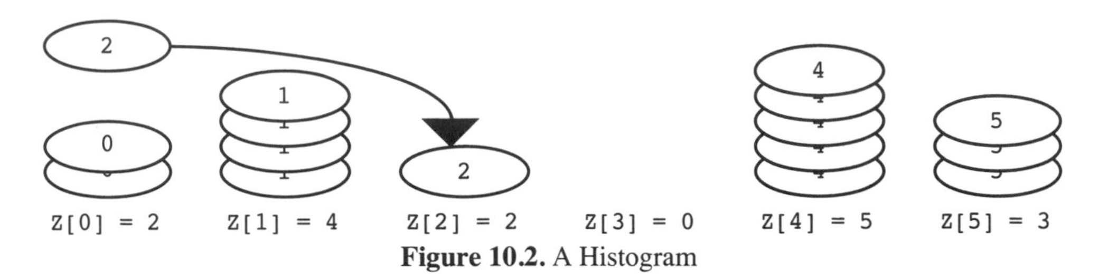

Figure 10.2. A Histogram

If the precision is 2 and the origin is 1, then if i is in accumulator B, Z(i) can be loaded into D with

```
LDX  #Z      ; Point X to Z
DECB         ; i-1 into B              (3)
ASLB         ; 2*(i - 1) into B
LDD  B,X     ; Z(i) into D
```

The origin is 1 for Z in the segments (2) and (3). It seems obvious by now that for assembly language or C programming, an origin of 0 has a distinct advantage because the DECB instruction can be eliminated from segments (2) and (3) if Z has an origin of 0. Unless stated otherwise, we will assume an origin of 0 for all of our indexable data structures. Accessing the elements of vectors with higher precision is straightforward and left to the problems at the end of the chapter.

A *histogram* is implemented with a vector data structure. In a histogram, there are, say, 20 counters, numbered zero through nineteen. Initially all counters are zero. A stream of numbers arrives, each between zero and nineteen. As each number i arrives, counter i is incremented. This vector of counts is the histogram.

Figure 10.2 illustrates the first six counts of the histogram Z. An item "2" arrives, so counter 2 should be incremented. In C, if the vector is *Z* and the number "2" is in i, then z[i]++; increments the counter for number "2"; and, in assembly language, if this number "2" is in index register X, then the instruction in (4) will increment the counter:

```
INC   Z,X      ; increment the Xth count          (4)
```

Histograms are useful in gathering statistics. We used them to "reverse engineer" a TV infrared remote control; the counts enabled us to determine how a "1" and a "0" were encoded as pulse widths and how commands were encoded into 1's and 0's. Note that the data structure is a vector. Counts are accessed in random order as items arrive.

A *list* is similar to a vector except that each element in the list may have a different precision. Lists are stored in memory in buffers just like vectors, successive elements in successive memory locations. Like a vector, there is an origin and a length and each element of the list can be accessed. However, you cannot access the ith element of a list by the simple arithmetic computation used for a vector. Consider the following example of a list L that consists of 30 bytes for a person's name (ASCII), followed by 4 bytes for his or her Social Security number (in C, an unsigned long), followed by a 45-byte address (ASCII), and another 4 bytes for the person's telephone number (an unsigned long). This list then has four elements, which we can label L0, L1, L2, and L3, and whose precisions are 30, 4, 45, and 4, respectively. In C, a list is conveniently handled by a struct. The list above can be represented by `struct{ char name[30]; long ss; char address[45]; long phone; } s;`. The `phone` element of struct `s` is indicated by the notation `s.phone`. The assembly-language implementation of lists is simple. Assuming that the label `L` is used for the address of the first byte of the list, we can load the jth byte of `L2` (the address) into A with the sequence

```
LDX  #L+34    ; Point X to L2                (5)
LDAA A,X      ; jth byte of L2 into A
```

where we have also assumed that the value of j is initially in accumulator A. While the segment (5) seems simple enough, remember that we have had to compute the proper offset to add to `L` in order to point X to L2. For simple lists, such as this example, this is not much of a problem, and the programmer may elect to do it "in his (or her) head." But the assembler can help. For our example list, we can create labels for the offsets to avoid remembering the sizes of each element.

```
NAME:    EQU  0            ; Name of person                (6)
SSN:     EQU  NAME+30      ; Social Security number
ADDRESS: EQU  SSN+4        ; Address of person
TN:      EQU  ADDRESS+45   ; Telephone number
NBYTES:  EQU  TN+4         ; Number of bytes in the list
```

In the following program segment, a telephone number is stored in accumulator D (high 16 bits) and index register X (low 16 bits). If the list L's telephone number matches this D:X, put the Social Security number in D:X, otherwise go to label NoMatch:

```
CMPD  L+TN      ; check high 16-bits of telephone number      (7)
BNE   NoMatch
CPX   L+TN+2    ; check low 16-bits of telephone number
BNE   NoMatch
LDD   L+SSN     ; get high 16-bits of Social Security number
LDX   L+SSN+2   ; get low 16-bits of Social Security number
```

Notice that with the `EQU` directives of (6) the program segment (7) becomes much more self-documenting. What makes this technique work is that it is easy to associate labels with attributes, particularly because the order of the list elements is usually unimportant. Notice that the `EQU` statements (6) not only let the programmer use labels for offsets, but also let the assembler calculate the offsets for the programmer. This same example will be continued for the description of a table, which is a vector of lists. However, we will first discuss an array, which is a bit simpler.

```
AR(0,0),   AR(0,1),   AR(0,2),...
AR(1,0),   AR(1,1),   AR(1,2),...
AR(2,0),   AR(2,1),   AR(2,2),...
  ...        ...        ...
```

Figure 10.3. An Array

A (two-dimensional) *array* is a vector whose elements are vectors, each of which has the same length and precision. For us, it suffices to consider an array as the usual two-dimensional matrix pattern of elements of the same precision, where, as before with vectors and lists, it is convenient to start indexing the rows and columns from 0. If we consider our array to be a vector of rows, the data structure is called a *row major* array. If we consider it to be a vector of columns, the data structure is called a *column major* array. In C, row-major order is used by arrays; a zero origin 1-byte precision two-dimensional array is declared as `char AR[5][5]`, and the ith row jth column element is designated `AR[i][j]`. Rows are kept together. For instance, `AR[0][4]` might be located in memory at location $833, `AR[1][0]` at location $834, and `AR[1][1]` at location $835. In assembly-language programming, two dimensional n by m arrays are declared as a vector of n times m elements, e.g., `AR DS n*m`. The address of AR(i,j) is given by

$$\text{address of AR(i,j)} = (i * 5) + j + \text{address of AR(0,0)} \tag{8}$$

Formula (8) can easily be modified for arrays with higher-precision elements. One uses `MUL` to compute array addresses. For instance, if the precision of each element of AR is two bytes, as if declared in C as *int* `AR[5][5]`, and if AR is the address of the first byte of the array, and further if i and j are in accumulators A and B, respectively, the following segment puts AR(i,j) into accumulator D.

```
PSHB                ; Save j
LDAB  #5            ; Number of columns into B
MUL                 ; i * 5 into D
ADDB  1,SP+
ADCA  #0            ; (i * 5) + j into D
ASLD                ; 2 * ((i * 5) +j) into D
XGDX                ; Put combined offset in X
LDD   AR,X          ; AR(i,j) into D
```

In this segment, multiplication by two for the contents of D is done by `ASLD`. Multiplication by powers of two can be done by repeating `ASLD`.

Consider a program that writes into `ZT` the transpose Z^T of a 5 by 5 matrix Z of 1-byte elements. The C procedure is shown below.

```
void ZTRANS( char Z[5][5], char ZT[5][5] ) {char i,j;
    for(i = 0; i < 5; i++)
        for(j = 0; j < 5; j++)
            ZT[i][j] = Z[j][i];
}
```

While this subroutine appears to pass its arguments by value, they are actually passed by name, because all vectors and arrays are passed by name. Figure 10.4 shows the assembly-language program that performs the same operation, but in an optimized way.

```
*   ZTRANS computes the transpose ZT of a 5 by 5 matrix Z of 1-byte elements.
*          CALLING SEQUENCE:
*          PSHX                      Address of the transpose matrix ZT
*          PSHY                      Address of the matrix Z
*          BSR    ZTRANS
*          LEAS   4,SP               Balance the stack
*
*          PARAMETERS
*
RA:        EQU    0                  ; Return address
ADDRZ:     EQU    RA+2               ; Address of Z
ADDRZT:    EQU    ADDRZ+2            ; Address of ZT
*
ZTRANS:    LDX    ADDRZ,SP           ; First row address into X
           LDY    ADDRZT,SP          ; First column address into Y
           LDAA   #5                 ; There are 5 columns in the matrix Z
*
STR1:      LDAB   #5                 ; There are 5 elements in a column of the matrix Z
*
STR2:      MOVB   5,X+,1,Y+          ; Transfer data and move pointers to next array element
           DBNE   B,STR2             ; Count down number of elements in a column
*
           LEAX   -24,X              ; Move to next element of row 0 (back up 25 - 1)
           DBNE   A,STR1             ; Count down number of columns
           RTS
```

Figure 10.4. Subroutine ZTRANS

A *table* is a vector of identically structured lists. For example, one might have a table of lists where each list is exactly like the list example just discussed, one for each person in the table. In C, a table of 100 telephone number and Social Security number lists can be represented by `struct{ char name[30]; long ss; char address[45]; long phone; } t[100];`. A search for a specific Social Security number *theSS*, putting the matching telephone number in *theTel*, is accomplished in the program main below:

```
main() { long theSS, theTel; int i;
  for(i = 0; i <1 00; i++)
      if(theSS == t[i].ss) break;
      theTel = t[i].phone;
}
```

We assume it finds a matching telephone number, which is left in accumulator D (low 16 bits) and index register X (high 16 bits) when we exit. In assembly language, index addressing can be used to access any particular list in the table and offsets can be used, as done earlier, to access any particular element of the list. For instance, the directive

```
NUM:      EQU   100            ; Number of lists in the table      (9)
TABLE:    DS    NUM*NBYTES     ; Allocation of table
```

creates a buffer for 100 of the lists defined by (6). The address of the first byte of the buffer is TABLE. The following program segment searches such a table for a certain telephone number, which is stored (in binary) in accumulator D (low 16-bits) and index register X (high 16 bits). We will assume it finds a matching telephone number, which is left in accumulator D (low 16-bits) and index register X (high 16 bits) when we exit.

```
LOOP:    CPX    TN,Y        ; check high 16-bits of telephone number of row Y
         BNE    NOMTC
         CPD    TN+2,Y      ; check low 16-bits of telephone number of row Y
         BEQ    MTCH
NOMTC:   LEAY   NBYTES,Y    ; skip to next list
         CPY    #TABLE+NUM*NBYTES ; at end of table?
         BNE    LOOP        ; if not, loop
MTCH:    LDD    SSN,Y       ; get high 16-bit Social Security number of row Y
         LDX    SSN+2,Y     ; get low 16-bit Social Security number of row Y
```

This discussion has examined indexable data structures. Each element of an indexable data structure can be accessed, and, furthermore, some form of indexing can be used for the access. The simple, but very useful, vector was easy to access because the address of the ith element, assuming a zero origin, is obtained by adding i*(precision) to the address of the vector. A list is like a vector but has fewer restrictions, in that elements can be of any precision. Arrays and tables are just mixtures of these two structures. These indexable structures are used often in a microcomputer like the 6812, because they are so easy to handle with its index addressing options and multiply instructions.

10.3 Sequential Data Structures

We now consider sequential data structures. The ubiquitous string, which you met earlier, and various deques, including the stack, are sequential structures. The key characteristic of sequential structures is that there is a current location, or top or bottom, to the structure, and access to the data in this structure is limited to this location.

Strings can be variable or constant. A buffer is used to hold a variable string, which, in particular, can have a variable length. In a program with string manipulations, the length of a particular string can change in the program (up to the size of its buffer) in contrast to a vector of, say, 2-byte numbers that has a constant length throughout most programs. In C, a global constant string is declared `char s[11] = "High there";`. Most strings in C are terminated by a null character ($0); the number of bytes allocated for a string generally must include this extra null character at the end. In assembly language, constant strings can be created in memory with assembler directives like

```
s:    DC.B      "This is a string"                (9)
```

In assembly language, a string's length can be ascertained in one of three ways: by also giving its length, either by knowing it implicitly or giving it in a variable, by terminating it with a special character such as null, carriage return, or $4 (end-of-text), or by setting the sign bit in the last, and only the last, character in the string. In C, a pointer is generally used to access the current location of a string. For instance, suppose we declare global `char *ptr;` and later we initialize `ptr = s;`. Then the pointer `ptr` can be used to access the current location. Alternatively, a numerical index can be used, declared as in *char i;*, and the index i can be intialized to zero, as in `i = 0;`, so that `s[i]` is the first character in the string `s`. In assembly language, this string (9) can appear in the program area as a constant to be displayed on a terminal or to be printed on a printer. It is often in the program area because it may be stored in ROM together with the program. To preserve position independence, the address of the string is generally put in an index register using program counter relative addressing as in

```
        LEAX      s,PCR
```

With this instruction, the address of the first character of **s** will be put into X regardless of where the ROM containing the program is placed in memory. A numerical index can also be used to read elements of a string in assembly language, as we did in C (see the problems at the end of the chapter).

Strings can be accessed at a current location, which can be moved as the string is accessed. In C, we can access this element and move to the next location using `*ptr++` or `*ptr--`. Alternatively, if we use an index to access the elements, we can increment or decrement an index, as in the expression `s[i++]` or `s[i--];` In assembly language, if the current location is a memory word whose address is in the index register X, the string can be accessed by instructions like `LDAA 0,X, LDAA 1,X+`, or `LDAA 1,X`.

In C, operations on a string can repeat until the null character is read at the end of the string, as in the statement `while(*ptr) f(*ptr++);`. If we are using indexes to access characters, this operation can be written as `while(s[i]) f(s[i++]);`. In assembly language, if a null-terminated string is used, the load instruction such as `LDAA 1,X+` will be followed by a branch instruction such as `BEQ END`, to terminate when the null character is read. If another special character such as a carriage return is used, then a `CMPA` instruction can be used to detect the end of the string. If a sign bit indicates the end of the string, then the `LDAA 1,X+` will be followed by a branch instruction such as `BMI END`, to terminate when the last character is read. The program generally has to strip off the sign bit before using the last character.

Strings of characters are frequently input and output from a terminal using a buffer. To discuss this, we first need to make some remarks about how single characters are input and output between a terminal and the MPU. In C, there is generally a procedure such as is given by the prototype `char inch();` (for "input character") to input characters from the keyboard. It waits for a key to be depressed at the terminal; and, when the key is depressed, the procedure returns to the calling routine with the ASCII code of the key depressed. There is also generally a procedure such as is given by the prototype `void outch(char c);` (for "output character") to output characters to the screen; it generally displays the ASCII contents of the seven low-order bits of *c*, but control characters, such as carriage return ($0D) and line feed ($0A), will move the screen cursor in the usual way. (The remaining ASCII characters are used for different purposes and are displayed differently on different terminals. These are usually input from the terminal keyboard by holding down a "control" key and pressing one of the other keys. These characters will not be needed in this discussion.) Implicit in the subroutine `OUTCH` is a segment of code that will wait until the previous character is displayed on the terminal before *c* is displayed. In assembly language, we might use two subroutines equivalent to these procedures; to input a character, execute subroutine `INCH`, which leaves the character in accumulator A, and to output a character, put it in accumulator A, and execute subroutine `OUTCH`. The exact code used in `INCH` and `OUTCH` are not important at this point.

In C, if `ptr` points to the beginning of a null-terminated string `s`, a statement `while(*ptr) outch(*ptr++);` outputs the string `s`, if `ptr` points to the beginning of a buffer `b`, a statement `do *ptr++ = c = inch(); while(c != '\r')` inputs characters to the buffer `b`, until a carriage return is received (the carriage return is written at the end of the string that is in `b`). In assembly language, we can input or output strings of ASCII characters almost as easily. For example, we could display the constant string (9) on the terminal beginning at the current cursor position with the segment

```
        LDAB    #16        ; Number of characters in STRING        (10)
        LDX     #STRING
LOOP:   LDAA    1,X+       ; Next character of STRING; into A
        JSR     OUTCH
        DBNE    B,LOOP
```

Program segment (10) assumes that the programmer will count the number of characters in the string. This can be avoided by adding

```
LENGTH:     EQU     *-STRING
```

after the definition of `STRING` and replacing `LDAB #16` with `LDAB #LENGTH` in the sequence (10). We can also input a string of characters from the terminal, terminated by an ASCII carriage return ($0D), with the program segment below. The string is stored in a buffer labeled `BUFFER` established with the directive

```
BUFFER:     DS      100
```

The string is entered with the program segment (11)

```
        LDX     #BUFFER    ; X -> buffer to hold the string         (11)
AGAIN:  JSR     INCH
        STAA    1,X+       ; Place character in buffer
        CMPA    #$0D       ; is the character input a carriage return?
        BNE     AGAIN
OUT:    RTS
```

Another type of sequential data structure is the deque, which we now discuss. A special case of the deque is the stack, which we studied extensively in Chapters 3 and 8. Our stacks have also been indexable, because the S register can be used as an index register as well as a stack pointer. Nevertheless, when these stack pointers are used only with push and pull instructions, they become true sequential structures.

A *deque* is a generalization of a stack. It is a data structure that contains elements of the same precision. There is a top element and a bottom element, and only the top and

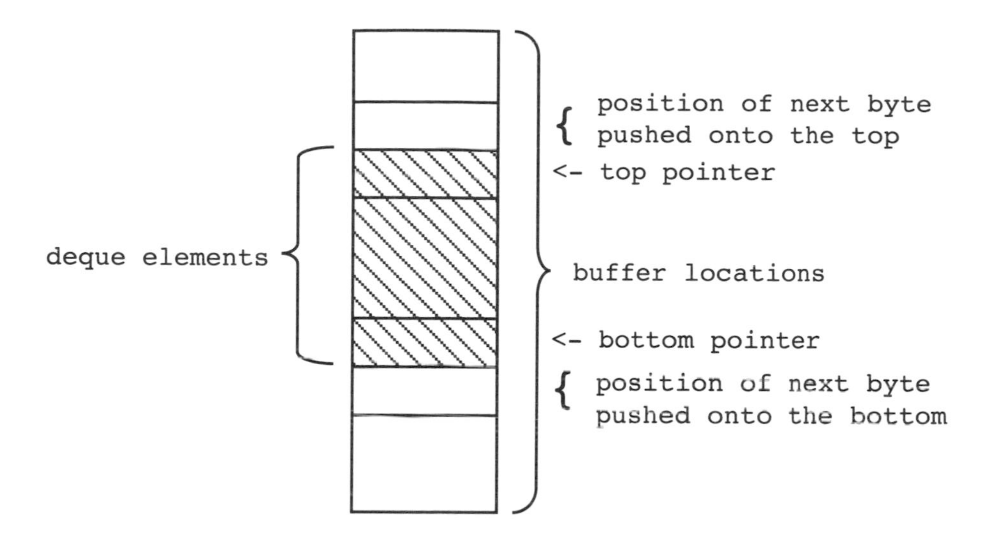

Figure 10.5. Deque Data Structure

bottom elements can be accessed. *Pushing* an element onto the top (or bottom) makes the old top (or bottom) the next-to-top (or next-to-bottom) element and the element pushed becomes the new top (or bottom) element. *Popping* or *pulling* an element reads the top (or bottom) element, removes it from the deque, and makes the former next-to-top (or next-to-bottom) element the new top (or bottom) element (see Figure 10.5). You start at some point in memory and allow bytes to be pushed or pulled from the bottom as well as the top.

In C, two pointers can be used, as a pointer was used in the string data structure, or else indexes can be used to read or write on the top or bottom of a deque, and a counter is used to detect overflow or underflow. We use indexes in this example and invite the reader to use pointers in an exercise at the end of the chapter.

The deque buffer is implemented as a 50-element global vector `deque,` and the indexes as global `unsigned chars` `top` and `bot` initialized to the first element of the `deque,` as in the C declaration

```
unsigned char deque [50], size,error, top, bot;
```

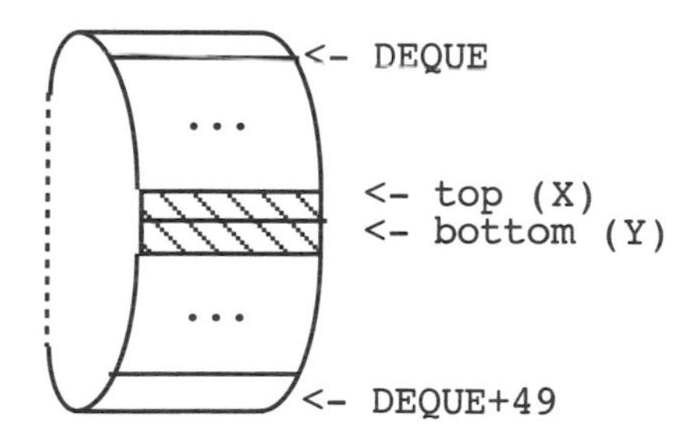

Figure 10.6. Buffer for a Deque Wrapped on a Drum

As words are pulled from top or bottom, more space is made available to push words on either the top or bottom. To take advantage of this, we think of the buffer as a ring or loop of words, so that the next word below the bottom of the buffer is the word on the top of the buffer (see Figure 10.6). That way, as words are pulled from the top, the memory locations can become available to store words pushed on the bottom as well as words pushed on the top, and vice versa. Then to push or pop data into or from the top or bottom of it, we can execute procedures:

```
void pstop(int item_to_push) {
 {if((size++)>=50)error=1;if(top==50)top=0;deque[top++]= item_to_push;}

int pltop()
   {if((--size) < 0)error=1;if(top == 0)top=50;return(deque[--top]);}

void psbot(int item_to_push) {
  if((size++)>=50)error=1; if(bot==0) bot=50;deque[--bot]=item_to_push;}

int plbot()
   {if((--size)<0) error=1; if(bot==50) bot=0;  return( deque[bot++]);}
```

In assembly language, a deque can use registers to point to its top and bottom elements. In our discussion, we will first assume that all of memory is available to store the deque elements, and then we will consider the more practical case where the deque is confined to a buffer rather than all of memory. We use register X to point to the top and Y to point to the bottom of the deque. If location L is where one wants the first possible push on the top to go, one initializes the top pointer with

```
            LDX         #L
```

A push from accumulator B onto the top of the deque then corresponds to

```
            STAB        1,X+
```

while a pull from the top into B corresponds to

```
            LDAB        1,-X
```

Just as we wrapped around a drum as shown in Figure 10.6 in C, we need to do the same in assembly language. When a byte is pushed into the bottom of the deque, it is actually put into the bottom byte of the buffer. The pointer is initialized to the top of the buffer, but upon the first push to the bottom of the deque, the pointer is moved to the bottom of the buffer. As an example, if we use a buffer with 50 bytes to hold the deque, we would have the directive

```
DEQUE:            DS                     50
        PSHTP:    CMPA   #50
                  LBEQ   ERROR        ; Go to error routine
                  INCA
                  CPX    #DEQUE+50    ; Pointer on top?
                  BNE    L1
                  LDX    #DEQUE       ; Move to bottom
        L1:       STAB   1,X+
                  RTS

        PLTP:     DECA
                  LBMI   ERROR        ; Go to error routine
                  CPX    #DEQUE       ; Pointer at bottom?
                  BNE    L2
                  LDX    #DEQUE+50    ; Move to top
        L2:       LDAB   1,-X
                  RTS
```

Figure 10.7. Subroutines for Pushing and Pulling B from the Top of the Deque

at the start of our program. If accumulator A contains the number of elements in the deque and if X and Y are the top and bottom pointers, we would initialize the deque with

```
CLRA                ; Initialize deque count to 0
LDX     #DEQUE      ; First push onto top into DEQUE
LDY     #DEQUE      ; First push onto bottom into DEQUE+49
```

Pushing and pulling bytes between B and the top of the deque could be done with the subroutines `PSHTP` and `PLTP`, shown in Figure 10.7. The index register X points to the top of the deque, while the index register Y points to the deque's bottom.

Similar subroutines can be written for pushing and pulling bytes between B and the bottom of the deque. In this example, if the first byte is pushed onto the top of the deque, it will go into location `DEQUE`, whereas, if pushed onto the bottom, it will go into location `DEQUE+49`. Accumulator A keeps count of the number of bytes in the deque and location `ERROR` is the beginning of the program segment that handles underflow and overflow in the deque.

Usually, you do not tie up two index registers and an accumulator to implement a deque as we have done above. The pointers to the top and bottom of the deque and the count of the number of elements in the deque can be kept in memory together with the buffer for the deque elements. The subroutines for this implementation are easy variations of those shown in Figure 10.7. (See the problems at the end of the chapter.)

A *queue* is a deque where elements can only be pushed on one end and pulled on the other. We can implement a queue exactly like a deque but now only allowing, say, pushing onto the top and pulling from the bottom. The queue is a far more common sequential structure than the deque because the queue models requests waiting to be serviced on a first-in first-out basis. Another very common variation of the deque, which is close to the queue structure, is the *shift register* or *first-in first-out buffer*. The shift register is a full deque that only takes pushes onto the top, and each push on the top is preceded by a pull from the bottom. If the buffer for the shift register holds N bytes, then, after N or more pushes, the bottom byte of the shift register is the first in among the current bytes in the shift register, the next-to-bottom is the second in among the current bytes, and so on. If all pushes into the shift register are from accumulator B, only one macro and one pointer are needed to implement this data structure. (See the problems at the end of the chapter.) Although these two sequential structures are more common than the deque, we have focused on the deque in our discussion to illustrate the differences between accessing these sequential structures' top and bottom with the 6812.

In this section we have studied the sequential structures that are commonly used in microcomputers: the string and the variations of the powerful deque, the stack, queue, and shift register. As you have seen, they are easy to implement on the 6812.

10.4 Linked List Structures

The last structure that we discuss is the powerful linked list structure. Because its careful definition is rather tedious and it is not as widely used in microcomputer systems as the structures discussed in the previous sections, we examine this structure in the context of a concrete example, a data sorting problem. Suppose that we have a string of ASCII letters, say,

```
t c x u a f b
```
(12)

where we want to print the letters of the string in alphabetical order.

We will store this string of letters in a data structure called a *tree*, using a linked list data structure to implement the tree. The reason that we do not want to store the letters in consecutive bytes of memory as, say, a vector, but in a linked list implementation of a tree, is that, stored as a vector, the time to find a particular letter grows linearly with the number of letters in the string. If the letters were files of data, searching for a particular file could take days if the number of files is large. Organized as a linked list implementation of a tree, the search time grows logarithmically with the number of files so that searching for that same file could be done in seconds. Large collections of data are typically stored in some manner to improve the time to search them, and the linked list implementation of a tree is a common way to do this.

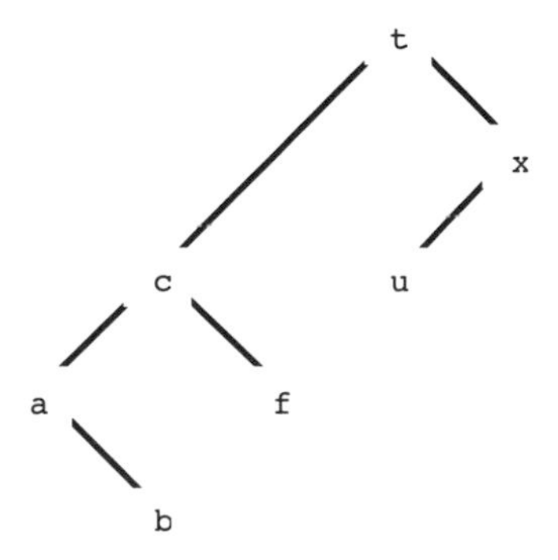

Figure 10.8. Picture of the Tree Representing the String (12)

We first describe the tree using the algorithm to generate its graph. The first letter in the string is put at the *root* of the tree (see Figure 10.8), while the second letter is put at the left or right *successor* node of the root, depending on whether it is alphabetically before or after the letter at the root. Successive letters begin at the root node, going left or right to successor nodes in the same way as the second letter until an empty node is found. The new letter is placed at this empty node. We recommend that you work through the characters in the string (12) above, and build up the tree shown in Figure 10.8 using the foregoing rule.

As you have just seen, it is fairly easy to generate the graph of the tree. We now describe how to store the tree in memory using a *linked list structure*. After each letter in the string, append two integers where the first is the string index of the letter for the left successor and the second is the string index of the letter for the right successor. (String indexes, as usual, begin at 0, so that 0 is the string index for t, 1 is the string index for c, and so on.) The symbol NS, which is $FF, indicates no successor of that type. (See Figure 10.9 for the linked list representation of the tree. The symbolic contents of each byte is shown rather than the usual hexadecimal contents.) This linked list structure contains identically structured lists, such as the first three bytes, t,1,2. The list index of each list is identical to the string index of the letter in the list so that the list t,1,2 is the 0th list in the linked list. Each list has three elements, a letter and two links. Although the elements of each list are the same precision in this example (each is one byte), the precisions are generally different, from one bit to hundreds of bits per element. The links in this example are equal to the indexes of the lists that contain the left and right successors. For the top, which is the 0th element and represents the root of the tree, the left successor of the root is the letter c, and the element that contains the letter c has index 1, so that the first link out of the 0th element is 1. The root's right successor is the letter x, and the element that contains x has index 2, so the second link of the 0th element is 2. You should verify that the other elements, which are identical in form to the 0th element, have the same relationship to the tree that the 0th element has.

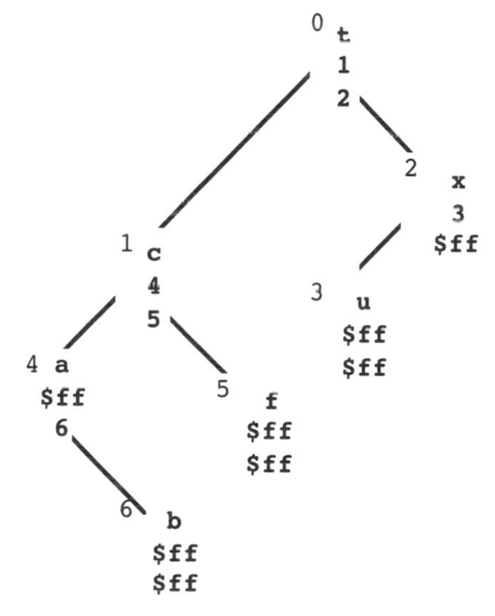

Figure 10.9. Linked List Representation of the Tree Shown in Figure 10.8

Once the linked list is formed, the tree can be scanned to print the letters in order by an algorithm pictured in Figure 10.10. The idea behind the algorithm is this. Starting at the root, wrap a cord around the outside of the tree and print each node's letter (except 0), as you pass under its *crotch*. Its crotch is the part between the branches to its successors or, if it does not have successors, the crotch is the part between where the successors would be connected. (Try this out on Figure 10.10.) Although a human being can visualize this easily, a computer has a hard time working with pictures. This algorithm can be implemented in a computer using the elegantly simple rule:

1. Process the tree at the left successor node. (13)
2. Print the letter.
3. Process the tree at the right successor node.

In processing the root node, you process the tree containing nodes c, a, b, and f first, then print the letter t, then process the tree containing x and u last. Before you print the letter t, you have to process the tree containing the nodes c, a, b, and f first, and that processing will result in printing some letters first. In processing the tree containing c, a, b, and f, you process the tree containing a and b, then print the letter c, then process the tree containing f. Again, before you print the letter c, you have to process the tree containing a and b first, and that will result in some printing. In processing the tree containing a and b, you process the "null" tree for the left successor node of letter a (you do nothing), then you print the letter a, then you process the tree containing b. In processing the tree containing b, you process the "null" tree, you print the letter b, then you process the "null" tree. After you print the letter c, you will process the tree containing f and then process the tree containing x and u after printing the letter t. Try this rule out on the tree, to see that it prints out the letters in alphabetical order.

The flowchart in Figure 10.11 shows the basic idea of the rule (13). The calling sequence sets `LINK` to 0 to process list 0 first. If `LINK` is $FF, nothing is done; otherwise, we process the left successor, print the letter, and process the right successor. Processing the left successor requires the subroutine to call itself, and processing the right successor requires the subroutine to call itself again so that this subroutine is recursive, as discussed in Chapter 5. (To read the flowchart of Figure 10.11, `RETURN` means to return to the place in the flowchart after the last execution of `SCAN`.)

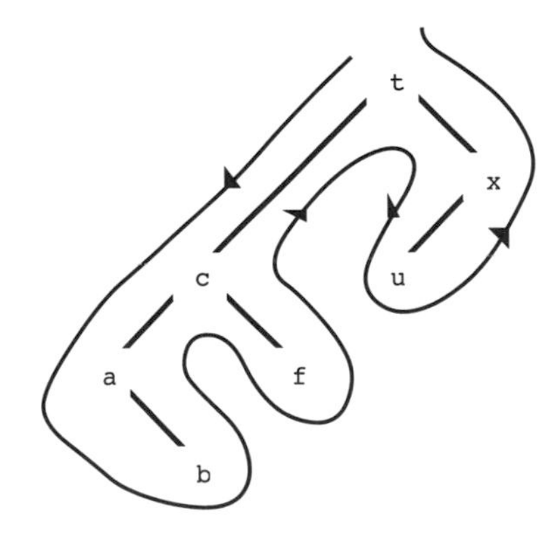

Figure 10.10. Path for Scanning the Tree

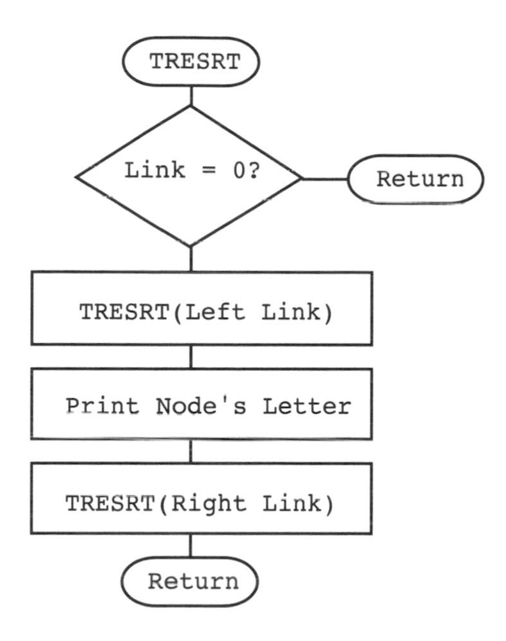

Figure 10.11. Flowchart for Scanning Tree

In C, the linkage can be by means of indexes into a table, which is a vector of `structs`. The following procedure is intially called as in `scan(0);`

```
typedef struct node{char c; unsigned char l,r; } node; node table[10];
void scan(unsigned char i)
  {if(i!=0xff) {scan(table[i].l);outch(table[i].c);scan(table[i].r); }}
```

A similar approach is used in assembly language, in the subroutine shown in Figure 10.12. It simply implements the flowchart, with some modifications to improve static efficiency. First, index register X points to the 0th list, so that `LINK` can be input as a parameter in accumulator B. This link value is multiplied by three to get the address of the character of the list. That address, with one added, gets to get the link to the left successor, and that address, with two added, gets the link to the right successor. The subroutine computes the value 3 * `LINK` and saves this value on the stack. In processing the left successor, the saved value is recalled, and one is added. The number at this location, relative to X, is put in B, and the subroutine is called. To print the letter, the saved value is recalled, and the character at that location is passed to the subroutine `OUTCH`, which causes the character to be printed. The saved value is pulled from the stack (because this is the last time it is needed), and two is added. The number at this location relative to X is passed in B as the subroutine is called again. A minor twist is used in the last call to the subroutine. Rather than doing it in the obvious way, with a `BSR SCAN` followed by an `RTS`, we simply do a `BRA SCAN`. The `BRA` will call the subroutine, but the return from that subroutine will return to the caller of this subroutine. This is a technique that you can always use to improve dynamic efficiency. You are invited, of course, to try out this little program.

```
* SUBROUTINE SCAN scans the linked list TREE from the left, putting out the
* characters in alphabetical order. The calling sequence below scans TREE
*
*         LDX     #TREE
*         CLRB                    ; Put LINK to 0
*         BSR     SCAN
*
SCAN:     CMPB    #$FF
          BEQ     L
          LDAA    #3
          MUL
          PSHB                    ; Save 3 * B on stack
          INCB                    ; 3 * (B) + 1 into B
          LDAB    B, X            ; Left successor link into B
          BSR     SCAN
          LDAA    0, SP           ; Recover 3 * B
          LDAA    A, X
          JSR     OUTCH           ; Put out next character
          PULB                    ; Recover 3 * B from stack, remove from stack
          ADDB    #2
          LDAB    B, X            ; Link to right successor into B
          BRA     SCAN
L:        RTS
```

Figure 10.12. Subroutine SCAN Using Indexes

The main idea of linked lists is that the list generally has an element that is the number of another list, or it has several elements that are numbers of other lists. The number, or *link,* allows the program to go from one list to a related list, such as the list representing a node to the list representing a successor of that node, by loading a register with the link element. The register is used to access the list. This is contrasted to a sequential search of consecutive rows of a table, which is a vector of lists. In a table, one

Location	Letter	Left	Right
0x800	t	0x803	0x806
0x803	c	0x80c	0x80f
0x806	x	0x809	0
0x809	u	0	0
0x80c	a	0	0x812
0x80f	f	0	0
0x812	b	0	0

Figure 10.13. Linked List Data Structure for SCAN

```
*     SUBROUTINE SCAN
*
* SCAN scans the linked list TREE from the left, putting out the characters in
* alphabetical order. The address of TREE is passed in X with the calling sequence
*
*           LDX     #TREE
*           BSR     SCAN
*
SCAN:       CPX     #0              ; If pointer is a "null" (0)
            BEQ     L               ; Exit without doing anything
            PSHX                    ; Save link
            LDX     1,X             ; Left successor link into X
            BSR     SCAN            ; Call this subroutine again
            LDAA    [0,SP]          ; Get character
            JSR     OUTCH           ; Put out next character
            PULX                    ; Pull pointer from stack
            LDX     3,X             ; Move pointer to right successor into B
            BRA     SCAN            ; Call this subroutine again
L:          RTS
```

Figure 10.14. Subroutine SCAN Using Address Pointers

usually accesses one list (row) after the list (row) above it was accessed. In a linked list, one can use any link from one list to go to another list. By providing appropriate links in the list, the programmer can easily implement an algorithm that requires going from list to list in a particular order. Linked lists generally store addresses rather than index numbers, to simplify the procedure and to avoid an artificial restriction on length. The linked list above can be stored as in Figure 10.13, and the procedure in Figure 10.14 can be used to read the list.

Compare the implementation of the tree structure above using a linked list with one using a simple table where the nodes are put down successively by levels. Not only are most offsets calculated to go from node to successor node, but gaps will be left in the table where the tree has no successors, and testing for the end of the search will be messy. Even constructing this table from the string will be difficult. However the linked list program can be written in a simple and logical form, using the power of the data structure to take care of many variations. In the example above, nodes that have no left successor or nodes that have no right successor are handled the same way as nodes that have two successors. While links can simplify the program, as we have just discussed, additional links can speed up a program by permitting direct access to lists that are linked via several link list-link list . . . -link list steps. Linked lists can simplify the program as well as speed it up, depending on how the designer uses them.

This final section briefly introduced the linked list structures. These are very useful in larger computers, although rarely seen in microcomputers, in our experience. Nevertheless, they are well known to be useful in artificial intelligence applications so that you may expect to see them used in the near future in robots and pattern recognition devices. You should read further material on these powerful structures.

10.5 Summary

This chapter has presented the data structures that are most commonly used in microcomputers. Although many programmers have written millions of lines of code without knowing about them, they help you to create order in a maze of possible ways to store data, they allow you to copy, or almost copy, the code needed to access the structures, and they allow you to save memory by using a better structure.

To show how your knowledge of data structures can save memory in a small computer, consider the storage and access of a mathematical array of ten rows by ten columns of 1-byte numbers, where 96 of the numbers are zero (this is called a sparse array). A natural way to store this data is in an array, but that array would take 100 memory locations, and most would contain zeros. It is more efficient to store the four numbers of the sparse array in a table that stores the nonzero elements of the array, where the first column of the table is the row number, the second is the column number, and the third is the data in that row and column. This could be done in only twelve bytes. Knowledge of data structures can enable you to make a program work in a limited amount of memory, which may not be possible otherwise.

This chapter only scratches the surface of this fertile area of study. If you study computer science, you will probably take a whole course on data structures, as well as meeting this material in other courses on database systems and compiler design. It is your best single course to take from the computer science area of study. Many textbooks are available for these courses, and you can use practically any of them to expand your comprehension of data structures. We suggest one of the earliest books, *Fundamental Algorithms. Vol. 1, The Art of Computer Programming,* 2nd ed. (D. Knuth, Addison-Wesley Publishing Co., Inc., Reading, Mass., 1973), for your reading.

From reading this chapter, you should be able to handle any form of the simple data structures that are likely to be met in microcomputer programming, and you should be able to handle the various types of sequential structures. You should also be able to recognize the linked list structures. But most important, you should be prepared to put some order in the way your programs handle data.

Do You Know These Terms?

See the end of chapter 1 for instructions.

data structure	base	deque	buffer
vector	histogram	push	tree
element	list	pop	root
component	array	pull	successor
index	row major	queue	linked list
precision	column major	shift register	structure
length	table	first-in	crotch
origin	string	first-out	link

PROBLEMS

1. What are the limitations on the precision and length of the vector Z that are accessed by program segments (2) and (3)? How would you change these segments for a vector Z with length 500?

2. Write a shortest subroutine `READV` that returns in index register Y (high 16 bits) and accumulators A and B (low 8 bits), the ith element of a zero-origin vector. For instance, V may be allocated as in `V DS M*N`, where V has precision `M` bytes (`M` ≤ 4) and cardinality `N` elements. To read V, the parameters can be passed on the stack with the calling sequence shown below. If `M` is 1, pass the result in ACCB; if 2, pass it in ACCD; if 3, pass it in Y (msb) and ACCB (lsb); and if 4, pass it in Y (msb) and ACCD (lsb).

```
MOVW    #V,2,-SP
MOVB    #M,1,-SP
MOVB    #i,1,-SP
JSR     READV
LEAS    4,SP
```

3. Write a shortest subroutine `WRITEV` that writes the `M` bytes (`M` ≤ 4) into the ith element of a one-origin vector. For instance, the data may be in index register Y (high 16 bits) and accumulators A and B (low 8 bits). V may be allocated as in `V DS M*N`, where `V` has precision `M` bytes (M ≤ 4) and cardinality `N` elements. To write the data into `V`, the parameters can be passed on the stack with the calling sequence as shown below. Pass returned results as in Problem 2.

```
MOVW    #V,2,-SP   ; Vector base address
MOVB    #M,1,-SP   ; Vector precision
MOVB    #i,1,-SP   ; Desired element number
JSR     WRITEV     ;
LEAS    4,SP       ; Balance stack
```

4. Consider the zero-origin vector of 32 bytes in locations 0 through 31. Assume that the bits in this vector are labeled 0 to 255 beginning with the first byte in the vector and going right to left within each particular byte. Write a subroutine `SETBIT` that will set the ith bit in this vector assuming that the value of i is passed on the stack with the calling sequence

```
MOVB   #58,1,-SP      ; Value of i into parameter
BSR    SETBIT
LEAS   1,SP           ; Balance stack
```

5. Write a subroutine `STRBIT` that will store the binary-valued variable `BIT` in the ith bit of the zero-origin vector in Problem 4. The value of `BIT` and `INDEX` can be passed on the stack with the calling sequence

```
        MOVB    BIT,1,-SP       ; Bit to be written
        MOVB    INDEX,1,-SP     ; Index to be written
        BSR     STRBIT
        LEAS    2,SP            ; Balance stack
```

6. Assume S[100] stores a 100-character (maximum) string that is accessed by an 8-bit index I that is initialized to zero. Write a shortest subroutine READ to return in accumulator A the elements one at a time, beginning with element zero.

7. Write a subroutine `DISPLY`, using `OUTCH`, whose input is in A, that will display a zero-origin vector with the structure of Problems 4 and 5 8 rows by 32 columns of 0's and 1's. The bits of the vector must be displayed left to right, 32 consecutive bits per row. You may assume that the address of the vector is passed after the call with the sequence

```
        BSR     DISPLY
        DC.W    VECTOR
```

8. Give two subroutines `PSHBT` and `PLBT`, to go with those of Figure 10.7, that will push and pull the contents of accumulator B on the bottom of the deque.

9. One would not usually tie up two index registers and an accumulator to implement a deque. Rewrite the two subroutines in Figure 10.7 and the initialization sequence to push and pull bytes from B and the top of the deque when the deque is stored in memory as

```
COUNT:  DS.B    1       ; Deque count
TPOINT: DS.W    1       ; Top pointer
BPOINT: DS.W    1       ; Bottom pointer
DEQUE:  DS.B    50      ; Buffer for the deque
```

Here, `COUNT` contains the number of elements in the deque, and `TPOINT` and `BPOINT` contain, respectively, the addresses of the top and bottom of the deque. The subroutines should not change any registers except B, which is changed only by the pull subroutine.

10. Assuming that the location of the deque and the error sequence are fixed in memory, how would you change the subroutines of Figure 10.7 so that the machine code generated is independent of the position of the subroutines? How would you change these subroutines if the size of the deque was increased to 400 bytes?

11. Do you see how you can avoid keeping a counter for the deque? For example, can you check for an empty or full deque without a counter? "Full" means the last element is now used up.

12. Assume that a 10-byte shift register is established in your program with

```
SHIFTR: DS      10      ; Buffer memory
POINT:  DS      2       ; Pointer to SHIFTR
```

Write a subroutine SHIFT to put a byte into the shift register from B and pull a byte out into A.

13. Write the shortest subroutines necessary to maintain five 8-element one-byte element queues, where each queue is in a buffer. Your implementation should include a branch to location `ERROR` if an overflow of the buffer to hold the strings occurs. The first queue is stored at label `Q1`; the second, at `Q2`, etc. Upon entering the subroutines, the address of the queue being used is in X and the data is passed in `ACCA`.

14. Write a subroutine `BUILD`, which is passed, by name, a string of ASCII lower-case letters terminated by a carriage return, to form the linked list shown in Figure 10.9.

15. What is the limitation on the number of characters in the tree for the subroutine of Figure 10.10? How would you change the subroutine to allow for 350 characters?

16. Write the subroutine `REVSCAN` that corresponds to `SCAN` but that now scans the tree from the right, printing the characters out in reverse alphabetical order.

17. Write a subroutine to add `M` 4-byte numbers that corresponds to the following header:

```
*               SUBROUTINE ADD4
*
* ADD4 adds the M 4-byte numbers pointed to by Z Placing
* the result in SUM. All parameters are passed on the
* stack with the sequence
*
*               LDAB    M           ; Value of M into B
*               LEAX    Z,PCR       ; Address of Z into X
*               LEAY    SUM, PCR    ; Address of SUM into Y
*               PSHY
*               PSHX
*               PSHB
*               BSR     ADD4
*               LEAS    5,SP        ; Balance stack
```

18. Write a position-independent reentrant subroutine to go with the header:

```
*
* SUBROUTINE INSERT inserts the string STG into string TEXT at the first
* occurrence of the ASCII letter SYMBOL. No insertion is made if SYMBOL does not
* occur in TEXT. Parameters are passed on the stack with the sequence
*
*               LDAA    SYMBOL          ; ASCII symbol into A
*               LDAB    LSTG            ; Length of STG into B
*               PSHD                    ; Push both parameters
*               MOVW    #STG,2,-SP      ; Push Address of STG
*               LDAB    LTEXT           ; Length of TEXT into B
*               PSHB                    ; Push parameter
*               MOVW    # TEXT,2,-SP    ; Push Address of TEXT
*               BSR     INSERT          ; Subroutine balances the stack
```

19. Write a shortest subroutine GET3 that puts three characters into B (first character) and Y (second and third character) pointed to by X, and moving X past them and a next space or carriage return. Write a shortest subroutine CHKEND that reads three characters using subroutine GET3, and checks for the characters "END". Assume the calling subroutine has not pushed anything on stack, so if "END" is read, return by pulling two bytes from the stack and executing rts, so as to terminate the calling subroutine, otherwise return by just executing rts as usual to return to the calling subroutine. These subroutines are to be used in problem 21.

20. Write a shortest subroutine FIND which searches a binary tree, pointed to by X for three letters, B:Y as returned by GET3 of problem 19. The binary tree nodes contain a 3-byte character string, a one-byte value, and two two-byte addresses, the first of which points to a left son, and the other points to a right son. If the three letters are found, return with X pointing to the beginning of the node, else if the letters are lower in the dictionary than the last node searached in the tree, X points to the left son field of the last node, but if higher, X points to the right son field of the last node. This subroutine is to be used in problem 21.

21. Write a 6812 assembler program using linked lists, able to assemble programs having the following specifications.

(a) The operations will be encoded in the left two bits of a one-byte opcode: LDA is 00, ADD (for ADDA) is 01, STA is 10, and SWI is 11. There will also be directives DCB and END.

(b) At most six lables can be used, each of which is exactly three uppercase letters long. Operands will be encoded in the right six bits of a one-byte opcode.

(c) Only direct addressing can be used with instructions, which will be coded in the right 6 bits of the instruction, and only hexadecimal numbers, beginning with $, can be used with the DCB directive.

(d) The input line will have a fixed format: label (3 characters), space, instruction mnemonic (three characters), space, address (3 characters or 2-digit hexadecimal number prefixed with a $) ending in a carriage return. There are no comments, and, if a label is missing, it is replaced by 3 spaces.

(e) The program will have from 1 to 10 lines, ending in an END directive.

(f) The source code has no errors (i.e., your assembler does not have to check errors). The origin is always zero.

Your assembler will have the source code prestored as a character array TEXT, 10 rows by 12 columns, and will generate an object code string OBJ up to 10 bytes long. No listing will be generated. The assembler should be able to at least assemble the following two programs, shown on the left and shown on the right.

```
        LDA   ABB
        ADD   BAB
        STA   BBA
        SWI
ABB     DCB   $01
BAB     DCB   $02
BBA     DCB   $00
        END
```

```
ALP     DCB   $01
GAM     DCB   $00
DEL     DCB   $04
BET     DCB   $03
        LDA   ALP
        ADD   BET
        ADD   DEL
        STA   GAM
        SWI
        END
```

A two-pass assembler is required, and labels and opcodes must be stored as linked lists. Use subroutines `GET3` and `CHKEND` (Problem 19) to input labels or opcode mnemonics, and `FIND` (Problem 20) to search both the symbol table and the mnemonics, in your assembler. Show the storage structure for your mnemonic's binary tree (it is preloaded) following the graph shown in Figure 10.15. On the first pass, just get the lengths of each instruction or directives and save the labels and their addresses in a linked list. End pass one when `END` is encountered. On pass two, put the opcodes and addresses in the sring OBJ.

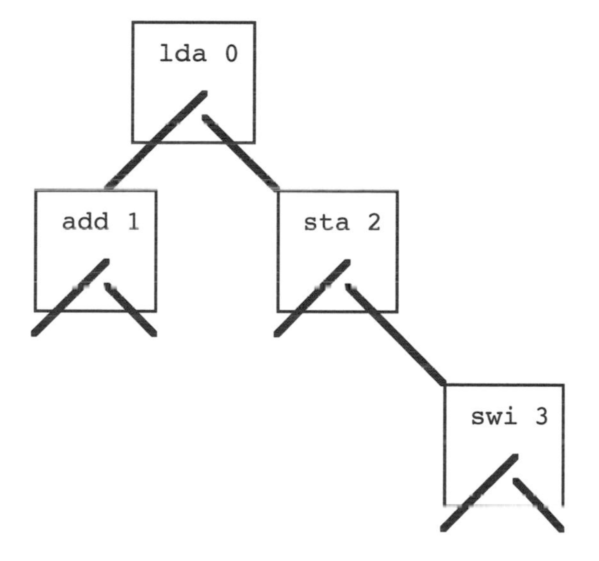

Figure 10.15. Graph of Linked List for Problem 10.21

The Adapt812 is connected to an M68HC12B32EVB board which is configured in POD mode, which in turn connects to a PC. We used this configuration to download and debug using HiWare, using the ASCIIMON target interface.

11

Input/Output

An *input routine is* a program segment that inputs words from the outside world into the computer, and an *output routine* is a program segment that does the reverse. It outputs words from the computer to the outside world. Clearly, a computer that does not have input and output routines, and the hardware to carry out these routines, would be useless regardless of its power to invert matrices or manipulate great quantities of data. Until now, we have implied that you should avoid knowing the details of these routines. Even though we have left the discussion of input and output until near the end of this book, it is really simple and should pose no problem to the reader.

In this chapter, we first describe how the basic input and output operations are implemented in hardware and executed in software, using simple ports available in both the 6812 'A4 and 'B32. We then discuss the use of buffers in input and output. To describe synchronization, we describe a greatly simplified timer interrupt mechanism, which is also available in both the 6812 'A4 and 'B32. We then discuss gadfly and interrupt mechanisms. Finally, we introduce D-to-A and A-to-D converters.

Upon concluding this chapter, you should understand how basic input and output operations are performed and be able to read and write input and output routines that use simple synchronization mechanisms.

11.1 Input and Output Devices

In this section, we introduce a simplified hardware model used to understand input-output routines. We also discuss simple input and output ports to provide enough background for the later sections of this chapter.

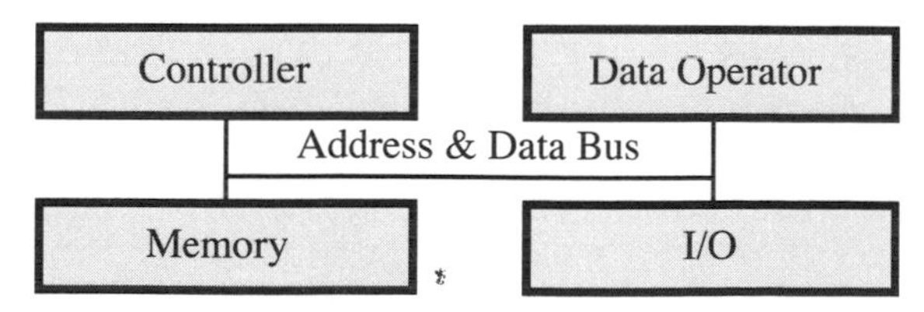

Figure 11.1. Simplified Diagram of a Microcomputer (Identical to Figure 1.1)

Recall from Chapter 1 that a computer is divided into its major components: the controller, data operator (arithmetic/logic unit), memory, and input-output unit (see Figure 11.1). Input and output instructions use the same address and data bus as load and store instructions with memory, but the action of input and output instructions on input-output hardware is a bit different than the action of load and store instructions on memory. In many microcontrollers, different instructions are used for memory reads or writes than for input or output operations, even though essentially the same pins are used for each instruction. However (see Figure 11.2), in microcontrollers such as the Motorola 6812, the load instruction, used to read data from memory such as words 1, 2, or 3, can also be used to input data such as from word 0, and the store instruction, used to store data in memory such as words 1, 2, or 3, can also be used to output data such as to word 0.

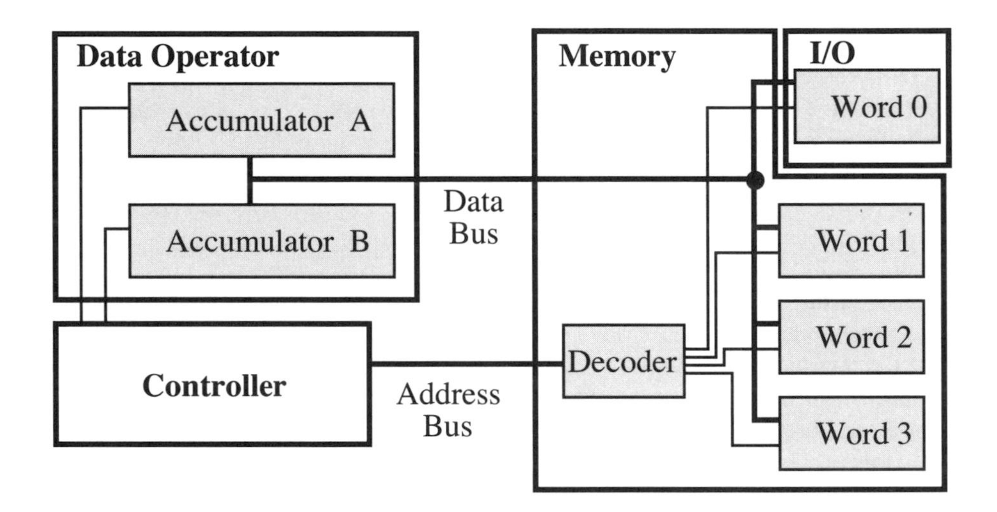

Figure 11.2. A Memory and Its Connection to the MPU (Compare to Figure 1.3)

Certain 8-bit or 16-bit memory locations are chosen to be *output ports* corresponding to a hardware component called an *output device,* which has *output lines* connected to the outside world. Certain 8-bit or 16-bit memory locations are chosen as *input ports,* corresponding to a hardware component called an *input device,* which has *input lines* coming from the outside world.

From the point of view of the I/O device, each address, data, or control line has a signal that is a (logical) one if the voltage is above a certain threshold level and a (logical) zero if the voltage is below that level. The voltages corresponding to a (logical) one and a (logical) zero are also termed *high signal* and *low signal*, respectively. The clock signal is alternately low and high repetitively in a square wave. The clock signal between high-to-low transitions is called a clock cycle. In each clock cycle, the microcontroller can read a word from an input port, such as word 3, by putting the address of the word to be read on the address bus and putting the read/write line to high throughout the clock cycle. At the clock cycle's end, the device will put data on input lines on the data bus, and the processor will copy the word on the data bus into some internal memory register. The microcontroller can also write a word into an output port, such as word 3, at a particular address in one clock cycle by putting the address on the address bus, putting the word to be written on the data bus, and making the signal low on the read/write line throughout the clock cycle. At the clock cycle's end, the microcontroller will write the word on the data bus into the device. The written data will become available on the device's output lines, until changed by another output instruction.

For example, whenever the microcontroller writes data into location $0000, such as in the instruction

```
STAA $00
```

the data written are put on the output lines of the device. This instruction can use the page-zero addressing mode because the address is at location zero. The simplest output port is writable but not readable: This "write-only memory" is usually a topic only for a computer scientist's joke collection, but it is a real possibility in an output port.

For example, location $0000 may be an input port, and a hardware input device will be built to input data from that port. Whenever the microcontroller reads data from location $0000, as in the instruction

```
LDAA $00
```

the signals on the input lines of the input device will be read into the microcontroller just like a word read in from memory. Note that an input operation "takes a snapshot" of the data fed into the input device at the end of the last clock cycle of the load instruction and is insensitive to the data values before or after that point in the last clock cycle. A final aspect is whether the port can be read from or written in, or both.

In a sense, the basic input and output devices trick the microcontrollers. The microcontroller thinks it is reading or writing a word in its memory at some address. However, the microcontroller designer has selected that address as an input or an output port and built hardware to input or output data that are read from or written into that port. By means of the hardware, the designer tricks the microcontrollers into inputting data when it reads a word at the address of an input port or into outputting data when it writes data into the word at the address of an output port.

One of the most common faulty assumptions in port architecture is that I/O ports are eight bits wide. For instance, in the 6812, the byte-wide `LDAB` instructions are used in I/O programs in many texts. There are a number of 16-bit I/O ports on I/O chips that are designed for 16-bit microcontrollers. But neither 8 nor 16 bits is a fundamental width. In this chapter, where we emphasize fundamentals, we avoid that assumption. Of course, if the port is 8 bits wide, the `LDAB` instruction can be used, and used in C by accessing a variable of type `char`. There are also 16-bit ports. They can be read by `LDD` instructions, or as an `int` variable in C or C++. A port can be 1 bit wide; if so, a 1-bit input port is read in bit 7; reading it will set the N condition code bit, which a `BMI` instruction easily tests. Many ports read or write ASCII data. ASCII data is 7 bits wide, not 8 bits wide. If you read a 10-bit analog-to-digital converter's output, you should read a 10-bit port.

11.2 Parallel Ports

The 'A4 and 'B32 have two parallel ports, shown in Figure 11.3. The description of each port and their special features and programming techniques are discussed in the first subsection. The second subsection describes an object-oriented class for these ports.

The 6812's parallel ports have a *direction port*. For port A, for each bit position, if the port A direction bit is zero, as it is after reset, the port bit is an input, otherwise if the port A direction bit is one, the port bit is a readable output bit. A direction port is an example of a *control port,* which is an output port that controls the device but doesn't send data outside it.

We illustrate the use of *PORTA* in assembly language first, and in C or C++ after that. To make *PORTA* an output port, we can write in assembly language:

```
LDAB  #$FF   ; generate all ones
STAB  $2     ; put them in direction bits for output
```

Then, any time after that, to output accumulator B to *PORTA* we can output accumulator B to port A by writing `STAB $0`. To make *PORTA* an input port, we put zeros in direction bits for input by executing `CLR $2`. Then, any time after that, to input *PORTA* into accumulator B we read *PORTA* into accumulator B by writing `LDAB $0`. It is possible to make some bits, for instance the rightmost three bits, readable output bits and the remaining bits input bits, as follows:

```
LDAB  #7     ; generate three ones in rightmost bits
STAB  $2     ; put them in direction bits for output
```

The instruction `STAB $0` writes the rightmost three bits into the readable output port bits. The instruction `LDAA $0` reads the left five bits as input port bits and the right three bits as readable output bits. A minor feature also occurs on writing the 3-bit word: The bits written where the direction is input are saved in a register in the device and appear on the pins if later the pins are made readable output port bits.

Ports can be identified by putting the `@` sign after port names, followed by their locations. These can be put in a header file that is `#included` in each program:

```
unsigned char PORTA@0, PORTB@1, DDRA@2, DDRB@3;
```

The equivalent operations in C or C++ are shown below. To make `PORTA` an output port, we can write

```
DDRA = 0xff;
```

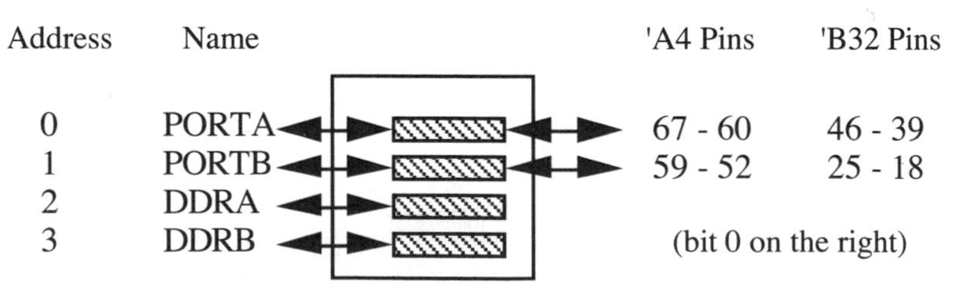

Address	Name	'A4 Pins	'B32 Pins
0	PORTA	67 - 60	46 - 39
1	PORTB	59 - 52	25 - 18
2	DDRA		
3	DDRB		

(bit 0 on the right)

Figure 11.3. Some 6812 Parallel I/O Ports

Note that *DDRA* is declared an `unsigned char` variable. Then, any time after that, to output a `char` variable `i` to `PORTA`, put

```
PORTA = i;
```

Note that `PORTA` is declared an `unsigned char` variable. To make `PORTA` an input port, we can write

```
DDRA = 0;
```

Then, any time after, to input `PORTA` into an `unsigned char` variable `i` we write

```
i = PORTA;
```

Generally, the direction port is written into before the port is used the first time and need not be written into again. However, one can change the direction port from time to time.

`PORTA` and `PORTB` together, and their direction ports `DDRA` and `DDRB` together, can be treated as a 16-bit port because they occupy consecutive locations. Therefore, they can be read from or written into using `LDD` and `STD` instructions. To make `PORTAB` an output port, we can write in assembly language:

```
LDD  #$FFFF   ; generate all ones
STD  $2       ; put them in direction bits for output
```

Then, any time after that, to output accumulator D to `PORTAB` we can write

```
STD  $0       ; output accumulator D
```

To make `PORTAB` an input port, we can write

```
CLR  $2       ; put zeros in high direction bits for input
CLR  $3       ; put zeros in low direction bits for input
```

Then, any time after that, to input `PORTAB` into accumulator D we can write

```
LDD  $0       ; read PORTA into accumulator D
```

Also, some of the 16 bits can be made input, and some can be output. In manner similar to how 8-bit ports are accessed in C, 16-bit ports can be declared in a header file that is `#included` in each program as follows:

```
int PORTAB@0, DDRAB@2;
```

To make `PORTA` and `PORTB` an output port, we can write: `DDRAB = 0xffff;`. Note that `DDRAB` is declared an `int` variable. Then, any time after that, to output an `int` variable `i`, high byte to `PORTA` and low byte to `PORTB`, we can write `PORTAB = i;`. Note that `PORTAB` is declared an `int` variable. To make `PORTA` and `PORTB` an input port, we write: `DDRAB = 0;`. Then, any time after that, to input `PORTA` (as high byte) and `PORTB` (as low byte) into an `int` variable `i` we can write `i = PORTAB;`. The ports A and B, or the combined port AB, can be made an input or output port and can be easily accessed in assembly language or in C.

As a simple example of the use of an input port, consider a home security system. See Figure 11.4a. Three switches, each attached to a different window, are normally closed. When any window opens, its switch opens, and the pull-up resistor makes PORTA's input high; otherwise the input is low. This signal is sensed in PORTA bit 0. The C program statement `if(PORTA & 1) alarm();` will execute procedure `alarm` if any switch is opened. It is optimally programmed into assembly language as follows:

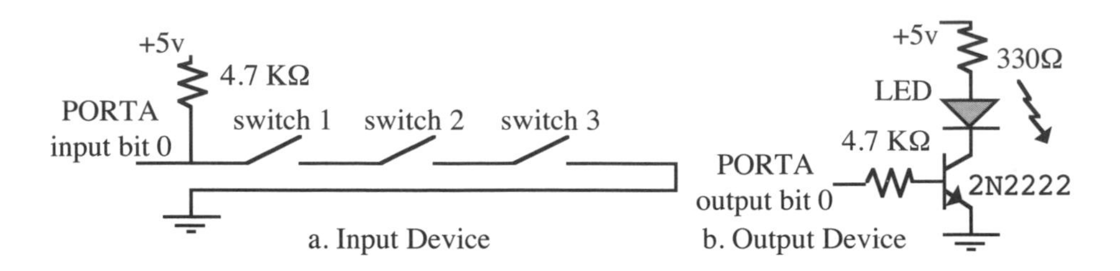

Figure 11.4. Simple Devices

```
BRCLR  0,#1,*+6     ; branch over BSR if bit one of PORTA is low
BSR    alarm        ; otherwise call the subroutine
```

As a simple example of the use of an output port, consider the light-emitting diode (LED) display shown in Figure 11.4b. When PORTA's output is high, current flows through the transistor and LED, being limited by the 330 Ω resistor. The C program statement `PORTA = 1;` will cause the LED to light up. It is optimally programmed into assembly language as follows:

```
LDAB #1      ; generate a one
STAB $0      ; output it to the port
```

11.3 Input and Output Software

Input or output of a single word is simple, but we often need to input or output a string of characters, an array of numbers, or a program consisting of many words. This section reviews how vectors can be used in these situations.

The simplest and one of the most common situations occurs when one is inputting or outputting a vector of bytes. To output a vector of bytes to PORTA, smallest indexed byte first, execute the following C procedure:

```
char buffer[0x10];
void main() { char i = -0x10;
     DDRA = 0xff; do PORTA = buffer[i + 0x10]; while(++i);
}
```

An optimized assembly language program segment for the body of this C procedure is

```
      LDD  #$FFF0        ; Put $FF in Accumulator A, -$10 in Accumulator B
      STAA $2            ; Put $FF in direction, to make it output
      LDX  #BUFFER+$10   ; Set index register X to just beyond end of vector
LOOP: LDAA B,X           ; Get an element of the vector
      STAA $0            ; Write it to the output port
      IBNE B,LOOP        ; For each element of the vector
```

Note the ease of indexing a vector in a `do while` loop statement. This operation, emptying data from a vector to an output port, is one of the most common of all I/O programming techniques.

Conversely, to input a vector of bytes, this time largest indexed byte first, with data on the input lines of the simple input device, execute the following C procedure:

```
char buffer[0x10];
void main() { char i = 0x10;
    DDRA = 0; do  buffer[i - 1] = PORTA; while(--i);
}
```

An optimized assembly-language program segment for the body of this C procedure is

```
      LDD   #$10          ; Generate 0 in Accumulator A, $10 in Accumulator B
      STAA   $2           ; Put 0 in direction, to make it input
      LDX   #BUFFER-1     ; Set index register X to base of vector
LOOP: LDAA   $0           ; Get a byte from the input port
      STAA   B,X          ; Write it into the vector, top element first
      DBNE B,LOOP         ; For each element of the vector
```

From the two examples above, the reader should be convinced that inputting to or outputting from a vector is a very simple operation in either C or assembly language.

We show a simple example of an output device, a toy traffic light. While this example is not the best way to control a traffic light, it serves to demonstrate output from a vector and other techniques discussed later in the chapter. Each of the six least significant bits of the output PORTA controls a pair of LED traffic lights (see Figure 11.5a). The north and south LEDs are wired in parallel, and the east and west LEDs are similarly paralleled and turned on if the transistor base input is HIGH (see Figure 11.5b, which is like Figure 11.4b). Making PORTA bit 5 HIGH turns on the red LEDs, bit 4 turns on the yellow LEDs, and bit 3 turns on the green LEDs, in north and south lanes. Making PORTA bit 2 HIGH turns on the red LEDs, bit 1 turns on the yellow LEDs, and bit 0 turns on the green LEDs, in east and west lanes. So making PORTA bits 5 and 0 HIGH would turn on the red north and south and green east and west LEDs.

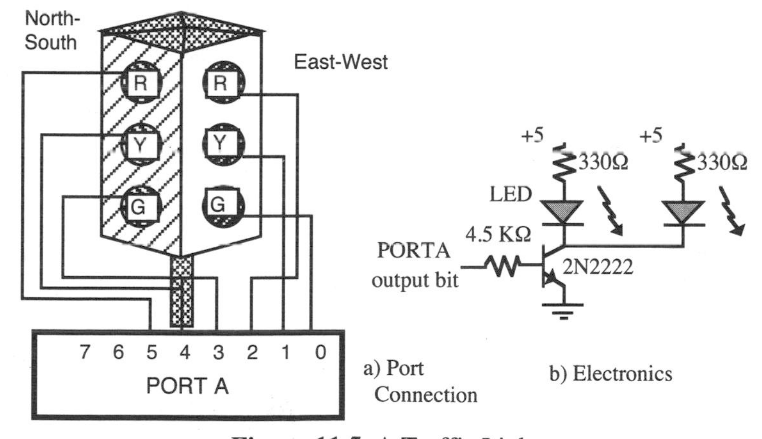

Figure 11.5. A Traffic Light

We will output a vector of light patterns to the LEDs, largest indexed element first. Because page-zero addressing is used for the `STAA $0` instruction, a byte is output to the LED display every eight clock cycles, which is every microsecond, much too fast for any type of useful display. But an arbitrary delay can be easily added in the loop to display each pattern as long as desired. To output an element every second, execute

```
char buffer[0x10];
void main() { char i = 0x10; long t;
    DDRA=0xff;do{PORTA=buffer[i-1];for(t=0;t<2666666;t++);}while(--i);
}
```

An optimized assembly language program segment for the body of this C procedure is

```
        MOVB  #$FF,$2       ; PORTA direction set for output
        LDAB  #$10          ; Vector Index initialized to high end
        LDX   #BUFFER-1     ; Base address of the vector
LOOP:   LDAA  B,X           ; Get an element out of the vector
        STAA  $0            ; Write it into the output port
        LDAA  #100          ; Execute outer loop 100 times
WT0:    LDY   #80000/3      ; Execute the inner loop 26666 times
WT1:    DBNE  Y,WT1         ; Inner loop takes 10 ms
        DBNE  A,WT0         ; Outer loop takes 1 second
        DBNE  B,LOOP        ; Output all elements of the vector
```

The inner loop, the single instruction, `WT1: DBNE Y,WT1`, takes three memory cycles. Because index register Y is initialized to 26666, this loop takes 80,000 memory cycles, which, for an 8 MHz 6812 clock, takes 10 ms. The outer loop, including the inner loop and `LDY #80000/3` and `DBNE A,WT0`, is executed 100 times, delaying very close to 1 second. So the program segment outputs a vector element each second. The vector `buffer` can therefore be initialized with appropriate bit pattern constants to produce the desired sequence of lighted LEDs. Figure 11.6a illustrates the general idea of a delay loop used to synchronize output from a vector.

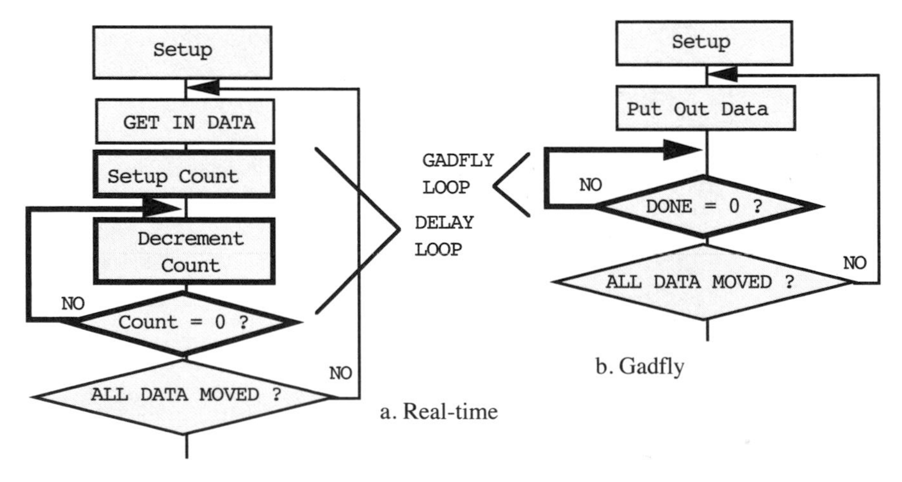

Figure 11.6. Flow Charts for Programmed I/O

Arrays or strings can be used with buffers to input or output many numbers in a similar way. Queues are implemented in buffers to input or output data when the program needs them in the same order that they are input or when the device needs them in the same order that they are provided by the program, but at different times than they are provided. Generally, data structures are used to maintain order in the data that are being transferred between the MPU and an input-output device.

11.4 Synchronization Hardware

The previous section illustrated the use of a delay loop to synchronize the output to a traffic light controller. In the next sections, we consider the use of gadfly loops and interrupts to synchronize input and output. These use hardware to indicate when I/O is to be done. The most commonly used hardware is an edge-triggered flip-flop that is set when an output is needed or when an input is available. In this section, we introduce the 'A4 or 'B32 basic counter/timer device to illustrate edge-triggered sensing of I/O status signals. (See Figure 11.7.) We also briefly discuss PORTT and its direction bits.

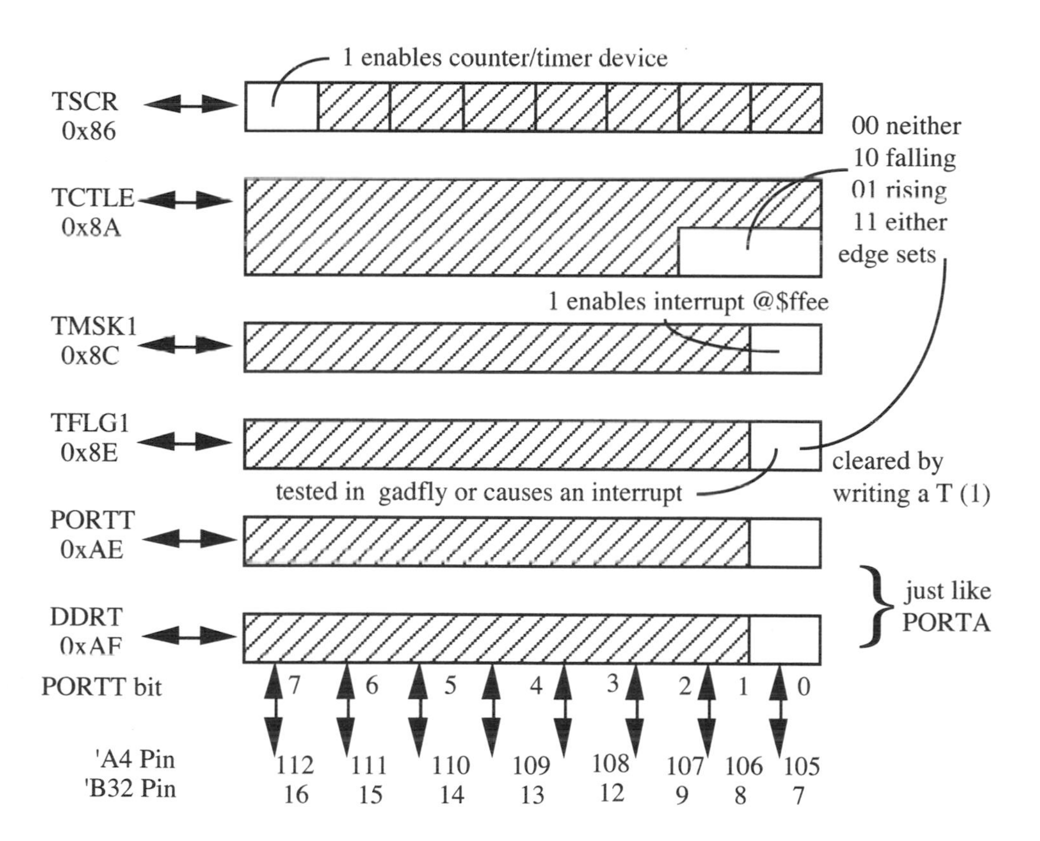

Figure 11.7. The Counter/Timer Subsystem

The counter/timer device is very flexible, and so it has many control bits that can be set or cleared to permit it to implement different functions. In this section, we use just one of these functions, to detect an edge of an input signal. Bits that we set or clear are shown in clear rectangles in Figure 11.7 and discussed below; unused bits are shown in rectangles with diagonal lines in them. We therefore merely initialize some control ports in a "fixed" way and do not discuss the other ports in this counter/timer device.

The counter/timer device is enabled when `TEN`, bit 7 of `TSCR`, is T (1). We illustrate our techniques using `PORTT` bit 0 (pin 105 in the 'A4, or pin 7 in the 'B32). Bits 1 and 0 of `TCTLE` determine which edges will be sensed; if that port's bit 1 is set, then a falling edge on `PORTT` bit 0 sets the flip-flop; if its bit 0 is set, a rising edge sets the flip-flop; and if both bits are set, either edge sets the flip-flop. Bit 0 of `TFLG1` reads this flip-flop. This bit can be tested in a gadfly loop, or if bit 0 of `TMSK1` is also set, it causes an interrupt vectored through 0xFFEE and 0xFFEF. Bit 0 of `TFLG1` *must* be cleared before it can be set sensed again; it is cleared by writing a T (1) into it.

For each bit position, `PORTT` can be used as a parallel I/O port whose direction is specified by the corresponding bit in `DDRT`. Even when a `PORTT` bit is used to detect an edge, the port can also be an input port to directly read the pin's signal.

11.5 Gadfly Synchronization

In the *gadfly synchronization* technique, the program continually "asks" one or more devices what they are doing (such as by continually testing the timer flag bit). This technique is named after the great philosopher, Socrates, who, in the Socratic method of teaching, kept asking the same question until he got the answer he wanted. Socrates was called the "gadfly of Athens" because he kept pestering the local politicians like a pesky little fly until they gave him the answer he wanted (regrettably, they also gave him some poison to drink). This bothering is usually implemented in a loop, called a *gadfly loop*, in which the microcomputer continually inputs the device state of one or more I/O systems until it detects DONE or an error condition in one of the systems. Gadfly synchronization is often called polled synchronization. However, polling means sampling different people with the same question—not bothering the same person with the same question. Polling is used in interrupt handlers discussed in the next section; in this text, we distinguish between a polling sequence and a gadfly loop.

A gadfly loop is illustrated by the flow charts shown in Figure 11.6b. The processor keeps testing a *status port,* which is set by the device when it is done with the request for the input, after which data can be read from the input port.

Gadfly synchronization generally requires more extensive initialization before the device can be used. The counter/timer control registers must be set up so that when data are to be output, a falling edge on `PORTT` bit 0 can set a flag bit. Bit 7 of the `TSCR` needs to be set to enable any operation in the counter/timer. Bits 1 and 0 of `TCTLE` must be set to 1 and 0 respectively, to indicate that a falling edge sets the `TFLG1` bit 0. This flag can accidentally become set before the first output operation occurs, so to clear it just in case it is set, \$01 should be written into the `TFLG1` port. The initialization of the counter/timer consists of the C statements: `TSCR = 0x80; TCTLE = 2; TFLG1 = 1;`. It is optimally compiled to assembly language as:

```
        LDAB   #$80           ; Put one in leftmost bit of accumulator B
        STAB   TSCR           ; Put it into the control register TSCR
        LDD    #$201          ; Generate 2 in Accumulator A, 1 in Accumulator B
        STAA   $8B            ; Write 2 in rightmost bit of control register TCTLE
        STAB   $8E            ; Write one in rightmost bit of control register TFLG1
```

In place of a delay loop, the gadfly loop itself is used whenever input or output is done: `do ; while((TFLG1 & 1) == 0); TFLG1 = 1;`. The loop waits until an edge sets the flag bit, and the next statement clears the flag bit. This is optimally compiled to assembly language as

```
L:      BRCLR  $8E,#1,L       ; Wait for one in TFLG1 bit 0
        LDAB   #1             ; Put one in rightmost bit of accumulator B
        STAB   $8E            ; Write it into control register TFLG1
```

We illustrate the use of gadfly synchronization for counting pulses from a Geiger counter. Each time a pulse occurs, `PORTT` bit 0 sees a rising edge. To count the pulses, execute the C procedure:

```
void main() { int i = 0;
    TSCR = 0x80; TFLG1 = 1;TCTLE = 1;
    do {while((TFLG1 & 1) == 0); TFLG1 = 1; i++; }while(1);
}
```

An optimized assembly language program segment for the body of the C procedure is

```
        LDAB    #$80          ; Put one in leftmost bit of accumulator B
        STAB    $86           ; Write it into control register TSCR
        LDD     #$102         ; Generate one in Accumulator A, 2 in Accumulator B
        STAA    $8E           ; Write one in rightmost bit of control register TFLG1
        STAB    $8B           ; Write 2 in rightmost bit of control register TCTLE
        CLRA                  ; Clear msbyte of local variable i
        CLRB                  ; Clear next byte of i
LOOP:   BRCLR   $8E,#1,LOOP   ; Wait for one in TFLG1 bit 0
        MOVB    #1,$8E        ; Write 1 into control register TFLG1 to clear bit 0
        ADDD    #1            ; Increment count
        BRA     LOOP          ; Loop forever
```

You can stop this program after it has counted the pulses and examine accumulator D.

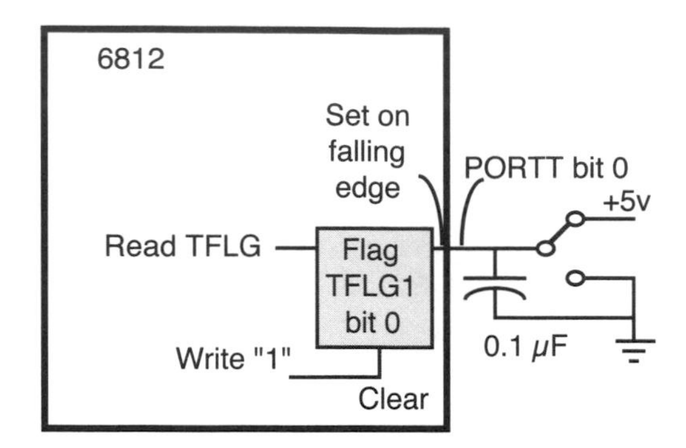

Figure 11.8. Status Port

We next illustrate the use of gadfly synchronization by implementing a variation of a traffic light controller, as it was described in Section 11.3. But each time another vector element is to be output, a switch is closed and PORTT bit 0 sees a falling edge. See Figure 11.8. Moving the switch to the bottom contact causes exactly one falling edge. The capacitor holds the voltage on the port input bit when the switch is in between the top and bottom contact. To output the vector, execute the C procedure:

```
char buffer[0x80];
void main() { char i = 0x80;
    DDRA = 0xFF;  TSCR = 0x80; TFLG1 = 1; TCTLE = 2;
    do {PORTA=buffer[i-1]; do ; while((TFLG1&1)==0); TFLG1=1;}while(--i);
}
```

Each time the `do while` loop is executed, it waits for the next edge to occur. The edge sets the flag, and the `do while` loop exits. The next statement clears this flag. An optimized assembly language program segment for the body of the C procedure is

```
        LDAB    #$80        ; Put one in leftmost bit of accumulator B
        STAB    $86         ; Write it into control register TSCR
        LDD     #$102       ; Generate one in Accumulator A, 2 in Accumulator B
        STAA    $8E         ; Write one in rightmost bit of control register TFLG1
        STAB    $8B         ; Write 2 in rightmost bit of control register TCTLE
        LDD     #$FF80      ; Generate two constants in Accumulators A and B
        STAA    $2          ; Make PORTA output
        LDX     #BUFFER-1   ; Get the address of the buffer into X
LOOP:   LDAA    B,X         ; Read a byte from the buffer
        STAA    $0          ; Output the data
L:      BRCLR   $8E,#1,L    ; Wait for one in TFLG1 bit 0, which occurs once/sec.
        LDAA    #1          ; Put one in rightmost bit of accumulator A
        STAA    $8E         ; Write it into control register TFLG1
        DBNE   B,LOOP       ; Count down and loop until all are output
```

The gadfly loop waits until a hardware-generated edge occurs, while the delay loop waits a prescribed number of instruction executions. In general, delay loops do not require as much hardware as gadfly loops, because a gadfly loop needs a status bit in hardware and a means to set it. However, gadfly synchonization can wait exactly as long as an I/O device needs, when hardware causes the edge to occur, while a delay loop is generally timed to provide a delay that is the worst case delay needed for the device.

11.6 Interrupt Synchronization

In this section, we consider interrupt hardware and software. Interrupt software can be tricky, so some companies actually have a policy never to use interrupts but instead to use the gadfly technique. At the other extreme, some designers use interrupts just because they are readily available in microcomputers like 6812 systems. We recommend using interrupts when necessary, but using simpler techniques whenever possible.

The *hardware* or *I/O interrupt* is an architectural feature that is very important to I/O interfacing. Basically, it is invoked when an I/O device needs service, either to move some more data into or out of the device or to detect an error condition. *Handling an interrupt* stops the program that is running, causes another program to be executed to service the interrupt, and then resumes the main program exactly where it left off. The program that services the interrupt (called an *interrupt handler* or *device handler*) is very much like a subroutine, and an interrupt can be thought of as an I/O device tricking the computer into executing a subroutine. An ordinary subroutine called from an interrupt handler is called an *interrupt service routine*. However, a handler or an interrupt service routine should not disturb the current program in any way. The interrupted program should get the same result no matter when the interrupt occurs.

I/O devices may request an interrupt in any memory cycle. However, the data operator usually has bits and pieces of information scattered around and is not prepared to stop the current instruction. Therefore, interrupts are always recognized at the end of the current instruction, when all the data are organized into accumulators and other registers (the machine state) that can be safely saved and restored. The time from when an I/O device requests an interrupt until data that it wants moved is moved, or the error condition is reported or fixed, is called the *latency time*. Fast I/O devices require low latency interrupt service. The lowest latency that can be guaranteed is limited to the duration of the longest instruction, because the I/O device could request an interrupt at the beginning of such an instruction's execution.

The condition code register, accumulators, program counter, and other registers in the controller and data operator, the machine state, and these nine bytes are saved and restored whenever an interrupt occurs. After completion of a handler, the last instruction executed is *return from interrupt* (RTI). It pulls the top nine bytes from the stack, replacing them in the registers the interrupt took them from.

Interrupt techniques can be used to let the I/O system interrupt the processor when it is DONE, so the processor can be doing useful work until it is interrupted. We first look at steps in an interrupt. Then we consider interrupt handlers and the accommodation of critical sections.

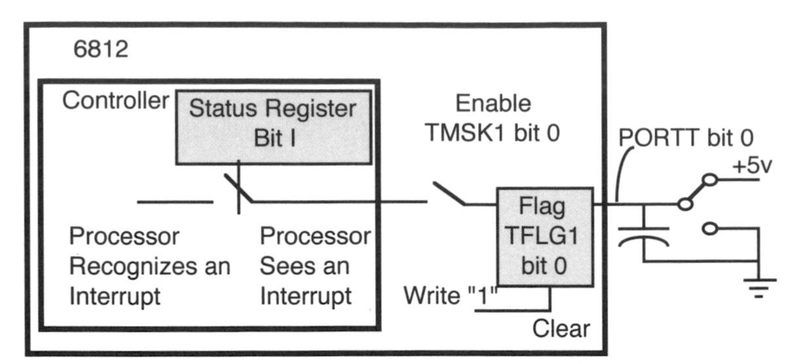

Figure 11.9. Interrupt Request Path

After port and counter/timer control registers are properly initialized, a falling edge on the counter/timer bit 0 can request an interrupt. See Figure 11.9. The six-step sequence of actions that lead to an interrupt and that service it are outlined below.

1. When the external hardware determines it needs service either to move some data into it or out of it or to report an error, we say the *device requests an interrupt*. This occurs when `PORTT` bit 0 falls.

2. If the `PORTT` bit 0 pin is an input (in `DDRT` bit 0) and had been assigned (in `TMSK1` bit 0) to sense interrupts, we say `PORTT` bit 0 *interrupts are enabled*.

3. If the microprocessor's condition code register's I bit is 0 we say the *microprocessor is enabled*. When I is 1, the *microprocessor is masked* (or the *microprocessor is disabled*). If a signal from the device is sent to the controller, we say the *microprocessor sees a request,* or a *request is pending,* and an interrupt will occur, as described below. (The bit I is also controlled by hardware in the next step.)

4. Most microcomputers cannot stop in the middle of an instruction. Therefore, if the microprocessor recognizes an interrupt, it *honors an interrupt* at the end of the current instruction. When the 6812 honors a counter/timer interrupt, it saves the registers and the program counter on the stack, sets the condition code register I bit, and loads the 16-bit word at 0xffee into the program counter to process this interrupt. Importantly, condition code bit I is set after the former I was saved on the stack.

5. Beginning at the address specified by 0xffee is a routine called the timer 0 *handler*. The handler is like a subroutine that performs the work requested by the device. It may move a word between the device and a buffer, or it may report or fix up an error. One of a handler's critically important but easy to overlook functions is that it must explicitly remove the cause of the interrupt (by negating the interrupt request) unless the hardware does that for you automatically. This is done by writing 1 into bit 0 of the `TFLG1` port.

6. When it is completed, the handler executes an `RTI` instruction; this restores the registers and program counter to resume the program where it left off.

Some points about the interrupt sequence must be stressed. As soon as it honors an interrupt seen on a line, the 6812, like most computers, sets the I condition code bit to prevent honoring another interrupt from the same device. If it didn't, the first instruction in the handler would be promptly interrupted—an infinite loop that will fill up the stack. You do not have to worry about returning it to its value before it was changed, because step 6 restores the program counter and the condition code register and its I bit to the values they had before the interrupt was honored. However, a handler can change that level using a `TFR`, `AND`, or `OR` to condition code instruction to permit honoring interrupts. Note that the I/O device is generally still asserting its interrupt request line because it doesn't know what is going on inside the microprocessor. If the `RTI` is executed or I is otherwise cleared, this same device will promptly interrupt the processor again and again—hanging up the machine. Before the handler executes `RTI` or changes I, it *must* remove the interrupt source! (Please excuse our frustration, but this is so simple yet so much of a problem.)

To handle the interrupts, we need to put the handler's address in the 16-bit word at 0xffee. The mechanism that puts an interrupt handler's address where the interrupt mechanism needs to find it is specific to each compiler. In a compiler that is specifically designed for I/O interfacing, all we need to do is write `interrupt 8` in front of the name of the procedure for counter/timer device 0 (a number different from 8 is used for other devices—see Table 11.1). This convention inserts the address of the handler into 0xffee, and further ends the procedure with an `RTI`. For other compilers a statement `*(int *)0xffee = (int)handler;` puts the address of the handler in the location that the hardware will use when an interrupt occurs. However, this location is in EEPROM in the 'A4 and flash in the 'B32, so special programming procedures are used.

There are two parts of the path that an interrupt request takes. In our example, the counter/timer flag is set if an edge occurs on its input. A switch in series with an input that can set this flag is called an *arm;* if it is closed the device is *armed,* and if it is opened, the device is *disarmed*. Any switch between the flag register and the 6812 controller is called an *enable;* if all such switches are closed the device is *enabled,* and if any are opened, the device is *disabled*. Arming a device lets it record a request and makes it possible to request an interrupt, either immediately if it is enabled or later if it is disabled. You disarm a device if you do not want to honor an interrupt now or later. But you disable an interrupt to postpone it. You disable an interrupt if you can't honor it now, but you may honor it later when interrupts are enabled.

Also, for gadfly synchronization, you arm the device so the flag register can become set when the device enters the DONE state, but you do not enable the interrupt because the program has to test the flag. If an interrupt did occur and was honored properly so as not to crash the computer, the gadfly loop wouldn't exit because the flag would be cleared in the handler before the gadfly loop could test it and exit its loop.

We illustrate the use of interrupt synchronization by implementing a variation of the Geiger counter that used gadfly synchronization. Each time a pulse occurs, `PORTT` bit 0 sees a rising edge, which causes an interrupt. To count the pulses, execute the C procedure

```
unsigned char i;
void main() {
    TSCR = 0x80; TTMSK1 = TFLG1 =  TCTLE = 1;  asm CLI
    do ; while(1);
}
void interrupt 8 hndlr(void){ TFLG1=1; /* clr reqst */ i++; /* inc count */}
```

An optimized assembly-language program segment for the body of the C procedure is

```
i:      DS.B    1        ; Global storage for 8-bit count i
        LDAB    #$80     ; Put one in leftmost bit of accumulator B
        STAB    $86      ; Write it into control register TSCR
        LDAA    #$1      ; Generate one in Accumulator A
        STAA    $8E      ; Write one in rightmost bit of control reg. TFLG1
        STAA    $8C      ; Write one in rightmost bit of control reg. TMSK1
        STAA    $8B      ; Write 1 in rightmost bit of control reg. TCTLE
        CLI              ; Enable interrupt
LOOP:   BRA     LOOP     ; loop forever
```

An optimized assembly-language program segment for the handler is

```
HNDLR: MOVB #1,TFLG1    ; Write one in rightmost bit of control register TFLG1
       INC  i           ; Increment count
       RTI              ; Return to interrupted routine
```

The main procedure above initializes the control registers and loops forever. The I condition code bit is generally clear after the 6812 is reset and must be cleared to enable interrupts to occur. A high-level language like C generally does not have a way to enable interrupts, except by inserting embedded assembly language. The statement `asm CLI` inserts `CLI` into the assembly-language program. Each time another vector element is to be output, a switch is closed, and `PORTT` bit 0 sees a rising edge. This causes an interrupt, and the handler is entered. This handler increments the count.

We further illustrate an interrupt-based traffic light controller that is essentially the same as the gadfly-based traffic light controller example. The main procedure initializes the control registers, waits for all elements to be output, and then disables the interrupt. Each time another vector element is to be output, a switch is closed, and `PORTT` bit 0 sees a falling edge. See Figure 11.9. This causes an interrupt, and the handler is entered. This handler outputs an element from the vector. To output the vector, execute

```
char buffer[0x80], i = 0x80;
void main() {
    DDRA = 0xFF;  TSCR = 0x80; TFLG1 = TMSK1 = 1; TCTLE = 2; asm CLI
    do ; while( i );  /* wait until all are output */ TMSK1 = 0; /* disable interrupt */
}
void interrupt 8 hndlr(void){TFLG1=1;PORTA=buffer[--i]; /* output data */}
```

An optimized assembly-language program segment for the body of the C procedure is shown below. For it and the subsequent handler program segment, we assume there is a global variable I that is initialized to $80, the number of elements in the buffer, and there is an $80-element buffer.

```
        LDAB  #$80      ; Put one in leftmost bit of Accumulator B
        STAB  $86       ; Write it into control register TSCR
        LDD   #$102     ; Generate one in Accumulator A, 2 in Accumulator B
        STAA  $8C       ; Write one in rightmost bit of control register TMSK1
        STAA  $8E       ; Write one in rightmost bit of control register TFLG1
        STAB  $8B       ; Write 2 in rightmost bit of control register TCTLE
        LDD   #$FF80    ; Generate two constants in Accumulators A and B
        STAA  $2        ; Make PORTA output
        CLI             ; Enable Interrupts
LOOP:   LDAB  i         ; Get index
        BNE   LOOP      ; Wait until all vector elements are output
        CLR   $8C       ; Disable interrupt
```

An optimized assembly-language program segment for the handler is

```
CCT0:   MOVB  #1,TFLC1  ; Write one in rightmost bit of control register TFLG1
        LDX   #BUFFER   ; Get the address of the buffer into X
        DEC   i         ; Decrement index
        LDAB  i         ; Get index
        LDAA  B,X       ; Read a byte from the buffer
        STAA  $0        ; Output the data
        RTI             ; Return to interrupted routine
```

This program is quite like the gadfly program in Section 11.5. However, the loop `do ; while( i );` can be replaced by a program that does useful work. Interrupts provide output whenever the device needs it, without wasting time in a delay or gadfly loop.

This raises an often-misunderstood point about interrupts. Gadfly has lower latency than interrupt synchronization. Gadfly does not have to save the registers and then initialize registers used in the handler. If, when using interrupt synchronization, you just waste time in a gadfly or wait loop, use gadfly synchronization to lower latency.

The 6812 has a number of I/O devices, and most of them have their own interrupt vector, shown in Table 11.1. The handler used in the previous example has its vector at $FFEE and $FFEF, for "Timer channel 0." When an edge occured on `PORTT` bit 0, the contents of this vector, the 16-bit data in $FFEE (high byte) and $FFEF (low byte), are put into the program counter, and the interrupt handler is then executed starting at this address. Other interrupt vectors of interest in this introductory discussion are the reset, SWI, and TRAP vectors. The reset vector at $FFFE and $FFFF is the address of the first instruction that is executed when a 6812 comes out of reset. Locations $FFF6 and $FFF7 usually contain the address of the monitor program, which is where you go when an SWI instruction is executed, or a BGND instruction is executed, but the background debug module is not set up to handle a monitor program. Locations $FFF8 and $FFF9 usually contain the address of the handler for illegal instructions, which can be used to emulate instructions not implemented in the 6812.

Table 11.1. Interrupt Vectors in the 6812 'A4

Interrupt Vector	Interrupt #	Name
$FFCE,CF	24	Key Wakeup H
$FFD0, D1	23	Key Wakeup J
$FFD2,D3	22	AtoD
$FFD4,D5	21	SCI1
$FFD6,D7	20	SCI0
$FFD8,D9	19	SPI Serial Transfer Complete
$FFDA,DB	18	Pulse Accumulator Input Edge
$FFDC,DD	17	Pulse Accumulator Overflow
$FFDE,DF	16	Timer Overflow
$FFE0,E1	15	Timer Channel 7
$FFE2,E3	14	Timer Channel 6
$FFE4,E5	13	Timer Channel 5
$FFE6,E7	12	Timer Channel 4
$FFE8,E9	11	Timer Channel 3
$FFEA,EB	10	Timer Channel 2
$FFEC,ED	9	Timer Channel 1
$FFEE,EF	8	Timer Channel 0
$FFF0,F1	7	Real Time Interrupt
$FFF2,F3	6	IRQ or Key Wakeup D
$FFF4,F5	5	XIRQ
$FFF6,F7	4	SWI
$FFF8,F9	3	Unimplemented Instruction
$FFFA,FB	2	COP Failure
$FFFC,FD	1	Clock Failure
$FFFE,FF	0	Reset

This section has introduced the interrupt and its implementation on the 6812. You have learned how interrupts work in hardware and how an assembly-language program can be written to handle the interrupt from the timer module 0, in particular for rising or falling edges on `PORTT` bit 0.

This section, together with the last section, has shown two commonly used alternative methods for synchronization. Gadfly synchronization actually provides lower latency than interrupt synchronization (that is, faster response to an edge signal), while interrupts let you do other work while waiting for an edge from an I/O device. Upon completion of these two sections, you should find either technique easy to use.

11.7 Analog-to-Digital and Digital-to-Analog Conversion

Throughout electrical engineering, microcontrollers interface with analog systems in which a voltage level represents a property like pressure or speed. The digital microcontroller uses an *analog-to-digital converter* (A-to-D) to convert analog voltages to digital numbers that it can process, and it uses a *digital-to-analog converter* (D-to-A)

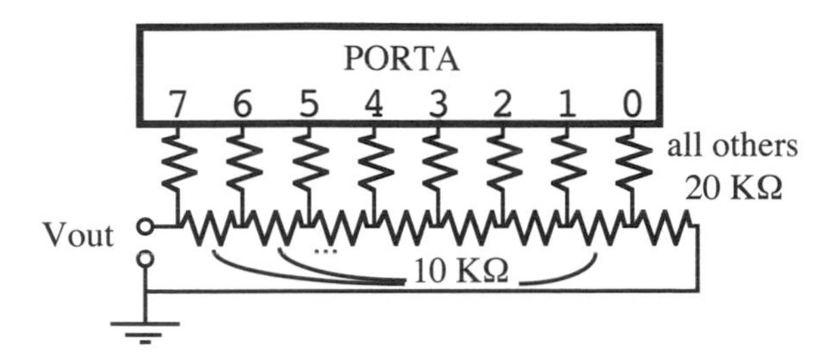

Figure 11.10. D-to-A Converter

to convert digital numbers that it has processed into analog voltages. This section illustrates A-to-D and D-to-A devices usable with the 'A4 and 'B32.

Both the 'A4 and 'B32 have built-in A-to-D converters. They do not have any built-in D-to-A converters, however, but this gives us an opportunity to build a D-to-A converter using a parallel port in order to better explain how it works. An A-to-D converter generally has within it a D-to-A converter. When we use the built-in A-to-D converter, we will refer to the D-to-A converter that we will build into a parallel port.

Figure 11.10 illustrates how an 8-bit *r-2r ladder* D-to-A converter can be implemented using Port A. The seven resistors on the bottom left of the figure have resistance r (10 KΩ), and all other resistors have resistance 2r (20 KΩ).

If the port bit output resistance is small compared to r (10 KΩ) and the microcontroller's supply voltage is 5.00 volts, then Vout is 5 * D / 256, where D is the binary number in port A. The reader can verify this using analog system analysis, which is simple enough but is outside the scope of this course.

In reality, the microcontroller's output port has significant resistance and is nonlinear; the input resistance of the Vout measuring instrument loads down this circuit, and the microcontroller's supply voltage is not 5.00 volts and has noise on it. So this is not a good D-to-A converter. An integrated circuit, such as a DAC-08, and an OP amp are used to implement an 8-bit r-2r ladder D-to-A converter. However, the latter does essentially the same thing as Figure 11.10. Furthermore we will find it instructive to measure the accuracy of the circuit shown in Figure 11.10 at the end of this section.

An A-to-D converter basically consists of a D-to-A converter and an *analog comparator*. The latter has two inputs and outputs a high signal only when one input is greater than the other input. By outputting different analog voltages and comparing these voltages to the input voltage, the input voltage can be determined.

The 'A4 and 'B32 both have an on-board A-to-D converter connected to input port `PORTAD`. Figure 11.11 shows the block diagram of this subsystem. Using voltages on pin VRH as high- and pin VRL as low-reference voltages, the analog voltage on inputs PAD0 can be converted to 8-bit digital values and put into registers `ADR0`. Initially, control bit `ADPU` must be set to apply power to this subsystem (100 μs are then needed for the voltages to become stable). The conversion is begun when control register `ATDCTL5` is written with a value 0. While `ATDSTAT0` bit 7 is 0, conversion is being done. Although four conversions are done, we will only use one result, which is read from port `ADR0` at location 0x70.

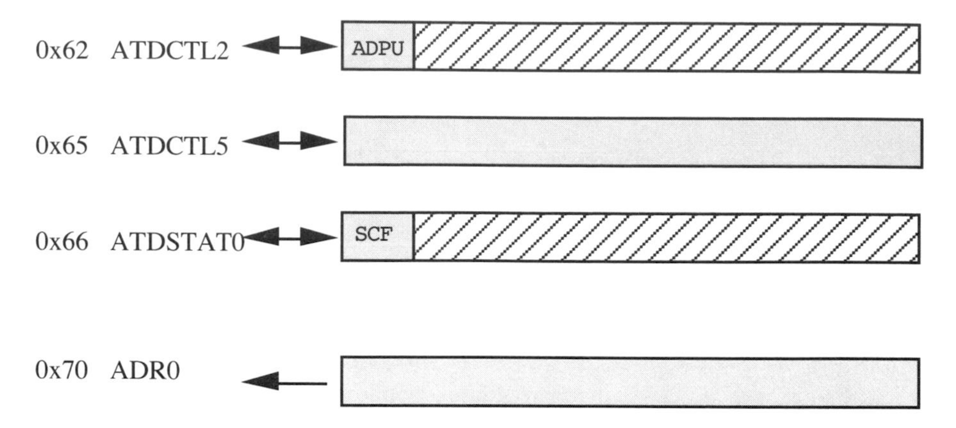

Figure 11.11. A–D Subsystem of the 'A4 or 'B32

Assuming VRH is 5 volts and VRL is ground, the following procedure converts the voltage (times 256/5) on `PORTAD` bit 0 into *i*:

```
main() { unsigned char i;
   ATDCTL2 = 0x80; for(i=1;i!=0;i++); /* wait for voltages to stabilize */
   ATDCTL5 = 0;  do ; while(!(ATDSTAT0 & 0x80)) ; i = ADR0;
}
```

An optimized assembly-language program segment for the body of the C procedure is shown below:

```
     LDD   #$80           ; Put one in leftmost bit of Accumulator B
     STAB  $62            ; Write $80 into control register ATDCTL2
     LDY   #800/3         ; Allow 100 μsec to elapse
L1:  DBNE  Y,L1           ; Wait for A-to-D device voltages to stabilize
     STAA  $65            ; Put zero in control register ATDCTL5 to start conversion
L2:  BRCLR $66,#$80,L2    ; Wait for one in bit SCF of the ATDSTAT0 port
     LDAB  $70            ; Get the converted value
```

If we connect the Vout from Figure 11.10 to PORTAD bit 0, the following C program will convert values from 0 to $FF and accumulate the largest error above and below the correct value:

```
unsigned char above, below;
main() { unsigned char i, j; DDRA = 0xff;
   ATDCTL2 = 0x80; for(i=1; i!=0; i++); /* wait for voltages to stabilize */
   for(i = 0; i != 0xff; i++){
      PORTA = i; ATDCTL5 = 0;  do ; while(!(ATDSTAT0 & 0x80)) ;
      if((ADR0 > i) && (( j = (ADR0 - i ) ) > above )) above = j;
      if((ADR0 < i) && (( j = ( i - ADR0) ) > below )) below = j;
   }
}
```

Writing an optimized assembly-language program for this C procedure is left as an exercise for the reader. But observe two aspects of this technique. The initialization of the A-to-D device, which is required only before the first time you use the device, writes a value in a device's control registers and waits 100 μsec for voltages to stabilize. Each time a conversion is needed, we write a value into the control register, wait in a gadfly loop until conversion is completed, and read the value from the ADR0 port.

We conducted an experiment that tied the output of Figure 11.10 to the AD port bit 0 input and ran the program shown above. The difference between the byte output from PORTA and the value read from port ADR0 was 4. Because the A-to-D converter is well-designed, we assume the D-to-A converter had an accuracy of a little less than 1%.

11.8 UART Protocol

The Universal Asynchronous Receiver-Transmitter (UART) is a module (integrated circuit) that supports a frame protocol to send up to eight-bit frames (characters). We call this the *UART protocol*. The UART frame format is shown in Figure 11.12. When a frame is not being sent, the signal is high. When a signal is to be sent, a *start bit*, which is a low, is sent for one bit time. The frame, from five to eight bits long, is then sent one bit per bit time, least-significant bit first. A parity bit may then be sent and may be generated so that the parity of the whole frame is always even (or always odd). To generate even parity, if the frame itself had an even number of ones already, a low parity bit is sent; otherwise a high bit is sent. Finally, one or more *stop bits* are sent. A stop bit is high and is indistinguishable from the high signal that is sent when no frame is being transmitted. In other words, if the frame has n stop bits (n = 1, 1½, or 2) this means the next frame must wait that long after the last frame bit or parity bit of the previous message has been sent before it can begin sending its start bit. However, it can wait longer than that.

The 'A4 and 'B32 have a UART-like device called the *Serial Communication Interfaces* (SCI0). The SCI0 device shown in Figure 11.13 is often used by the debugger. When not so used, it can be available for serial communication in an experiment or project. We describe its data, baud rate generator, and control and status ports. Then we will show how the SCI can be used in a gadfly synchronization interface.

The SCI has, at the same port address, a pair of data registers that are connected to shift registers. Eight bits of the data written at `SC0DRL` (0xc7) are put into the shift register and shifted out, and eight bits of the data shifted into the receive shift register can be read at `SC0DRL` (0xc7). Observe that, though they are at the same address, they are different ports. Reading the address reads the input port; writing writes the output port.

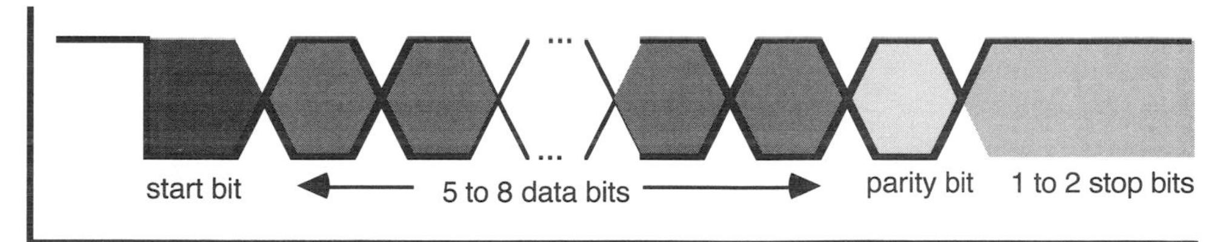

Figure 11.12. Frame Format for UART Signals

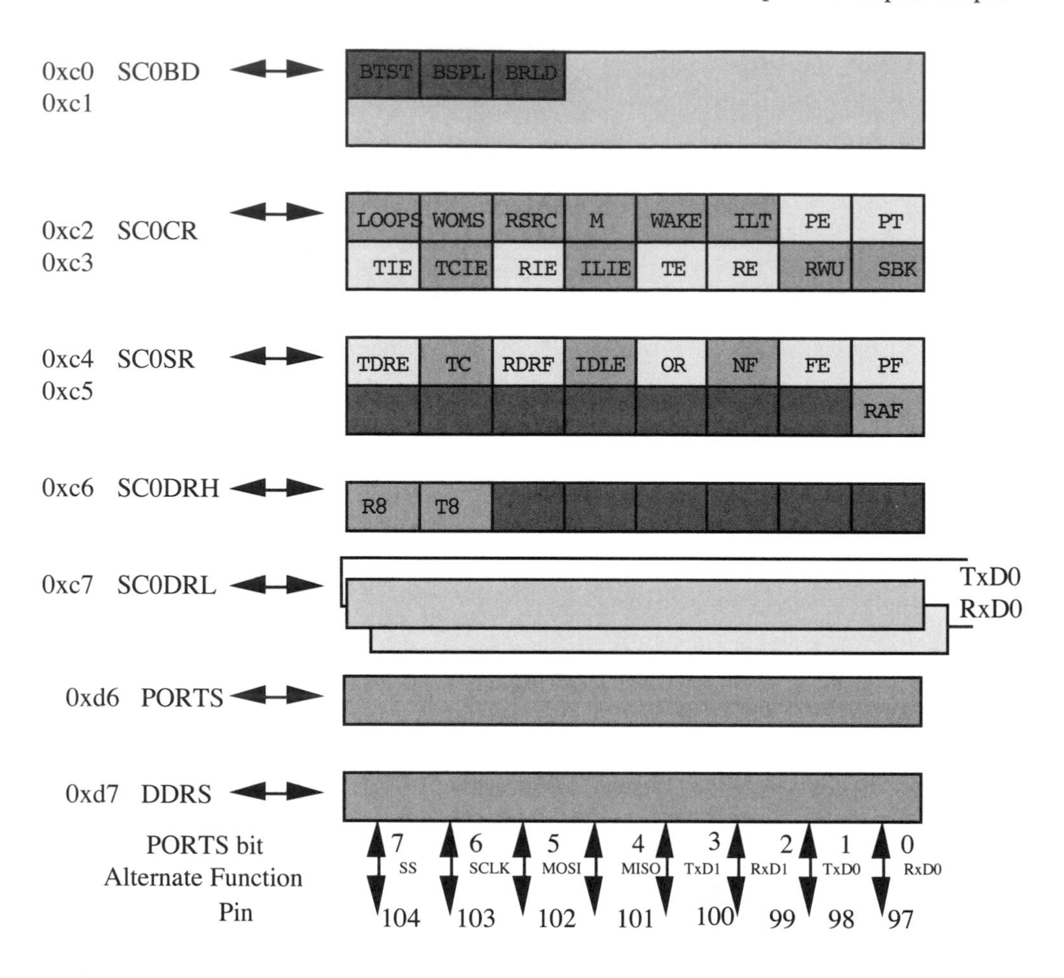

Figure 11.13. 6812 Serial Communication Interface

The clock rate is established by the 12-bit `SC0BD` port (0xc0). The number put in this port is the clock going to the SCI (Figure 11.13) divided by 16 times the desired baud rate. For example, to get 9600 baud, put 52 into the `SC0BD` port.

The 16-bit control port, `SC0CR`, at 0xc2 has parity enable `PE` and parity type `PT` to establish the parity, transmitter interrupt enable `TIE` and receiver interrupt enable `RIE` to enable interrupts, and transmitter enable `TE` and receiver enable `RE` to enable the device. The 16-bit status port at 0xc4 indicates what is happening in the transmitter and receiver. `TDRE` is T (1) if the transmit data register is empty; it is set when data are moved from the data register to the shift register and is cleared by a read of the status port followed by a write into the data port. The remaining status bits are for the receiver. `RDRF` is T (1) if the receive data register is full because a frame has been received. Receive error conditions are indicated by `OR`, set when the receiver overruns (that is, a word has to be moved from the input shift register before the previously input word is read from the data register) `FE`; T (1) if there is a framing error (that is, a stop bit is expected but the line is low; and `PE`, T (1) if there is a parity error.

Again, we offer a C program to send and receive serial data using gadfly synchronization; we leave the assembly-language program as an exercise for the reader. In *main*, the SCI is initialized for gadfly synchronization of 9600 baud and eight data bits without parity. Reading status and data registers twice clears the `RDRF` flag. The `put` procedure gadflies on transmitter data register empty (`TDRE`); when it is empty, `put` outputs its argument. The `get` procedure gadflies on the receive data register full (`RDRF`); when the receive register is full, `get` returns the data in data port `SC0DRL`. These procedures are the `get` and `put` procedures we have mentioned many times in this book:

```
enum{ PE=0x200, PT=0x100, TIE=0x80, RIE=0x20, TE=8, RE=4 };

enum{ TDRE = -32768, RDRF =0x2000, OR = 8, FE = 2, PF = 1 };

void main() { char i;
    SC0BD = 52; /*9600 baud*/ SC0CR = TE + RE; /* enable Xmt, Rcv devices */
    i=SC0SR; i=SC0DRL; i=SC0SR; i=SC0DRL;/* clear RDRF */
    put(0x55); i = get();
}

put(char d) { while( ( SC0SR & TDRE ) == 0) ; SC0DRL = d; }

char get() { while( ( SC0SR & RDRF ) == 0) ; return SC0DRL; }
```

An interrupt-based C program below will send a character vector and receive serial data into a buffer; we again leave the assembly-language program as an exercise for the reader. In `main`, the SCI is initialized for interrupt synchronization at 4800 baud and eight data bits without parity. Reading status and data registers twice clears the `RDRF` flag. The interrupt handler outputs a byte from the character string whenever a transmitter data register empty (`TDRE`) bit is set. The handler also moves a byte from the SCI0 data port whenever receive data register full (`RDRF`) bit is asserted. In this simplified example, we do not worry about overrunning the input buffer or output character string.

```
enum{ PE = 0x200, PT = 0x100, TIE = 0x80, RIE = 0x20, TE=8, RE=4 };

enum{ TDRE = -32768, RDRF =0x2000, OR = 8, FE = 2, PF = 1 };

char outString[] = "Hi there\r", inString[10], *in, *out;

void main() { char i;
    SC0BD = 104; /*4800 baud*/ SC0CR = TE + RE + RIE + TIE; // en. int.
    i=SC0SR; i=SC0DRL; i=SC0SR; i=SC0DRL;/* clear RDRF */
    in = inString; out = outString; asm CLI
    do ; while(1);
}

void interrupt 20 hndlr(void){
    if( SC0SR & TDRE ) SC0DRL = *out++;
    if( SC0SR & RDRF )  *in++ = SC0DRL;
}
```

11.9 Summary and Further Reading

This chapter has introduced you to input-output programming, a somewhat obscure area because many texts, magazine articles, and courses define input-output to be beyond their scope while they concentrate on some other topic. But it is not obscure. This chapter showed that the basic notion of input-output in a microcontroller is really a minor extension of reading and writing to memory. The input-output integrated circuit was studied. Synchronization mechanisms were discussed, and the gadfly and interrupt techniques were detailed. The `get` and `put` subroutines were then shown and their operation explained. You can now see how simple input-output programming can be. We then looked at simple examples of the use of the A-to-D and D-to-A converters and the SCI device used for serial I/O. We looked at the last example to understand the way that the `get` and `put` subroutines work, which have been mentioned many times earlier in this book. The method of outputting several words one after another to the same device was discussed. This method, using buffers, was seen to be very straightforward.

If input-output programming interests you, we recommend the following books. Of course, we recommend the accompanying textbook, *Single and Multiple Chip Microcontroller Interfacing for the Motorolla 68HC12* (G. Jack Lipovski). It emphasizes the software used to control devices, using the 6812 and chips in the 6800 family for concrete examples, experiments, and problems. Two books by P. Garrett, *Analog I/O Design* and *Analog Systems for Microprocessors and Minicomputers* (both published by Reston Publishing Co., Inc., Reston, Va., 1981 and 1978, respectively), give excellent discussions of operational amplifiers and filters used in input-output devices and also discuss the characteristics of transducers and measurement hardware.

Do You Know These Terms?

See the end of chapter 1 for instructions.

input routine
output routine
output port
output device
output line
input port
input device
input line
high signal
low signal
direction port
control port
gadfly synchronization
gadfly loop
status port
hardware interrupt
I/O interrupt
handling an interrupt
interrupt handler
device handler
interrupt service routine
latency time
return from interrupt
device requests an interrupt
interrupts are enabled
microprocessor is enabled
microprocessor is masked
microprocessor is disabled
microprocessor sees a request
request is pending
honors an interrupt
handler
arm
armed
disarmed
enable
enabled
disabled
analog-to-digital converter
digital-to-analog converter
r-2r ladder
analog comparator
UART protocol
start bit
stop bit
Serial Communication Interface

PROBLEMS

1. The least significant bit of a port, as shown in Figure 11.4, is connected as input to a push-button switch. The signal from the switch is high if the button is not pressed and low if it is pressed. However, when it is pressed or released, it "bounces," that is, it changes value rapidly from high to low to high to low . . . for about 10 milliseconds. Write a program segment that will loop until the button is down for at least 5 milliseconds without the signal changing value. (This is called a debouncer.)

2. Draw a diagram similar to that shown in Figure 11.4 in which three switches, normally open, are connected in parallel, such that PORT A bit 0 is normally high, becoming low when any switch is closed. Comment on why the circuit in Figure 11.4 is better than this circuit for a security system. (*Hint*: Consider ways to thwart the alarm.)

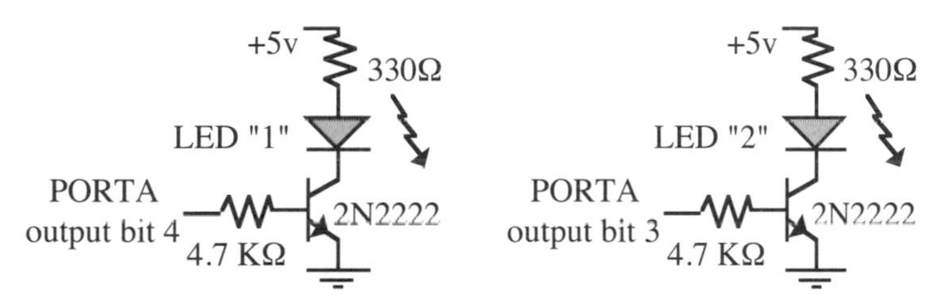

Figure 11.14. Parallel Output Port of the 'A4 or 'B32

3. Write a shortest program to flash the two LEDs in Figure 11.14. Only one LED should be lit at a time, and each LED should be lit for 1 second. Initially, the LED connected to PORTA bit 4 should be lit for 1 second. Then the LED connected to PORTA bit 3 should be lit for 1 second, and so on. This cycle should be repeated indefinitely.

4. Write a shortest program to use the three switches in Figure 11.15 in a game. Initially all switches are open. When the first switch is closed, print its switch number on the terminal using the OUTCH subroutine.

5. Write a shortest program to record the sequence of pressing the three switches in Figure 11.15 ten times. Only one switch should be pressed at a time. When it is pressed, it is closed multiple times, but within 5 milliseconds. A 10-element, 8-bit element vector INPUTS is to store input key strokes. Each time any switch is pressed, the number of the switch (1 on the left, 2 in the middle, and 3 on the right) is to be put into the next vector element.

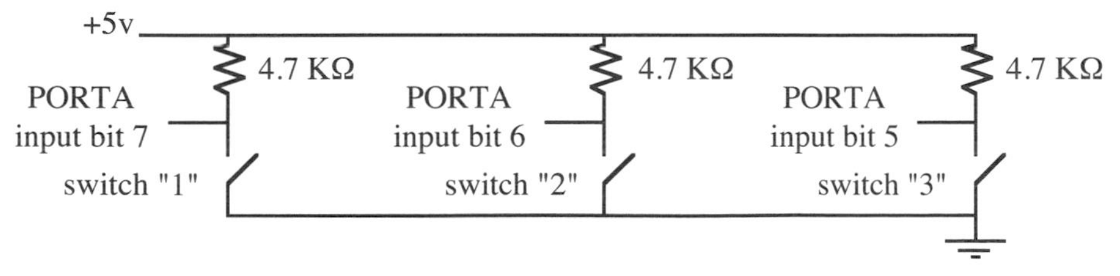

Figure 11.15. Parallel Input Port of the 'A4 or 'B32

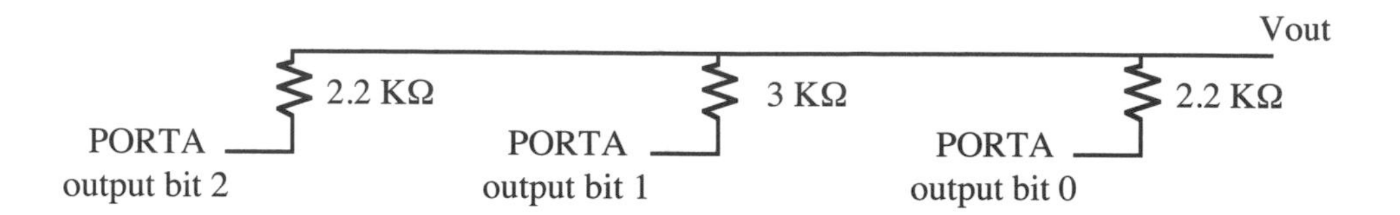

Figure 11.16. Parallel D-to-A Output Port of the 'A4 or 'B32

6. Write a keyless entry system, using the output port shown in Figure 11.14 and the input port shown in Figure 11.15. When the keys 1, 3, and 2 are pressed, LED 1 lights up; otherwise, after three keys are pressed that are not in the sequence, LED 2 lights.

7. Write a shortest program to use the switches to play the first four notes of the tune, "The Eyes of Texas," using a parallel port to output a staircase approximation to a sine wave. The first four notes are C F C G. A Johnson counter is simulated in software using output port A bits 2 to 0, essentially as shown in Figure 11.16. A 4-bit Johnson counter has sequence of bit values from Figure 11.17, where H is high (1) and L is low (0). The left three bits of this bit vector are connected to output port A bits 2 to 0 respectively, as shown in Figure 11.16. The note C has a frequency of 523 Hz or a period of 1912 μsec and is to be played when the left push-button switch is closed; the note F has a frequency of 698 Hz or a period of 1432 μsec and is to be played when the middle push-button switch is closed; and the note G has a frequency of 783 Hz or a period of 1277 μsec and is to be played when the right push-button switch is closed. Write a shortest program to play the notes when the switches are pressed.

8. Draw the staircase approximation to a sine wave generated by the program in Problem 7 for a period of 1 millisecond, showing the voltage levels to two decimal places, assuming the port outputs 0 or 5 volts.

9. Write the contents of vector **BUFFER** so the model traffic light in Figure 11.5 will have the north-south red LED on and east-west green LED on for 10 seconds, the north-south red LED on and east-west yellow LED on for 2 seconds, the north-south green LED on and east-west red LED on for 8 seconds, and the north-south yellow LED on and east-west red LED on for 2 seconds, using the program listed below that figure.

0 0 0 0
1 0 0 0
1 1 0 0
1 1 1 0
1 1 1 1
0 1 1 1
0 0 1 1
0 0 0 1

Figure 11.17. Johnson Counter Sequence

10. Write a shortest assembly-language program to output data from a buffer using autoincrement addressing to Port B each time Port A bit 7 rises from low to high. Instead of using `LDAA B,X STAA 1`, use `LDAA 1,X+ STAA 1` to transfer a byte from input to buffer. The program outputs ten bytes from vector `BUFFER` and exits.

11. Write a shortest assembly-language program to input data, from Port B each time Port A bit 7 rises from low to high, to a buffer using autoincrement addressing. Instead of using `LDAA 1 STAA B,X,` use `LDAA 1 STAA 1,X+` to transfer a byte from the input to the buffer. The program should store ten such bytes in vector BUFFER and then exit.

12. In Figure 11.7, if the two least significant bits of port `TCTLE`, which is the 16-bit port at $8a, are 10 instead of 01, the flag, which is the least significant bit of `TFLG1` and is the 8-bit port at $8e, is set on the rising edge. Write a shortest program segment to initialize the device and a shortest program segment to gadfly until a rising edge occurs on `PORTT` bit 0.

13. In Figure 11.7, if the two least significant bits of port `TCTLE`, which is the 16-bit port at $8a, are 11 instead of 01, the flag, which is the least significant bit of `TFLG1` and is the 8-bit port at $8e, is set on both the rising and falling edge. Write a shortest program segment to initialize the device and a shortest program segment to gadfly until an edge of either type occurs on `PORTT` bit 0.

14. Write a shortest program to initialize the devices and input a byte from `PORTA` to the 10-element `BUFFER` in an interrupt handler each time a rising edge occurs on `PORTT` bit 0.

15. Write a shortest program to initialize the devices and input a byte from `PORTA` to the 10-element `BUFFER` in an interrupt handler each time a rising or a falling edge occurs on `PORTT` bit 0.

16. Explain why interrupts are useful when unexpected requests are made from a device but are actually slower than gadfly routines when expected requests are made from a device.

17. Give two concrete examples of devices that will require each of the following synchronization mechanisms, so that they should only use that mechanism and no other, and give reasons for your choice: real-time, gadfly, and interrupt. For example, a microcontroller in an electric stapler, that generates a pulse to engage a solenoid, should use real-time synchronization, because the microcontroller is doing nothing else, and this is the least costly approach, requiring minimal hardware.

18. An A-to-D converter consists of (1) a D-to-A converter that outputs data through `PORTA` in Figure 11.10, which is converted to an analog voltage Vref, and (2) an analog comparator that compares Vref to an input voltage Vin, inputting a high (1) in `PORTB` bit 7 if Vref < Vin, otherwise inputting a low (0). Write a shortest subroutine `RAMP` that begins outputting Vref = 0, increments Vref until it just exceeds Vin, and returns the byte it last output in `PORTA`. This is called a ramp converter.

19. An A-to-D converter is connected as in Problem 18. Write the shortest subroutine `SUCCESS` that begins outputting Vref = 2.5 and recursively determines whether Vin is in the upper or lower half of the range, reducing the range by one half each time it determines which half of the range Vin is in. SUCCESS returns the final value output on `PORTA`. This divide-and-conquer algorithm is like binary number division; it is called successive approximation conversion.

20. For the compiled C procedure in Figure 11.18, encircle and label the following C statements and assembly-language program segments: (1) declaration of *global* variables; (2) declaration of *I/O* ports; (3) declaration, allocation, and deallocation of *local* variables; (4) the *initialization* ritual; (5) *for loop* control statements to output *str;* (6) *while loop* control statements to input until a carriage return; (7) *gadfly* loops; (8) *input* operation; and (9) *output* operation. Next to each circle, write the label shown above in *italics* or a number enclosed in parentheses () to identify which part the code corresponds to. If one of the above appears in two parts, circle each part.

21. For the compiled C procedure in Figure 11.19, encircle and label the following C and assembler program segments: (1) declaration of *global* variables; (2) declaration of *I/O* ports; (3) declaration of *local* variables; (4) the *initialization* ritual; (5) the *loop* control statements to try each value from 1 to 128; (6) the *gadfly* loop; (7) the *input* operation; (8) the *output* operation; (9) *calculation* and collection of results; and (10) an infinite loop to *stop* the program. Next to each circle, write the label shown above in *italics* or a number enclosed in parentheses () to identify which part the code corresponds to.

```
* volatile int SC1BD@0xC8;volatile char SC1CR@0xCB,SC1SR@0xCC,SC1DR@0xCF;
SC1BD:      EQU    $C8
SC1CR:      EQU    $CB
SC1SR:      EQU    $CC
SC1DR:      EQU    $CF
            ORG    $800

* char j; const char str[10] = "Hi There\r";
j:          DS.B   1
str:        DC.B "Hi There\r"

* void put(char d) { while( ( SC1SR & 0x80 ) == 0) ; SC1DR = d; }

put:        BRCLR  SC1SR,$80,put
            STAA   SC1DR
            RTS

* char get() { while( ( SC1SR & 0x20 ) == 0) ; return SC1DR; }
```

```
get:        BRCLR  SC1SR,$20,get
            LDAB   SC1DR
            RTS

* void main() { char i;

            PSHA

* SC1BD = 52; /*9600 baud*/ SC1CR = 0xC; /* enable Xmt, Rcv devices */

            MOVW  #52,SC1BD
            MOVB  #$C,SC1CR

* i = SC1SR; i = SC1DR; i = SC1SR; i = SC1DR; /* clear RDRF */

            LDAB  SC1SR
            LDAB  SC1DR
            LDAB  SC1SR
            LDAB  SC1DR

* for(i = 0; i < 9; i++) put(str[i]);

            CLR   0,SP
            LDX   #str
l1:         LDAB  B,X
            BSR   put
            INC   0,SP
            LDAB  0,SP
            CMPB  #9
            BLT   l1

* do j = get(); while( j != '\r');

l2:         BSR   get
            STAB  j
            CMPB  #13
            BNE   l2
            PULA
            RTS
```

Figure 11.18. Program with Disassembly

```
unsigned char hi = 0, lo = 0;
HI:           DC.B    0
LO:           DC.B    0

unsigned char
    PORTB@1,DDRB@3,ATDCTL2@0x62,ATDCTL5@0x65,ATDSTAT0@0x66, ADR3@0x76;

PORTB:        EQU     1
DDRB:         EQU     3
ATDCTL2:      EQU     $62
ATDCTL5:      EQU     $65
ATDSTAT0:     EQU     $66
ADR3:         EQU     $76

void main() { unsigned char i, j, k; DDRB = 0xff;

              ORG     $850
MAIN:         LDD     #$FF80
              STAA    DDRB                ;DDRB=0XFF

        ATDCTL2 = 0x80; for(j = 1; j != 0x80; j++) ;

              STAB    ATDCTL2             ;ATDCTL2=0x80
              LDY     #$800               ;for(j=1; j!=0x80; j++)
LOOP1:        DBNE    Y, LOOP1

        for(i = 0; i != 0xff; i++){ PORTB = i; ATDCTL5 = 3;

              LDD     #03
LOOP2:        STAA    PORTB               ;PORTB = i
              STAB    ATDCTL5             ;ATDCTL5 = 3

              do ; while(!(ATDSTAT0 & 0x80)) ; k = ADR3;

              PSHA
LOOP3:        BRCLR   ATDSTAT0, #$80, LOOP3
;do-while(!ATDSTAT0 & 0x80)

              k = ADR3;

              LDAB    ADR3                ;k = ADR3

              if((k > i) && (( j = (k - i ) ) > hi )) hi = j;
```

```
            SUBB    0,SP            ;j=k-i
            BEQ     LOOP4
            BLO     LOOP5
            CMPB    HI              ;j>hi?
            BLS     LOOP4
            STAB    HI              ;hi=j
            BRA     LOOP4

            if((k < i) && (( j = ( i - k) ) > lo )) lo = j;

LOOP5:      NEGB                    ;j=i-k
            CMPB    LO
            BLS     LOOP4           ;j>lo?
            STAB    LO

            for(i = 0; i != 0xff; i++){

LOOP4:      PULA
            INCA                    ;i++
            CMPA    #$FF            ;j!=0x
            BNE     LOOP2
LOOP6:      BRA     LOOP6           ;do-while(1)
```

Figure 11.19. Another Program with Disassembly

This board from Axiom Manufacturing has an MC68HC912B32, 64K external static RAM and AX-BDM-12 Debug 12 debugger. It is a full-function platform for developing products using the MC68HC912B32.

12

Other Microcontrollers

The microcomputer is a powerful tool, as we have learned in the preceding chapters. But the microcomputer is more than just one type of computer. There is a wide variety of microcomputers with different capacities and features that make them suitable for different applications. This chapter gives you some idea of this variety and the applications for which particular microcomputers are useful. To keep our discussion within the scope of this book, we examine microcomputers that are related to the 6812. This is particularly convenient for us since the 6812 is in the middle of the Motorola family of microcomputers, so that being thoroughly familiar with the 6812 makes it fairly easy to learn the other microcomputers in this family. This discussion of Motorola microcomputers will also help you with microcomputers designed by other companies.

This chapter has two themes. The first consists of a discussion of Motorola microcomputers that are simpler than the 6812, which include the 6811, the 6808, and the 6805. The second theme is an overview of the 68300, 500, and M·CORE series, machines more powerful than the 6812. After you read this chapter, you should be able to write simple programs for the 6811, 6808, and 6805 and their variants. You should be able to answer such questions as: Where should a 32-bit, 16-bit, or 8-bit microcomputer be used? You should be able to approach a microcomputer designed along quite different lines than the Motorola family, with some idea of what to look for. You should appreciate the capacities of different microcomputers, and you should be able to pick a microcomputer that is suitable for a given application.

Although the 6805 came first historically, we will treat the 8-bit microcomputers in order of their similarity to the 6812. We begin with a discussion of the 6811 followed by the 6808 and 6805. The final sections cover the 68300, 500, and M·CORE series, and observe the suitability of each microcomputer for different applications.

12.1 The 6811

The 6811, the immediate predecessor of the 6812, was designed for a specific application, automotive control, and then made available for other applications. The 6812 is upward compatible to the 6811. The register set for the 6811 is identical to that of the 6812. The

```
* SUBROUTINE DOTPRD
* LOCAL VARIABLES
TERM:     EQU  0
* PARAMETERS
RADDR:    EQU  2
LOCV:     EQU  4
LOCW:     EQU  6
DOTPRD:   PSHX             ; Allocate locals
          TSX              ; SP+1 -> X (Note: SP -> first free byte)
          LDAA LOCV,X
          LDAB LOCW,X
          MUL
          STD  TERM,X      ; Copy first term to local variables
          LDAA LOCV+1,X
          LDAB LOCW+1,X
          MUL
          ADDD TERM,X      ; Dot product into D
          PULX             ; Deallocate local variables
          RTS
```

Figure 12.1. A 6811 Dot Product Subroutine

6811 lacks some of the instructions and many of the addressing modes of the 6812 (see Table 12.1). You will be concerned mostly with the absence of stack index addressing.

Before we examine the differences, we should emphasize the similarities. The instruction sets are so similar that many of the programs in earlier chapters can be used in the 6811. The example that follows, the subroutine DOTPRD of Chapter 6 with the parameters passed on the stack, shows how similar the 6811 is to the 6812.

In Figure 12.1, bold type shows differences from the 6812 example, which are in allocating and indexing the stack. The 6811 has immediate, page-zero, 16-bit direct, and inherent addressing, exactly as in the 6812. The relative address mode is available only as an 8-bit relative address in BRA type of instructions, which behave exactly like their counterparts in the 6812. Indirect, postincrement, predecrement, and program counter relative addressing modes are missing in LDAA and similar instructions, and the index mode is only available in the form where an unsigned 8-bit offset in the instruction is added to the 16-bit index register X or Y. This program had to use the X register to access the local variables and parameters on the stack because there is no indexing mode that uses the stack pointer SP. The stack pointer actually points to the first free byte below the top of the stack; the TSX instruction puts SP+1 in X, so the top of the stack is at 0,X. The LEAS instruction is absent from the 6811, so we allocate using PSHX, or DES, or temporarily putting SP into accumulator D and using SUBD; and we deallocate using PULX, or INS, or temporarily putting SP into accumulator D and using ADDD. Also, instructions that read, modify, and write the same word in memory, such as INC COUNT, may use only 16-bit direct or 8-bit unsigned offset indexed addressing. All 6811 address arithmetic is unsigned, so that if X contains $1000, then LDAA $FF,X loads accumulator A from locations $1000 to $10FF. Programs using the 16-bit index addressing mode of the 6812, such as LDAA $1000,X have to be modified too, because the 6811 has only an 8-bit offset. You often have to calculate the address explicitly as the effective address is calculated within the 6812 instruction, put this effective address in the X register, and use the instruction `LDAA 0,X.` Also, index arithmetic can be done in accumulator D; the 6811 has the instruction `XGDX` to move the result to and from X.

The 6812 `LBRA` and other 16-bit branch instructions, which simplify the writing of position-independent code, are missing in the 6811. Writing position-independent code is tedious. However, except for this capability, the 6811 can get the effect of long branch

Addressing Modes		Moves				
Implied	SWI	LDAA	STAA	TAB	PSHA	CLRA
Register	INCA	LDAB	STAB	TBA	PULA	CLRB
Immediate	LDAA #1	LDD	STD	TAP	PSHB	CLR
Page 0	LDAA ALPHA	LDX	STX	TPA	PULB	TSTA
Direct	LDAA ALPHA	LDY	STY	TSX	PSHX	TSTB
Index	LDAA 5,X	LDS	STS	TXS	PULX	TST
	LDAA 3,Y		XGDX	TSY	PSHY	
Page Relative	BRA ALPHA		XGDY	TYS	PULY	

Arithmetic Instructions	
ADDA, ADDB, ABA	INCA, INCB,INC
ADCA, ADCB	DECA, DECB,DEC
SUBA, SUBB, SBA	NEGA, NEGB,NEG
SBCA, SBCB	DAA, MUL, ABX, ABY
CMPA, CMPB, CBA	FDIV, IDIV

Logic Instructions	Edit Instructions
EORA, EORB	ASLA,ASLB,ASL
ORAA, ORAB	ASRA, ASRB, ASR
ANDA, ANDB	LSLA, LSLB, LSL
BITA, BITB	LSRA, LSRB, LSR
COMA, COMB, COM	ROLA, ROLB, ROL
SEC, SEI, SEV	RORA, RORB, ROR
CLC, CLI, CLV	LSLD,ASLD,LSRD
BSET, BCLR	

Control Instructions
JMP , BRA , BRN, NOP, JSR, BSR, RTS, RTI, SWI, STOP, WAI
BEQ, BNE, BMI , BPL, BCS, BCC, BVS , BVC, BRSET, BRCLR
BGT, BGE, BEQ, BLE, BLT, BHI, BHS, BEQ, BLS , BLO

Table 12.1. Instruction Set and Addressing Modes of the 6811

instructions. For example, the 6812 instruction LBCC can be replaced by the 6811 sequence

```
BCS *+5
JMP L2
```

The 6812 extended arithmetic instructions _ EMUL, EMULS, EDIV, EDIVS, IDIVS, and EMACS _ are not 6811 instructions. These have to be implemented as subroutines. Other arithmetic instructions, MUL, FDIV, and IDIV, are quite a bit slower in the 6811. Fuzzy logic instructions MEM, REV, REVW, WAV, and similar instructions—ETBL, MAXA, MAXM, MINA, MINM, EMAXD, EMAXM, EMIND, and EMINM—are not 6811 instructions; they are implemented as subroutines or macros. 6812 control instructions—CALL, DBNE, DBEQ, IBNE, IBEQ, TBNE, and TBEQ—are not 6811 instructions, so a modified strategy is used to control loops and effect conditional expressions. Finally, the 6811 does not have the MOVB, MOVM, PSHC, PULC, and SEX instructions and does not have the full capabilities of TFR and EXG instructions.

With these modifications, you can rewrite a program written for the 6812. Try a few programs. Scan through the earlier chapters, and pick programs you have already written. Rewrite them for the 6811. It is not too hard. However, we caution you that each computer has its strong points, and writing a good program by adapting a program from another computer for a new computer does not take full advantage of the strong points of the latter. In the 6811, for instance, ABX is one byte, takes three cycles, and adds B as an unsigned 8-bit number; its equivalent in the 6812, LEAX B,X, is two bytes, takes two cycles, and adds B as a signed 8-bit number. The 6811 should use programs where an unsigned index is added to the X register, in preference to those programs where a signed index is added to X. You have to be more careful in the 6811 to organize your data to use only positive offsets from X. For instance, you may select to implement a second stack in the same direction as the hardware stack, to avoid using a negative offset to a stack that moves in opposite direction to the hardware stack that shares its buffer space.

Your 6811 programs can be tested using the debugger MCUez or HiWare, which can be implemented for the 6811 using a PC. We strongly recommend that you try your programs on such a system to get a feeling for the 6811.

12.2 The 6808

The 6808 is upward compatible from the 6805 (described in the next section), which it is intended to replace, while having the essential capabilities that are needed to implement code generated by a C compiler. Moreover, the 6808 is likely to replace the 6805 in newer designs and is more similar to the 6812, so we study it before the 6805. We will show how you might program these microcomputers by comparing them to the 6811 and 6812.

Maintaining compatibility will be discussed after the 6805 is presented. We focus on the differences between the 6812 and the 6808, so that you can adapt your 6812 programs to the 6808. Registers for the 6808 are like those in the 6811, but there is neither accumulator B nor index register Y (see Figure 12.2, and see its condition codes in Figure 12.3).

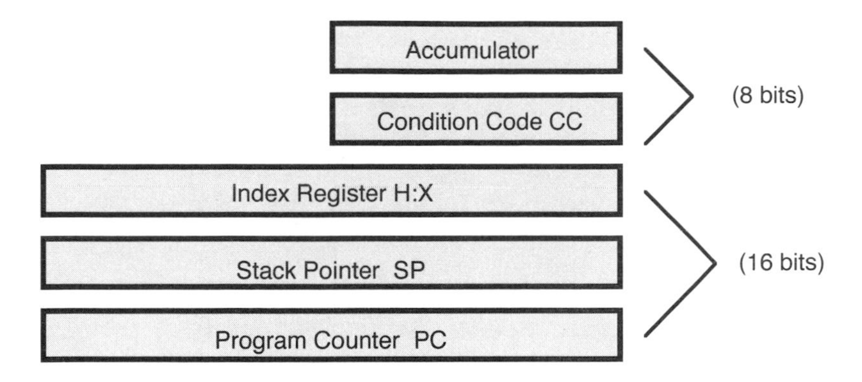

Figure 12.2. Registers of the 6808

The 6808 addressing modes, except indexed, are the same as those in the 6812. The 16-bit index register HX is used as a pointer register without offset, as an index register with an unsigned 8-bit offset, and as an index register with a 16-bit offset. The stack pointer SP is used as a 16-bit register with an unsigned 8-bit or 16-bit offset. The MOV (byte move) and `CBEQ` (compare and branch if equal) can use either pointer post-increment or post-increment with 8-bit unsigned offset.

The instructions dealing with accumulator D in the 6811 and 6812 generally can be replaced with those dealing with the 8-bit accumulator A in the 6808, at some loss in efficiency. The index register HX is treated as a pair of registers, register H and register X, just as accumulator D is accumulator A and accumulator B in the 6812. X can also be compared, incremented, decremented, shifted, or used in `MUL`, like an accumulator.

The instruction set of the 6808 and its addressing modes appear in Table 12.2. New instructions, not in the 6812 or 6811, are discussed below. These include `MOV`, `RSP`, `CPX`, `CPHX`, `AIX`, `AIS`, `NSA`, `CBEQ`, `DBNZ`, `BHCC`, `BHCS`, `BIH`, and `BIL`. Also the `BSET`, `BCLR`, `BRSET`, and `BRCLR` instructions use a bit position rather than a mask.

The 6808 `MOV` instruction is like the 6812 `MOVB` instruction but is restricted to a source that may be immediate, page-zero, or autoincrement addressed and a destination that is page-zero addressed or else a page zero source and destination that is autoincrement addressed. It is especially useful for I/O that are ports on page zero.

RSP, needed for upward compatibility to the 6805, writes $FF into register X but leaves register H unmodified. However, `TXS` and `TSX` move the index register HX to and from the stack pointer SP (`TSX` increments and `TXS` decrements the value moved), so the stack pointer can be set up in index register HX and transferred into SP. Because SP points to the first free word below the stack, byte 0,SP shouldn't be read or written.

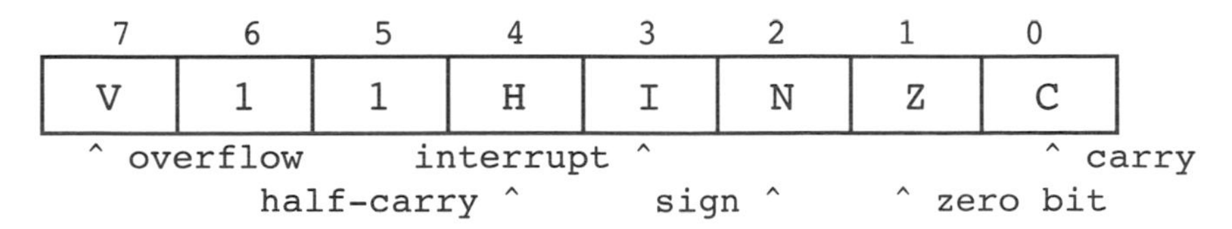

Figure 12.3. Bits in the 6808 Condition Code Register

Addressing Modes	
Implied	SWI
Register	INCA
Immediate	LDA #1
Page 0	LDA ALPHA
Direct	LDA ALPHA
Index	LDA ,X LDA 5,X LDA $1234,X
Page Relative	BRA ALPHA

Moves			
LDA	TSX	PSHA	CLRA
LDHX	TXS	PULA	CLRX
LDX	TAP	PSHH	CLR
STA	TPA	PULH	TSTA
STHX	TAX	PSHX	TSTX
STX	TXA	PULX	TST
MOV	RSP		

Arithmetic Instructions		
ADD, ADC	INCA, INCX, INC	AIS, AIX , CPHX
SUB, SBC	DECA, DECX, DEC	
CMP, CPX, CPHX	NEGA, NEGX, NEG	
CBEQ	DAA, MUL, DIV	

Logic Instructions	Edit Instructions
EOR, ORA, ANDA, BIT	ASLA,ASLX,ASL
COM, COMX, COM	ASRA, ASRX, ASR
SEC, SEI, CLC, CLI	LSLA, LSLX, LSL
BSET, BCLR	LSRA, LSRX, LSR
	ROLA, ROLX, ROL
	RORA, RORX, ROR, NSA

Control Instructions

Table 12.2. Instruction Set and Addressing Modes of the 6808

While CPX compares only the low byte of the index register, CPHX compares both bytes of register HX. AIX adds a signed 8-bit constant to register HX to access vectors. AIS does the same to the SP, to allocate and deallocate local variables.

The instruction NSA exchanges the high and low nibbles in the accumulator; it can be used to edit BCD numbers. DIV divides X into H (high byte) and A (low byte), putting the remainder in H and the quotient in A. The MUL instruction multiplies A by X, putting the 16-bit product in X:A.

The CBEQA and CBEQX instructions compare the accumulator or the X register to an immediate operand. The CBEQ instruction compares the accumulator to a byte addressed using page-zero, autoincrement, or stack pointer unsigned 8-bit offset index addressing. A special CBEQ instruction compares the accumulator to a byte addressed using an unsigned 8-bit offset index addressing using register HX and then increments HX. The DBNZA and DBNZX instructions decrement the accumulator or the X register and branch if nonzero. The DBNZ instruction decrements a byte addressed using page-zero or unsigned 8-bit offset index addressing using the index register HX, the stack pointer SP, or HX without an offset. `DBNZA` and `DBNZX` are similar to the 6812 `DBNE` using accumulator A, but `DBNZ` using a stack-indexed address is especially suited to using a C local variable for a loop counter. For instance, the following program clears a 16-byte vector whose base address is initially in HX; although the accumulator can be better used as a loop count, using a local variable shows this special 6808 mechanism:

```
        LDAA #$10        ; generate loop count
        PSHA             ; save it on the stack
LOOP:   CLR  ,X          ; clear word pointed to by HX (no offset)
        AIX  #1          ; increment pointer
        DBNZ 0,SP,LOOP   ; count down
        PULA             ; restore the stack
```

The branch instructions have a means to test the half-carry condition code bit H, which are `BCHS` and `BHCC`; a means to test the interrupt request pin IRQ, which are `BIH` and `BIL`; and the interrupt request mask, which are `BMC` and `BMS`. Finally, the `BSET`, `BCLR`, `BRSET`, and `BRCLR` instructions use a binary bit number to indicate which bit is set, cleared, or tested, rather than the bit mask used in the 6811 and 6812. For instance, the instruction `BSET 3,$10` will set bit 3 in word $10.

```
* SUBROUTINE DOTPRD
* LOCAL VARIABLES
TERM:    EQU  1            ; Note: location 0,SP is first free byte above stack
NBYTES:  EQU  2
* PARAMETERS
RADDR:   EQU  3            ; Return address
LOCV:    EQU  5            ; Vector V passed by value
LOCW:    EQU  7            ; Vector W passed by value
DOTPRD:  AIS  #-NBYTES     ; Allocation for local variables
         LDAA LOCV,SP      ; Get V(0)
         LDX  LOCW,SP      ; Get W(0)
         MUL
         STX  TERM,SP      ; High byte to local variable
         STAA TERM+1,SP    ; Low byte to local variable
         LDAA LOCV+1,SP    ; Get V(1)
         LDX  LOCW+1,SP    ; Get W(1)
         MUL
         ADD  TERM+1,SP    ; Add low byte
         STAA TERM+1,SP    ; Store low byte
         TXA               ; Get high byte
         ADC  TERM,SP      ; Add high byte
         TAX               ; Put high byte in X
         LDAA TERM+1,SP    ; Get low byte
         AIS  #NBYTES      ; Deallocate local variables
         RTS
```

Figure 12.4. A 6808 Dot Product Subroutine

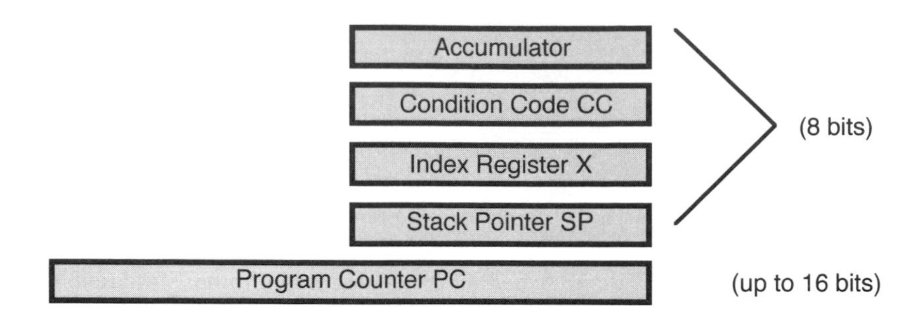

Figure 12.5. The Register Set of the 6805

Figure 12.4, the DOTPRD subroutine with stack parameters from Chapter 6, shows some similarity between the 6808 and the 6812. Differences, shown in bold type, are in allocating local variables and implementing 16-bit arithmetic.

With these "patches," you can refer to the *CPU08 Central Processor Unit Reference Manual* to learn how to program it without much difficulty, although you might have a certain amount of frustration after being accustomed to the 6812.

12.3 The 6805

The very inexpensive single-chip 6805 was designed for simple control applications that utilize bit manipulation and data structures but require simple instructions and only a few registers. Thus the 6805 has many addressing modes and bit manipulation instructions, even though it has few registers and few complex instructions.

Figure 12.5 shows the 6805 register set. Compared to the 6808, the stack pointer and index register are only eight bits long. Figure 12.6 shows the 6805 condition code register. Compared to the 6808, there is no V bit to indicate a two's-complement overflow. Finally, Table 12.3 displays the 6805 addressing modes and instruction set. Compared to the 6808, the main difference is there are no push or pull instructions or stack index addressing. The 6805 stack is designed to hold only a subroutine return address and hold the machine state during an interrupt. There is no effective way to pass parameters on the stack or allocate and access local variables on the stack. Also, because there is no V condition code bit, there are no conditional branches for signed arithmetic; and because there is no HX register, there are no instructions for that register. And 6808 special instructions, `DAA`, `MOV`, `NSA`, `DIV`, `DBNZ`, and `CBEQ` are missing.

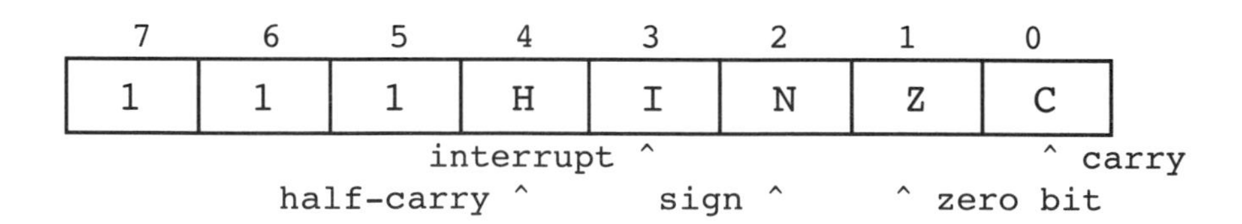

Figure 12.6. Bits in the 6805 Condition Code Register

Addressing Modes		Moves		
Implied	SWI	LDA	TAX	CLRA
Register	INCA	LDX	TXA	CLRX
Immediate	LDA #1	STA	RSP	CLR
Page 0	LDA ALPHA	STX		TSTA
Direct	LDA ALPHA			TSTX
Index	LDA X LDA 5,X LDA $1234,X			TST
Page Relative	BRA ALPHA			

Arithmetic Instructions	
ADD, ADC	INCA, INCX, INC
SUB, SBC	DECA, DECX, DEC
CMP, CPX	NEGA, NEGX, NEG
	MUL

Logic Instructions	Edit Instructions
EOR, ORA, AND, BIT	ASLA, ASLX, ASL
COMA COMX, COM	ASRA, ASRX, ASR
SEC, SEI, CLC, CLI	LSLA, LSLX, LSL
BSET, BCLR	LSRA, LSRX, LSR
	ROLA, ROLX, ROL
	RORA, RORX, ROR

Control Instructions
JMP , BRA , BRN, NOP, JSR, BSR, RTS, RTI, SWI, STOP, WAIT
BEQ, BNE, BMI, BPL, BCS, BCC, BHCS, BHCC, BVS , BVC, BIH, BIL
BHI, BHS, BLS , BLO, BMC, BMS, BRSET, BRCLR

Table 12.3. Instruction Set and Addressing Modes of the 6805

Figure 12.7 is a 6805 example of the subroutine DOTPRD in Chapter 6 that passes parameters as global variables, because the 6805 stack can't effectively pass parameters. Also, local variables are not allocated or deallocated, because the stack can't effectively hold local variables. Otherwise this subroutine is very similar to the 6808 DOTPRD.

Returning to the 6808, which is upward compatible to the 6805, the assembly language as well as machine code instructions that are in both will execute the same, except that the 6808 V condition code is modified as it is in the 6812. If the H register is not used, it is initialized to zero, and index addressing is the same in both machines. In fact, the H register is not stacked when SWI is executed or when an interrupt occurs. An interrupt handler has to save and restore H using an explicit PSHH or PULH instruction.

We focus here on programming this microcontroller, but we are intrigued by the 6805 hardware and spellbound by the possibilities of applying it. With this coverage and the information in a data sheet for the 6805, you should be able to write short programs for that very-low-cost microcomputer.

```
* SUBROUTINE DOTPRD
* GLOBAL VARIABLES USED IN SUBROUTINE AND AS PARAMETERS
TERM:    RMB  2
LOCV:    RMB  2
LOCW     RMB  2
LOCDP:   RMB  2
*
DOTPRD:  LDAA LOCV
         LDX  LOCW
         MUL
         STAA TERM+1        ; Copy first term low byte to local variables
         STX  TERM          ; Copy first term high byte to local variables
         LDAA LOCV+1
         LDX  LOCW+1
         MUL
         ADD  TERM+1        ; Add first term low byte to product
         STAA LDCDP+1       ; Copy first term low byte to out. param.
         TXA                ; Move high byte to accumulator
         ADC  TERM          ; Add first term high byte to product
         STAA LDCDP         ; Copy first term high byte to out. param.
         RTS
```

Figure 12.7. A 6805 Dot Product Subroutine

12.4 The 68300 Series

The preceding sections covered microcomputers that are less powerful than the 6812. We now present an overview of the 68300 series of microcomputers (the 68332, 68340) to convey an understanding of the strengths and weaknesses of these microcomputers in particular and of similar 16-bit microcomputers in general. The next section will similarly introduce the 500 and M·CORE series of RISC microcontrollers. However, in these two sections, we will at best be able to prepare you to write a few programs, similar to those written for the 6812, for these microcomputers. There is much more to these computers than we can discuss in the short section we can allot to each computer.

The register set for the 68300 series features seventeen 32-bit registers, a 32-bit program counter, and a 16-bit status register (see Figure 12.8). The eight data registers are functionally equivalent to the accumulators in the 6812, and the nine address registers are similar to the index registers.

The low byte of the status register is similar to the 6812 condition code register, having the familiar N, Z, V, and C condition code bits and a new condition code bit X, which is very similar to the carry bit C. Bits X and C differ in that C is changed by many instructions and is tested by conditional branch instructions, while X is changed only by a few arithmetic instructions and is used as the carry input to multiple-precision arithmetic operations. Having two carry bits, X and C, avoids some dilemmas in the design of the computer that are inherent in simpler computers such as the 6812. This

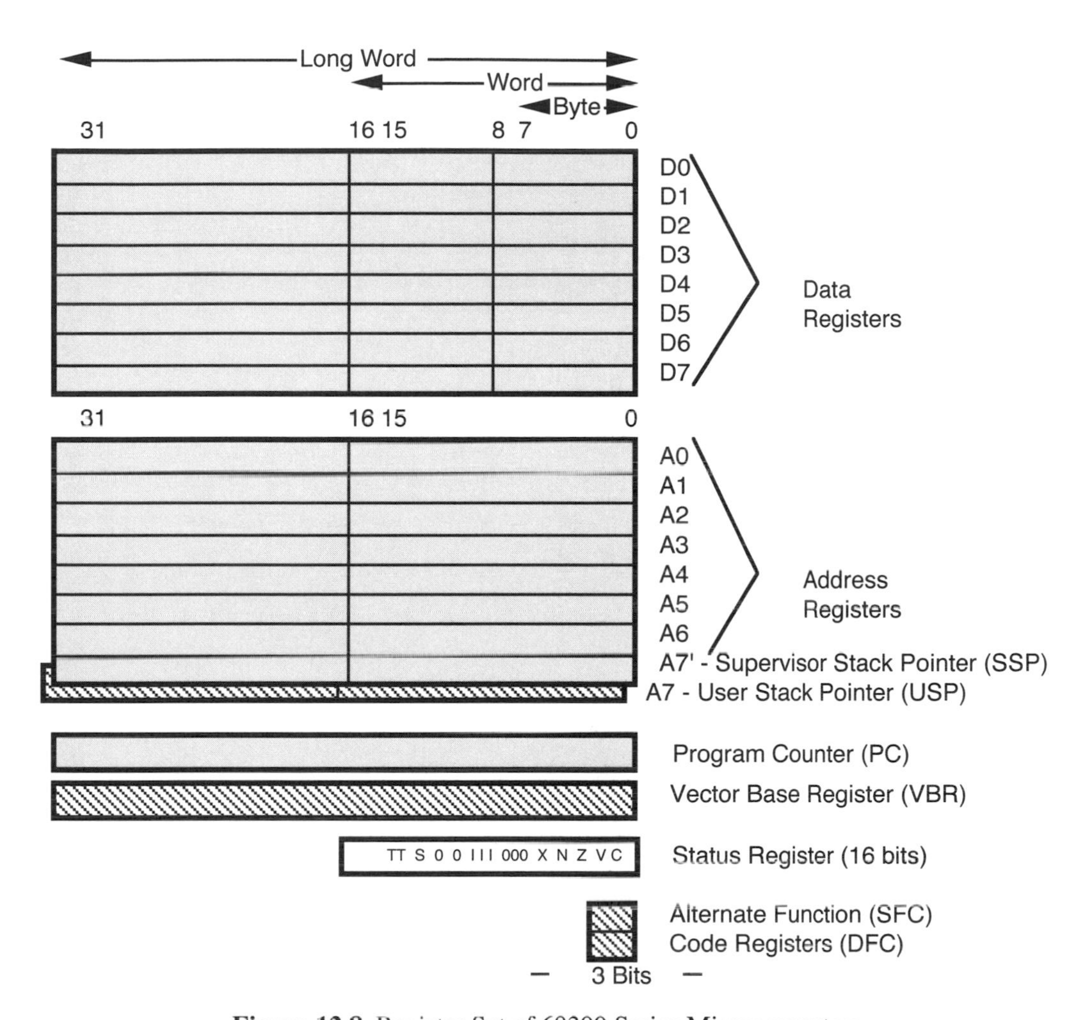

Figure 12.8. Register Set of 68300 Series Microcomputers

allows X to be set specifically for multiple-precision arithmetic and lets C be set for more instructions (such as `MOVE`) to facilitate testing using instructions (such as `BLS`). The high byte of the status register contains a bit, S, that distinguishes the mode as user or system. When it is set, the program uses the system stack pointer whenever it uses address register 7; and when it is clear, the program uses the user stack pointer whenever it uses address register 7. Further, several instructions can only be executed when the program is in the system mode (S = 1), and hardware can be built so that some memory or I/O devices may be accessed only when the program is in the system mode. This permits the writing of secure operating systems that can have multiple users in a time-sharing system, so that the users cannot accidentally or maliciously damage each other.

The 68300 series memory is organized as shown in Figure 12.9a. The 16-bit-wide memory is actually addressed as an 8-bit memory, so that a 16-bit word (the unit of memory read or written as a whole) is logically two consecutive locations. Instructions

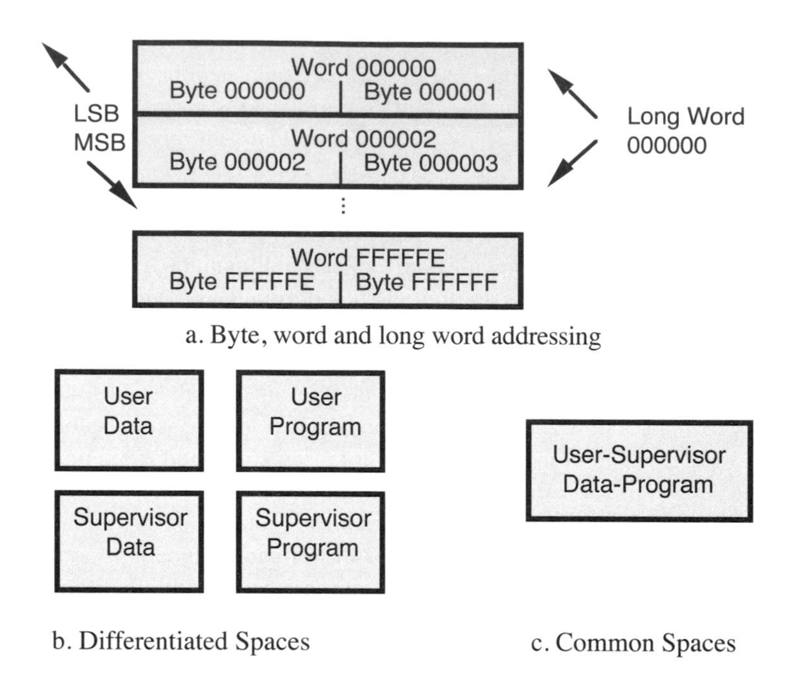

Figure 12.9. Memory Spaces in the 68300 Series

can read or write a byte (the mnemonic ends in `.B` for byte), a 16-bit word (these end in `.W` for word), or two consecutive words (these end in `.L` for long). If the suffix `.B`, `.W`, or `.L` is omitted, it is generally assumed to be a word (`.W`) instruction, unless such an option is not available. Word and long accesses must be aligned with memory so their addresses are even numbers; byte accesses using even addresses will read or write the high byte; and those with odd addresses will access the low byte of a word. This is consistent with the 6812 convention that puts the most significant byte at the lower-numbered address. Bits are numbered from right (0) to left (7) in a byte exactly as the 6812. If the hardware is so designed, access in the supervisor mode can access different memory than in the user mode, and fetching instructions can be done in different memories than reading or writing data, as shown in Figure 12.9b. Otherwise all memory can be the same regardless of whether it is accessed in supervisor mode or user mode, fetched from program space, or memorized or recalled from data space, as shown in Figure 12.9c.

The instruction set and the addressing modes are shown in Tables 12.4 to 12.6. You may observe the general `MOVE` instruction, which has variations for moving one byte (`MOVE.B`), a word (`MOVE.W`), or two words (`MOVE.L`). The source is always the first (left) operand, and the destination is the second (right) operand. Any addressing mode may be used with the source or destination. This general instruction is equivalent to the 6812 `LDAA`, `LDD`, `LDX`, and so on, and `STAA`, `STD`, `STX`, and so on, as well as the `TFR A,B` and `TFR D,X` instructions. It also includes a capability to move directly from memory to memory without storing the moved word in a register. In fact, there are over 12,000 different combinations of addressing modes that give different move instructions. There are similar byte, word, and long modes for arithmetic, logical, and edit instructions.

Mode (General)	Example
Long Direct	`MOVE 0x123456,D1`
Word Direct	`MOVE 0x1234,D1`
Implied	`RTS`
Register	`MOVE D0,D1`
Immediate	`MOVE # 0x1234,D1`
Pointer	`MOVE (A0),D1`
Autoincrement	`MOVE (A0)+,D1`
Autodecrement	`MOVE -(A0),D1`
Index	`MOVE (0x1234,A0),D1`
Double Indexed	`MOVE (0x12,A0,D0.W*2),D1`
"	`MOVE (0x1234,A0,D0.L*8),D1`
"	`MOVE (0x12345678,A0,D0.L*8),D1`
Short Relative	`BRA.S ALPHA`
Word Relative	`BRA.W ALPHA`
Long Relative	`BRA.L ALPHA`
Rel. Indexed	`MOVE 0x1234(PC),D1`
	`MOVE (0x12,PC,D0.L*8),D1`
	`MOVE (0x1234,PC,D0.L*8),D1`
	`MOVE (0x12345678,PC,D0.L*8),D1`

Table 12.4. Addressing Modes for the 68300 Series

The addressing modes listed in Table 12.4 can be used with almost all instructions. We will note those that are essentially the same as 6812 modes first, and then we will examine those that are significantly different.

Moving a byte, or word, into a 32-bit data register will result in replacing only the low byte, or word, in the register, leaving the other bits of the register untouched. Moving a word to an address register results in filling the high sixteen bits with the sign bit of the word that was moved, and moving a byte to or from an address register is not permitted. Immediate addressing can provide 16- or 32-bit operands. (Long) direct addressing uses a 32-bit address and can therefore address any word in memory. Pointer, postincrement, and predecrement are the same as in the 6812; a postincrement read of a word increments the pointer by 2, and a postincrement read of a long word increments the pointer by 4. Predecrement works similarly and write works similarly. Your experience with the 6812 should facilitate learning these modes.

The remaining modes consistently use sign extension to expand 8- or 16-bit instruction offsets to 32 bits before using them in address calculations. (Short) direct addressing is somewhat like direct page addressing, requiring a short 16-bit instruction offset; but, using sign extension, it can access locations 0 to \$7FFF and \$FFFF8000 to \$FFFFFFFF. Similarly, index addressing uses sign extension, so if A0 contains \$10000000, A0 index addressing accesses locations \$0FFF8000 to \$10007FFFF.

In Tables 12.5 and 12.6, Dn is any data register; An or Am is an address register; Xn is any An or Dn; nnnnnnnn is any 32-bit number; nnnn is any 16-bit number; nn is any 8-bit number; and n is a number 0 to 7. rrrr is any 16-bit offset for which PC+rrrr = L1; and rr is any 8-bit offset for which PC + rr = L1. Rc—SFC, DFC, VBR, or USP—

Instruction	Operands	Size
MOVE	<ea>,<ea> (see figure 1-7d)	LWB
	<ea>,SR/CCR or SR/CCR,<ea>	W
MOVEA	<ea>,An	LW > L
MOVE	USP,An or An,USP	L
MOVEQ	#<data>,Dn -128≤ data ≤ 127	B > L
MOVEC	Xc,Xn or Xn,Xc	L
MOVES	Rn,<ea> or <ea>,Xn	LWB
EXG	Xn,Xn	L
MOVEP	Dn,(d16,An)	LW
	(d16,An),Dn	LW
MOVEM	list,<ea. or <ea>,list	LW
LEA	<ea>,An	L
PEA	<ea>	L
CLR	<ea>	LWB
TST	<ea>	LWB
BTST	Dn,<ea> or #<data>,<ea>	LB
Scc	<ea>	B
ADD	<ea>,Dn or <ea>,An or Dn,<ea> or #data,<ea>	LWB
ADDA	<ea>,An	LW
ADDI	#<data>,<ea>	LWB
ADDQ	#data,<ea> 1 ≤ data ≤ 8	LWB
ADDX	Dn,Dn or -(An),-(An)	LWB
ABCD	Dn,Dn or -(An),-(An)	B
LEA	d(An),Am or d(PC),Am	L
SUB	<ea>,Dn or <ea>,An or Dn,<ea> or #data,<ea>	LWB
SUBA	<ea>,An	LW
SUBI	#<data>,<ea>	LWB
SUBQ	#data,<ea> 1 ≤ data ≤ 8	LWB
SUBX	Dn,Dn or -(An),-(An)	LWB
SBCD	Dn,Dn or -(An),-(An)	B
CMP	<ea1>,Dn or <ea>,An or #<data>,<ea>	LWB
CMPA	<ea>,An	LW
CMPI	#<data>,<ea>	LWB
CMPM	(An)+,(An)+	LWB
CHK	<ea>,Dn	LW
CHK2	<ea>,Xn	LWB
CMP2	<ea>,Xn	LWB
NEG	<ea>	LWB
NEGX	<ea>	LWB
NBCD	<ea>	B
DIVU/DIVS	<ea>,Dn	L/W > W:W
	<ea>,Dn	L/L > L
	<ea>,Dn	D/L > L:L
DIVSL	<ea>,Dr:Dq	L/L > L:L
MULU/MULS	<ea>,Dn	WxW > L
	<ea>,Dr:Dq	LxL >L
	<ea>,Dh:Dl	LxL > D

Table 12.5. Move and Arithmetic Instructions for the 68300 Series

Logical		
AND	<ea>,Dn or Dn,<ea>	LWB
ANDI	#<data>,<ea> or #<data>,CCR/SR	LWB
BCLR	Dn,<ea> or #<data>,<ea>	LB
OR	<ea>,Dn or Dn,<ea>	LWB
ORI	#<data>,<ea> or #<data>,CCR/SR	LWB
BSET	Dn,<ea> or #<data>,<ea>	LB
TAS	<ea>	B
EOR	<ea>,Dn or Dn,<ea>	LWB
EORI	#<data>,<ea> or #<data>,CCR/SR	LWB
BCHG	Dn,<ea> or #<data>,<ea>	LB
NOT	<ea>	LWB

Edit		
ASL	Dn,Dn or #data>,Dn	LWB
	<ea>	W
ASR	Dn,Dn or #data>,Dn	LWB
	<ea>	W
LSL	Dn,Dn or #data>,Dn	LWB
	<ea>	W
LSR	Dn,Dn or #data>,Dn	LWB
	<ea>	W
ROL	Dn,Dn or #data>,Dn	LWB
	<ea>	W
ROR	Dn,Dn or #data>,Dn	LWB
	<ea>	W
ROXL	Dn,Dn or #data>,Dn	LWB
	<ea>	W
ROXR	Dn,Dn or #data>,Dn	LWB
	<ea>	W
SWAP	Dn	L
EXT	Dn	B > W or W > L
EXTB	Dn	B > L

Control			
Conditional 2'S Complement	Conditionals (General)	Unconditional & Subroutine	Conditional Simple
BGT <label>	Bcc <label>	JMP <ea>	BEQ <label>
BGE <label>	DBcc Dn,<label>	BRA <label>	BNE <label>
BEQ <label>	TRAPV	BRN <label>	BMI <label>
BLE <label>	TRAPcc	NOP	BPL <label>
BLT <label>	TRAPcc #<data>	JSR <ea>	BCS <label>
Conditional Unsigned	Interrupt	BSR <label>	BCC <label>
BHI <label>	BGND	RTS	BVS <label>
BHS <label>	BKPT #<data>	RTR	BVC <label>
BEQ <label>	TRAP #<data>	RTD #<d>	
BLS <label>	RTE	LINK An,<d>	
BLO <label>	LPSTOP #<data>	UNLK An	
	RESET		

Table 12.6. Other Instructions for the 68300 Series

is a list of data and/or address registers; #<data> is an immediate operand; <ea> is an addressing mode; <label> is a label on a program statement; and cc is a condition code and value.

The 68300 effective address can be the sum of three values. The sum of a general register (which is any address or data register), an address register, and a signed 16-bit offset is used as the effective address in base index addressing.

Several special move instructions are provided. A `MOVE` instruction can move data to or from the status register (although the user can only access the low byte using `MOVE.B`), and the user stack pointer can be set while in the system state using a special `MOVE`. An `EXG` instruction permits exchanging the bits in the data or address registers. The instruction `MOVEM`, for move multiple, is a generalized `PSHX` or `PULX` instruction. Registers to be pushed or pulled are specified by separators "/," meaning AND, and "–," meaning TO. The instruction `MOVEM D0/D1/A0,-(A7)` pushes D0, Dl, and A0 onto the user's stack (or system stack if in the system mode). However, any address register may be used in lieu of A7, so that the user may create many stacks or queues and use this instruction with them. `MOVEA` is a variation of `MOVE` that moves to an address register and that does not affect the condition codes, `MOVEQ` is a short version of `MOVE` immediate using an 8-bit signed immediate operand, and `MOVEP` is a `MOVE` that can be used to move data to an 8-bit I/O device that might be designed for the 6812.

Other instructions from the move class include the `LEA` instruction, which works just like `LEAX` in the 6812; `PEA`, which pushes this effective address on the stack; and the familiar `TST` and `CLR` instructions. The `LINK` and `UNLINK` instructions are designed to simplify allocation and deallocation of local variables using the stack marker, as discussed in Chapter 6. The instruction `LINK A0` will push A0 onto the stack, put the resulting stack address into A0, and add (negative 10) to the stack pointer to allocate ten bytes. The instruction `UNLINK A0` deallocates by reversing this procedure, copying A0 into the stack pointer and then pulling A0 from the stack.

Arithmetic instructions are again similar to 6812 arithmetic instructions. As with `MOVE` instructions, `ADD`, `SUB`, and `CMP` have byte, word, and long forms, and `ADDA` and `SUBA` are similar to `MOVEA`. A memory-to-memory compare `CMPM` uses preincrement addressing to permit efficient comparison of strings. There are no `INC` or `DEC` instructions. Rather `ADDQ` can add 1 to 8, and `SUBQ` can subtract 1 to 8, from any register. These instructions are generalized `INC` and `DEC` instructions. Multiple-precision arithmetic uses `ADDX`, `SUBX`, and `NEGX` in the same way that ADC is used in the 6812, except that the Z bit is not set, only cleared if the result is nonzero and only predecrement addressing is used. The handling of Z facilitates multiple-precision tests for a zero number. Decimal arithmetic uses `ABCD`, `SBCD`, and `NBCD` and is designed to work like multiple-precision binary arithmetic such as `ADDX`. However, only bytes can be operated on in these instructions. A special compare instruction `CHK` is used to check addresses. For example, `CHK D0,#$1000` will allow the program to continue if D0 is between 0 and 1000; otherwise, it will jump to an error routine much as the `SWI` instruction does in the 6812. Finally, this machine has multiply and divide instructions for signed and unsigned 16-bit operands that produce 32-bit results. Logic instructions are again very familiar. We have `AND`, `OR`, and `EOR` as in the 6812. As with `ADD`, the instructions `AND` and `OR` can operate on a data register and memory word, putting the result in the memory word. We have `BCLR`, `BSET`, and `BTST` as in the 6805 and also a

BCHG instruction that inverts a bit. Moreover, the chosen bit can be specified either by an immediate operand or by the value in a data register. The S*** group of instructions copies a condition code bit, or a combination of them that can be used in a branch instruction, into a byte in memory. For example, `SEQ $100` copies the Z bit into all the bits of byte $100. The test and set instruction (`TAS`) is useful for some forms of multiprocessing. It sets the condition codes as in `TST`, based on the initial value of a byte, and then sets the byte's most significant bit.

Edit instructions include the standard shifts, with some modifications. All shifts that shift the contents of a data register can be executed many times in one instruction. The instruction `ASL.W #3,D0` will shift the low byte of `D0` three times, as in the 6812 sequence

```
ASLD
ASLD
ASLD
```

The number of shifts can be specified as an immediate operand or can be the number in a data register. However, when shifting memory words, an instruction can shift only one bit. Also, `ROL` and `ROR` are circular shifts of the 8- , 16-, or 32-bit numbers that do not shift through the X bit; `ROXL` and `ROXR` are 9-, 17-, or 33-bit shifts that shift through the X bit as the `ROL` and `ROR` instructions shift through the C bit in the 6812. `EXT` is a sign extend instruction like the 6812 instruction `SEX`, and `SWAP` exchanges the low and high words in the same data register.

Control instructions include the familiar conditional branch (B*** S), branch (`BRA.S`), branch to subroutine (`BSR.S`), long branch (`BRA.L`), conditional long branch (`B***.L`), long branch to subroutine (`BSR.L`), jump (`JMP`), and jump to subroutine (`JSR`) instructions, as well as the `NOP`, `RTS`, and `RTE` (equivalent to the 6812 `RTI`). The instruction `RTR` is like `RTS`, which also restores the condition codes. Special instructions `STOP` and `RESET` permit halting the processor to wait for an interrupt and resetting the I/O devices.

The decrement and branch group of instructions permits decrementing a counter and simultaneously checking a condition code to exit a loop when the desired value of the condition code is met. The condition code specified by the instruction is first tested, and if true, the next instruction below this instruction is begun. If the condition is false, the counter is decremented, and, if –1, the next instruction is executed; otherwise, the branch is taken. The sequence

```
L1:  CLR.B (A0)+
     DBF   D0,L1
```

will execute the pair of instructions n + 1 times, where n is the number in `D0`. This powerful instruction allows one to construct fast program segments to move or search a block of memory. Moreover, in the 68300 series and such short loops are detected, and, when they occur, the two instructions are kept inside the MPU so that the opcodes need not be fetched after the first time, so that these loops run very fast.

We now consider a few simple programs that illustrate the 68300 series instruction set. The first is the familiar program that moves a block of 10 words from SRC to DST. This program shows the way to specify the byte, word, or long form of most instructions, and it shows the powerful decrement and branch instruction.

Let us now look at the overused inner product subroutine, passing the parameters in the in-line argument list by value, in Figure 12.10. Although this method of passing parameters is not the best for the 68300 series because it has plenty of registers to pass parameters by registers, it illustrates the use of data and address registers and is used by C and C++. The LINK and UNLINK instructions are generally used to duplicate the stack pointer in another register, commonly A6. Register A6 accesses parameters with positive offsets and local variables with negative offsets. But, with an abundance of registers, the 68300 series can often save intermediate results in registers rather than on the stack, as in this example. So this example doesn't use any local variables.

Whereas we can see that the 68300 series is superior to the microcontrollers discussed earlier in this chapter for 16-bit and 32-bit arithmetic, this example shows they have a little difficulty in dealing with 8-bit data. The multiply instructions do not have an 8-bit by 8-bit multiply such as `MULU.W LOCW+1(A6),D0`. Moreover, just using an instruction like `MOVE.B LOCV(A6),D0` to bring in the operand and then using `MULU.W D1,D0` leaves the high-order bits of D0 unmodified. Bits 15 to 8 will be used in the `MULU.W D1,D0` instruction. So these bits must be cleared unless they are known to be clear already. The instruction `CLR.L D0` will take care of this. We didn't have to worry about such a case in the microcontrollers discussed earlier in this chapter. The 68330 series, designed for 16-bit and 32-bit arithmetic, is often no better than the 6812, 6811, 6808, and 6805 microcontrollers for operating on 8-bit data and may even be less efficient for some operations than those microcontrollers are.

With this short example, some of the flavor of the 68300 series can be seen. The machine offers a very large address space, over 16 megabytes, and 17 data and address registers. They offer superior performance for 16-bit and 32-bit arithmetic, and a more extensive instruction set than the microcontrollers discussed earlier in this chapter. However, especially in handling 8-bit data, those microcontrollers discussed earlier in this chapter may well exhibit comparable or even superior performance.

```
* SUBROUTINE DOTPRD
* PARAMETERS
LOCV:   EQU  4
LOCW:   EQU  6
DOTPRD: LINK    A6,0            ; Allocate no locals; put stack frame in A6
        CLR.L   D0              ; Clear out high bits of D0
        CLR.L   D1              ; Clear out high bits of D1
        CLR.L   D2              ; Clear out high bits of D2
        MOVE.B  LOCV(A6),D0     ; Get V(0)
        MOVE.B  LOCW(A6),D1     ; Get W(0)
        MULU.W  D1,D0           ; Multiply V(0) by W(0), result in D0
        MOVE.B  LOCV+1(A6),D1   ; Get V(1)
        MOVE.B  LOCW+1(A6),D2   ; Get W(1)
        MULU.W  D2,D1           ; Multiply V(1) by W(1)
        ADD.W   D1,D0           ; Dot product into D0
        UNLINK  A6              ; Deallocate local variables
        RTS
```

Figure 12.10. A 68300 Series Dot Product Subroutine

12.5 The 500 Series

The 500 series of microcomputers are *Reduced Instruction Set Computers* (RISC) that differs from *Complex Instruction Set Computers* (CISC) discussed heretofore. Its registers are shown in Figure 12.11, and its instruction set is given in Table 12.7.

A RISC computer trades off control complexity for additional general purpose registers. See Figure 12.11. The 500 series has 32 32-bit registers that can be used as address or as integer data registers are used in the 68300 series. Additionally, 32 64-bit floating-point registers each can hold a double-precision floating-point number. Finally, there is a link register that holds a subroutine return address, a count register that holds a loop counter, a condition register that holds codes for conditional branching, a floating-point status and control register, and an integer exception register.

The RISC architecture has very simple move instructions with limited addressing modes. The load instruction mnemonics (Table 12.7) are parsed as illustrated by `lhau r3,10(r4)`. The "`l`" means load; the "h" means half-word, which is 16 bits; "`u`" means unsigned (fill with zero bits) to load general register r3; and the effective address is the sum of the instruction's offset 10 and general register 4. The last general register, r4 in this example, may be register 0, in which case the value 0 is used in place of it. This permits a page-zero addressing mode but uses a sign-extended offset. If an "x" is appended, a second general register is used in place of the offset in the address calculation. The instruction `lhaux r3,r4,r5` loads general register 3, as an unsigned number, with the half-word, at the effective address which is the sum of general register 4 and general register 5. In place of "`l`" the letters "`st`" cause the register to be stored, in place of "`h`" the letter "`b`" causes eight bits to be moved or the letter "`w`" causes 16 bits to be moved; and in place of "`u`" the letter "`a`" causes the number to be sign extended when it is loaded. The letter "`r`" in `lhbrx` indicates byte reversal; a 16-bit word is loaded, but the two bytes in it are reversed as they are loaded. The load multiple instruction `lmw r3,10(r5)` loads the registers from r3 to r31 with memory data starting from the effective address, which is the contents of register 5 plus ten. The load string instruction `lswx` similarly loads string data into registers, but with more complexity, which we skip in this introductory treatment. There are corresponding integer store instructions and similar load and store instructions for floating-point and for special-purpose registers.

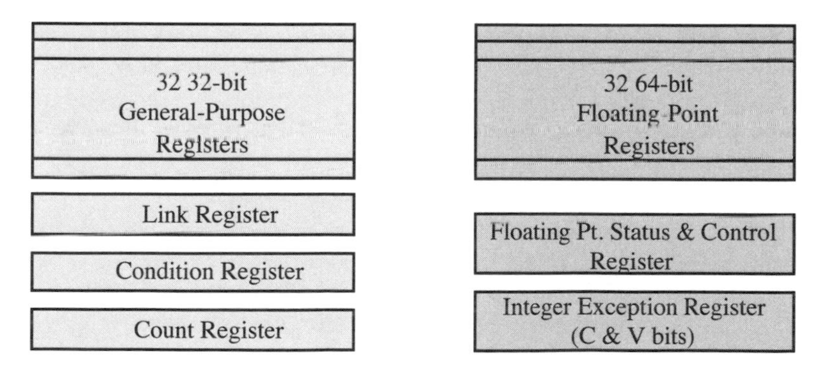

Figure 12.11. User Registers for the 500 Series

Move	Arithmetic
lbz lbzx lbzu lbzux lhz lhzx lhzu lhzux lha lhax lhau lhaux lwz lwzx lwzu lwzux stb stbx stbu stbux sth sthx sthu sthux stw stwx stwu stwux lhbrx lwbrx sthbrx stwbrx lmw stmw lswi lswx stswi stswx	addi addis add addo addic addc addco adde addeo addme addmeo addze addzeo subf subfo subfic subfc subfco subfe subfeo subfme subfmeo subfze subfzeo mulli mullw mullwo mulhw mulhwu divw divwo divwu divwuo neg nego cmpi cmp cmpli cmpl
Logical	**Floating-point Arithmetic**
andi andis and andc ori oris or orc xori xoris xor nand nor eqv	fadd fads fsub fsubs fmul fmuls fdiv fdivs fmadd fmadds fmsub fmsubs fnmadd fnmadds fnmsubfnmsubs
Edit	
rlwinm rlwnm rlwimi slw srw srawi sraw extsb extsh	frsp fctiw fctiwz fcmpu fcmpo fmr fneg fabs fnabs
Control	**Floating-point Move**
b ba bl bla (blr) bc bca bcl bcla bclr bclrl bcctr bcctrl sc rfi two tw	lfs lfsx lfsu lfsux lfd lfdx lfdu lfdux stfs stfsx stfsux stfd stfdx stfdu stdux stfix
Special	
cntlzw crand cror crxor crnand crnor crequ crandc crorc mcrf eieio isync lwarx stwcx sync icbi	mffs mcrf mtfsfi mtfsf mtfsb0 mtfsb1 mtcrf mcrxr mfcr mfmsr mtspr mfspr

Table 12.7. Instructions for the 500 Series

The RISC architecture uses three-register-address instructions for all arithmetic, logical, and edit instructions. There are also immediate operand arithmetic instructions, but no arithmetic instructions that access memory. For instance, the add instruction `add r3,r4,r5` puts the sum in r3 of r4 plus r5. Letters appended to the mnemonic enable other results of addition to be recorded or used. If a letter "c" is appended, the carry bit in the integer exception register is loaded with the carry out of the addition; if a letter "o" is appended, the overflow in the integer exception register is loaded with a two's-complement overflow status bit; and if a period (.) is appended, the condition code registers are updated for conditional branching. An appended letter "e" adds the previous carry bit into the sum, in the manner of the ADC instruction, and another appended "m" adds a minus 1 to it. An appended letter "i," as in `addi r5,r6,7`, indicates that the sum put in r5 is source general register, r6 plus a 16-bit signed immediate operand, 7. But if the source register is r0, the constant 0 is used instead of the contents of the source register; this add immediate instruction is thus used to load immediate data into a general register. Also, if a letter "s" is appended, as in `addis`, the immediate value can be added to the high 16 bits of the destination general purpose register rather than the low-order 16 bits.

There are subtract instructions analogous to the add instructions. The letter "f" in their mnemonics, as in `subf r2,r3,r4`, just means subtract r3 "from" r4, putting the result in r2. The multiply instruction can multiply a register's high word or low word as signed or unsigned numbers, and a register's word value can be divided by another register's word value as a signed or unsigned number. A register can be ANDed, ORed, or exclusive-ORed with an immediate value (in the register's left 16 bits or right 16 bits) or another register, and source values can be complemented before being operated on. The edit instructions include the logical shift left instruction `slw`, logical shift right instruction `srw`, where the amount shifted is an immediate operand or the contents of another general purpose register, and the sign extension instructions `extsb` and `extsh`. The novel `rlwinm` edit instruction rotates the register contents left, and also generates a mask that is ANDed with the result of stripping off some of the bits. It and similar instructions can efficiently extract and insert bit fields in a `struct`.

The control instructions differ from the instruction sets discussed heretofore in the way conditions are recorded and tested and the way a return address is saved and restored.

The conditional branches `bc`, `bca`, `bcl`, and `bcla` test bits in the 32-bit condition register. The branch address is a signed page-zero address (`ba` and `bla`) or a relative address (`b` and `bl`). Four bits are typically used for each condition. Within each four-bit set, the leftmost bit indicates less than; the next bit, greater than; the next bit, equal to; and the last bit, indicating that the numbers are not able to be computed (an overflow occurred) or compared (they are unordered). There are eight sets of these 4-bit conditions. The leftmost set reflects integer arithmetic condition codes, modified by an arithmetic instruction if a period is put at the end of the opcode mnemonic. A subsequent conditional branch instruction reacts to the arithmetic instruction's result condition. The second leftmost set reflects floating-point exceptions. The other six sets are updated by compare instructions; some of its instruction bits designate which set in the condition register is updated. In effect, the conditional branch instruction contains a bit number for a bit in this condition register, and the branch takes place if the selected bit is true. There are also similar unconditional branch instructions `b`, `ba`, `bl`, and `bla`.

```
* SUBROUTINE DOTPRD
DOTPRD: MULU    r26,r27,r28    ; Multiply V(0) by W(0), result in r26
        MULU    r31,r29,r30    ; Multiply V(1) by W(1), result in r31
        ADD     r31,r31,r26    ; Dot product into r31
        BLR                    ; Return link register to program counter
```

Figure 12.12. A 500 Series Dot Product Subroutine

Branch instructions (`bl`, `bla`, `bcl`, and `bcla`) save the address of the next instruction in the link register (Table 12.7). This link register can later be moved into the program counter to return from the subroutine by a `blr` instruction (which can also be conditional). If a subroutine is to call another subroutine, the contents of the link register must be saved. While there is no hardware stack on which return addresses are automatically saved, the programmer can implement a software stack, using any register as a software stack pointer. The link register can be saved on this stack just after the subroutine is entered and can be restored into the link register when the subroutine is about to be exited by a `blr` instruction. It should be noted that this overhead, of saving the return address on a stack and restoring it, is only needed if the subroutine calls subroutines. It should be further noted that if a subroutine is called ten times, the code to save and restore the return address is put just once in the subroutine, not in each occurrence where the subroutine is called. Thus, even though hardware does less work in a RISC processor than a conventional CISC processor, these tradeoffs can be justifiable.

The 500 series has a count register able to be used in decrement and count instructions. This register can also be used like the link register to save return addresses.

To conclude this section, in Figure 12.12 we illustrate the overused inner product subroutine, passing the parameters in registers by value. High-number registers are easy to load and unload, using load multiple and store multiple instructions. Therefore we assume that the vector elements V[0] and V[1] are in general registers 27 and 28, the vector elements W[0] and W[1] are in general registers 29 and 30, and the inner product result is left in register 31. So this example doesn't use any local variables.

12.6 The M·CORE Series

Motorola has recently introduced the M·CORE series of RISC microcontrollers. Memory in an M·CORE processor is organized similarly to the memory in the 3xx family. This processor has user and supervisor modes. The user mode has 16 general-purpose registers, a program counter PC, and a carry bit C. The supervisor mode has this set, and an alternate set, of these general purpose registers, and 13 special-purpose status and control registers. See Figure 12.13. We will mainly discuss the user mode in this overview.

The instructions are described in Table 12.8. `LD.B` can load any general-purpose register (GPR) with a byte from memory at an effective address that is the sum of a GPR and a 4-bit unsigned constant multiplied by the data size. `LD.H` similarly loads a 16-bit word, and `LD.W` loads a 32-bit word. These load instructions load the right bits of the GPR, filling the other bits with zeros. Similarly, `ST.B`, `ST.H`, and `ST.W` store a GPR into memory using the effective address discussed for `LD.B`. The `LRW` instruction can load a GPR with a 32-bit word at an effective address that is the program counter PC added to an 8-bit offset; `MOV` can move any GPR to any GPR; and `MOVI` can write a 7-bit unsigned immediate number into a GPR. `MTCR` can move any GPR to any special register, and `MFCR` can move any special register to any GPR. `LDM` permits a group of GPRs, from a register number designated by an operand, to register r15, using register R0 as a pointer; and `STM` stores such a group of registers. Similarly `LDQ` and `STQ` load and store GPR registers r4 to r7 using any GPR as a pointer.

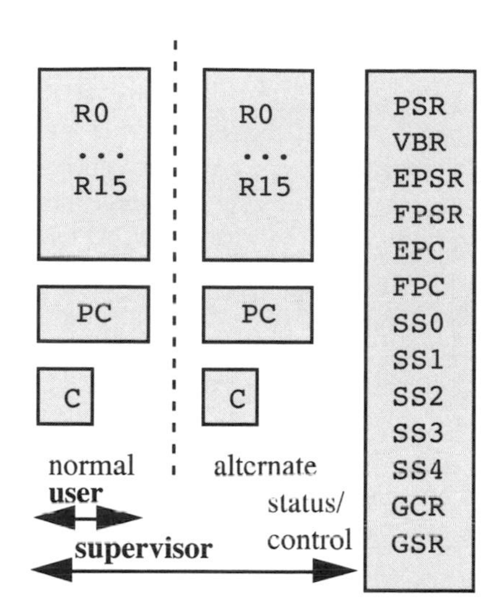

Figure 12.13. M·CORE Register Set

One of the interesting features of this microcontroller is that it has only one condition code bit, which is affected by only a few of the instructions. Generally, the result of an operation is put in a general-purpose register; a compare instruction is executed to put the result of this result into C; and an instruction tests this bit to branch to another location. Instruction `MOVT` (and `MOVF`) transfers any GPR to any GPR if the C bit is true (false). Instruction `MOVC` (or `MOVC`) transfers C (or C complemented) to the low-order bit of any GPR, filling the other bits with zero. `TSTNBZ` sets C if none of the four bytes of the designated GPR is zero; otherwise it clears C. The clear instruction clears if the C bit is true (`CLRT`) or if it is false (`CLRF`).

M·CORE arithmetic instructions include addition (`ADDU`) to add a GPC to a GPC, add immediate (`ADDI`), add with carry (`ADDC`), similar subtract (`SUBU`), subtract immediate (`SUBI`), subtract with carry (`SUBC`), reverse subtract (`RSUB`), and reverse subtract immediate (`RSUBI`), wherein the subtrahends are reversed.

Compare instructions set the C bit if one GPR is higher or the same as another (`CMPHS`), if one GPR is less than another (`CMPLT`), if a GPR is less than an immediate operand (`CMPLTI`), if a GPR is not equal to another GPR (`CMPNE`), or to an immediate operand (`CMPNEI`). `CMPLT` and `CMPLTI` are signed compares; the others are unsigned compares.

The arithmetic instructions include unsigned multiplication (`MULT`) and division (`DIVU`) and signed division (`DIVS`). Multiplication multiplies one GPR by another and puts the product in a 32-bit product in the first GPR. Division always divides any GPR by `r1`, putting the quotient in the GPR.

The increment instruction adds one to a GPR if the C bit is true (`INCT`) or if it is false (`INCF`), and the decrement instruction subtracts one from a GPR if the C bit is true (`DECT`) or if it is false (`DECF`). Other decrement instructions subtract one from a GPR and set the C bit if the result is greater than zero (`DECGT`), if the result is less than zero (`DECLT`), or if the result is not zero (`DECNE`).

Move	Arithmetic
LD.[B,H,W] ST.[B,H,W] MOV MOVI LRW MOVT MOVF LDM STM LDQ STQ MFCR MTCR MVC MVCV CLRF CLRT TSTNBZ	ADDC ADDI ADDU RSUB RSUBI SUBC SUBI SUBU MULT DIVS DIVU CMPHS CMPLT CMPLTI CMPNE CMPNEI DECF DECT INCF INCT DECGT DECLT DECNE ABS FF1 IXH IXW
Logical	**Edit**
AND ANDI ANDN OR XOR NOT BCLRI BSETI BTSTI BMASKI BGENI BGENR TST	ASR ASRC ASRI LSL LSR LSLC LSRC LSLI LSRI BREV ROTLI XSR SEXTB SEXTH ZEXTB ZEXTH XTRB0 XTRB1 XTRB2 XTRB3
Control	
BR BF BT JMP JMPI LOOPT BSR JSR JSRI TRAP RTE RFI BKPT WAIT DOZE STOP SYNC	

Table 12.8. Instructions for the M·CORE Series

M·CORE has some unusual arithmetic instructions. The ABS instruction gets the absolute value of a GPR's data. The FF1 instruction finds the first one of a bit pattern in a GPR. This instruction is useful in emulating floating-point arithmetic (alternatively a floating-point hardware accelerator can be put on the chip to more quickly execute floating-point instructions). Index instructions IXH and IXW add one GPR to another, multiplying one of the addends by two or four. These are useful in calculating an address of an element of a vector.

M·CORE logical instructions include AND to AND, a GPC to a GPC; ANDI to AND, an immediate operand to a GPC; and ANDN to AND, the negative of a GPC to a GPC. TST sets C if the AND of two designated GPR is nonzero; otherwise it clears C. It has an OR instruction to OR a GPC to a GPC, an XOR instruction to exclusive-OR a GPC to a GPC, and an instruction, NOT, to complement all bits in a GPC. Further logical instructions are BCLRI, which clears a bit of a GPR specified by an immediate operand; BSETI, which sets a bit of a GPR specified by an immediate operand; and BTSTI, which puts a bit of a GPR, specified by an immediate operand, into the C bit. The instruction BGENI sets a bit of a GPR specified by an immediate operand and clears all other bits, and BGENR sets a bit of a GPR selected by another GPR. BMASKI sets all bits of a GPR to the right of a bit selected by an immediate operand.

M·CORE edit instructions include `ASR` to shift a GPR right arithmetically a number of bits specified by a GPR; `ASRC` to shift a GPR right arithmetically one bit, putting the bit shifted out into C; `ASRI` to shift a GPR right arithmetically a number of bits specified by an immediate operand; and similar instructions `LSL`, `LSR`, `LSRC`, `LSLI`, `LSLRI`, and `ROTLI`. The last instruction is a circular shift of a GPR. The `BREV` instruction reverses the bits in a GPR. `XSR` shifts a GPR one bit right, putting the bit shifted out into the C bit. `SEXTB` sign extends a GPR from 8 to 32 bits, `SEXTH` sign extends a GPR from 16 to 32 bits, `ZEXTB` zero extends a GPR from 8 to 32 bits, and `ZEXTH` zero extends a GPR from 16 to 32 bits. XTRB0 extracts byte zero (the LSbyte) of any GPR to GPR register 1. XTRB1 similarly extracts byte one of any GPR, XTRB2 extracts byte two, and XTRB3 extracts byte three.

Addition and subtraction are unsigned, there being no V condition code bit needed for a signed overflow check. But because data moved into a GPR can be sign-extended using `SEXTB` or `SEXTH`, and addition and subtraction are 32-bit operations, a 32-bit signed overflow is unlikely. Before a store instruction such as `ST.B` or `ST.H`, the high bits, which are not stored, can be checked to see if they are all zeros or all ones.

The reader should observe that the M·CORE architecture has unusually extensive logic and edit instructions. These instructions are valuable for I/O operations. However, there are comparatively fewer arithmetic and move instructions in this RISC processor.

M·CORE control instructions include `BR` to branch to a relative address using an 11-bit relative address, `BRT` to branch if C is true, `BRF` to branch if C is false. `JMP` copies a GPR into the PC, and `JMPI` jumps indirectly to an address at a word specified by an 8-bit displacement. If the C bit is 1, the `LOOPT` instruction decrements a GPR and branches backwards up to 15 instructions to implement a loop. Otherwise it decrements the GPR and continues to execute the instruction below it.

`JSR` saves the PC in GPR register 15 and copies a GPR into the PC, and `JSRI` saves the PC in GPR register 15 and jumps indirectly to an address at a word specified by an 8-bit displacement. TRAP, having a 2-bit immediate operand, effects an exception like an interrupt, through an address stored at 0x40 to 0x4f. TRAP and other exceptions and interrupts save the processor status register (special purpose register 0) and return address in status/control registers. The instruction `RTE` returns from an exception, and `RFI` returns from a fast interrupt, restoring the saved PC and processor status register.

The instruction `BKPT` causes a breakpoint exception. It can be used to stop a program so that the debugger can examine memory or registers and resume. `WAIT` causes the processor to enter low-power wait mode in which all peripherals continue to

```
*
* SUBROUTINE DOTPRD: a JSR/BSR instruction puts the return address into R15
*
DOTPRD: MUL     r1,r3         ; Multiply V(0) by W(0), result in r1
        MUL     r2,r4         ; Multiply V(1) by W(1), result in r2
        ADD     r1,r2         ; Dot product into r1
        JMP     r15           ; Return R15 to program counter
```

Figure 12.14. An M·CORE Dot Product Subroutine

run, and `DOZE` causes the processor to enter low-power doze mode in which some peripherals continue to run. `STOP` causes the processor to enter low-power stop mode. The MMC2001, a first implementation of the M·CORE family, uses 40 milliamps at 3.3 volts. Both wait and doze modes have current drain of 3 milliamps, and the stop mode has current drain of only 60 microamps. `SYNC` causes the processor to suspend, fetching new instructions until all previously fetched instructions complete execution.

To conclude this section, in Figure 12.14, we illustrate the overused inner product subroutine again, passing the parameters in registers by value. Upon input, GPR r1 has v[0], r2 has v[1], r3 has w[0], and r4 has w[1], and upon exit, r1 contains the result.

12.7 Selecting a Microcontroller for an Application

Suppose you are designing a product that will have a microprocessor in it. Which one should you use? You have to look at many different alternatives, such as the ones we looked at in this chapter and similar microcomputers made by other companies. You should not select one with which you are very familiar, such as the 6812, or one that you are overwhelmed with, such as the 500 series, unless you have good reason to select it. You have to analyze the needs of the application to pick the most suitable microcomputer. Smaller computers are less costly, and larger computers make it easier to write large programs. However, many of the techniques are the same as those you have already learned, passing parameters, handling local variables, writing clear programs, and testing them. You are prepared to learn to read the 68300 series, 500 series, or M·CORE family programs. However, the greater size and complexity of these microcomputers requires longer to master all of their peculiarities than smaller microcomputers, to enable you to fluently read their programs.

Generally, the larger the microprocessor, the easier it is to write large programs. The 68300 series, 500 series, and M·CORE family have more capabilities to handle high-level languages, such as C or C++, and have an instruction set that allows assembly-language programs to be written that can handle fairly complex operations in short fashion (such as the `LINK` instruction). It is easy to say that the larger the microprocessor, the better, and to select the largest one you can get. But consider some other aspects.

The smaller microcomputers such as the 6805 are very inexpensive. A version of this 6805 sells for only 50 cents. You can build a fully functioning microcomputer using just the 6805 and a couple of resistors and capacitors. The 68300 series requires external SRAM and ROM to make a working computer. The cost of the integrated circuits, the printed circuit board, and the testing needed to get the board working make the 68300 series system more than an order of magnitude more expensive than the 6805 system. An M·CORE microcontroller, running with a 32 MHz clock, can require more than a two-layer printed circuit board. Multilayer boards are significantly more expensive than one- or two-layer boards. This can make a big difference to the cost of your product, especially if you intend to make thousands of copies of the product.

The trend toward networking should be observed. If you divide your problem in half, each half may fit on a smaller microcomputer. We once read a news article that claimed the Boeing 767 jet had over a 1000 microcomputers scattered throughout the wing tip, landing gear, and cockpit to control the plane. The distributed computer system saved wire and thus weight. Offices are using personal computers so that each person has a microcomputer dedicated to his or her work rather than time-sharing a large computer. Small jobs or small parts of a larger job should be put on small computers.

The main criterion for selecting a microprocessor (within a family such as the Motorola family described here) is the size of the program stored in it. The microcomputer should be able to store the programs and data, with a little to spare to allow for correcting errors or adding features. That is, as the program and data approach the maximum size of the microprocessor, the cost of programming rises very sharply, because squeezing a few extra instructions in will require moving subroutines around and cause errors to propagate as assumptions are forgotten and violated. The 6805 and its successor, the 6808, are the best choices when the program size is about 4K bytes. The 6811, and its successor, the 6812, are better when the size is above that but less than about 32K bytes. Generally the successors, the 6808 and 6812, should be used on new designs, but the older 6805 and 6811 may be less costly and fully adequate. The 68300 series, 500 series, and M·CORE are the best choices when the program size exceeds 64K bytes.

Other criteria include the requirements for I/O and speed. All the microcontrollers have some peripherals in the MPU chip. If the application needs more than the chip has, the advantage of that chip relative to memory size may be overshadowed by the extra cost of peripherals needed for the application. Speed can be a factor. Especially in communication systems and control of electronic systems, the fastest microprocessor may be needed. However, speed is often overrated. In most systems having I/O, the microprocessor will spend much, if not most, of its time waiting for I/O. The faster microprocessor will spend more time waiting. If you can select faster I/O (such as a Winchester disk in place of a floppy disk) the overall performance of the system will be much better than if you spend a great deal more money on the microprocessor.

A final and often overwhelming criterion is available software. Your company may have been using the 6805 for years and may therefore have millions of lines of code for it. This may force you to select the 6805 or the upward compatible 6808, even though the 6811 or 6812 may be indicated due to memory size, I/O, or speed requirements. Often, the availability of operating systems and high-level languages selects the microprocessor. The Z80 microprocessor from Zilog and the 8080-8085 microprocessors from Intel run the popular Microsoft operating system, which will support a very wide range of languages and other programs for business data processing.

This section has pointed to the need to consider different microprocessors. You should be able to select a microprocessor for an application and defend your selection. You should extend your understanding and appreciation of microprocessors made by other manufacturers.

12.8. Summary

This chapter has examined other microcomputers related to the 6812. There are smaller microcomputers, such as the 6805 and the 6811, that are ideal for controlling appliances and small systems; and there are larger microcomputers, such as the 68300 series, that are excellent for larger programs. Moreover, having learned to program the 6812, you are well prepared to learn the languages for the 6805 and the 68300 series. It is rather like learning a second foreign language after you have learned the first. Although you may err by mixing up the languages, you should find the second easier to learn because you have been through the experience with the first language. After learning these languages for the Motorola family, you should be prepared to learn the languages for other microcomputers and become a multilingual programmer.

This text has taken you through the world of microcomputer programming. You have learned how the microcomputer actually works at the level of abstraction that lets you use it wisely. You have learned the instruction set and addressing modes of a good microcomputer and have used them to learn good techniques for handling subroutine parameters, local variables, data structures, arithmetic operations, and input-output. You are prepared to program small microcomputers such as the 6805, which will be used in nooks and crannies all over; and you have learned a little about programming the 68300 series, which will introduce you to programming larger computers. But that should be no problem. A computer is still a computer, whether small or large, and programming it is essentially the same.

Do You Know These Terms?

See the end of chapter 1 for instructions.

Reduced Instruction Set Computer (RISC)
Complex Instruction Set Computer (CISC)

PROBLEMS

1. Write an 6811 subroutine `DOTPRD` that passes parameters after the call as that subroutine was written in Figure 6.25. It should be reentrant, position independent, and as short as possible.

2. Write a shortest 6811 subroutine `SRCH` that finds a string of ASCII characters in a text. The label `STRNG` is the address of the first letter of the string, `STLEN` is the length of the string, `TXT` is the address of the first letter of the text, `TXLEN` is the length of the text, and the subroutine will exit with C = 1 and the address of the first occurrence of the first letter of the string in the text in X, if it is found, or C = 0 if it is not found.

3. Write a position-independent reentrant 6811 subroutine `QUAD` that evaluates the quadratic function $ax^2 + bx + c$, where unsigned 8-bit arguments a, b, c, and x are passed on the stack from low to high addresses respectively, named `PARA`, `PARB`, `PARC`, and `PARX`, and the output is returned in register B. In order to demonstrate local variables, as part of your subroutine, store ax^2 in an 8-bit local variable on the stack. Write a calling sequence that writes 1, 2, 3, and 4 into `PARA`, `PARB`, `PARC`, and `PARX`, calls `QUAD`, and moves the result to global variable `ANSWER`.

4. Write a shortest position-independent reentrant 6811 subroutine PAR that computes the parallel resistance of two resistors R1 and R2, where unsigned 8-bit arguments are passed on the stack and named R1 and R2, and the output is returned in register B. In order to demonstrate local variables, as part of your subroutine, store R1 times R2 in a 16-bit local variable on the stack. Write a calling sequence that writes 100 into R1 and R2, calls `PAR`, and moves the result to global variable `ANSWER`.

5. Write a shortest reentrant 6808 `SWI` interrupt handler `AAX` that will add A to X.

6. Write a shortest reentrant 6808 `SWI` interrupt handler `EMUL` that will multiply A by HX, putting the result in HX, exactly as the 6812 `EMUL` works. Ignore CC bits.

7. Write a position-independent reentrant 6808 subroutine `QUAD` that evaluates the quadratic function $ax^2 + bx + c$, where unsigned 8-bit arguments a, b, c, and x are passed on the stack from low to high addresses respectively, named `PARA`, `PARB`, `PARC`, and `PARX`, and the output is returned in register A. In order to demonstrate local variables, as part of your subroutine, store ax^2 in an 8-bit local variable on the stack. Write a calling sequence that writes 1, 2, 3, and 4 into `PARA`, `PARB`, `PARC`, and `PARX`, calls `QUAD`, and moves the result to global variable `ANSWER`.

8. Write a shortest position-independent reentrant 6808 subroutine `PAR` that computes the parallel resistance of two resistors `R1` and `R2`, where unsigned 8-bit arguments are passed on the stack, and named `R1` and `R2`, and the output is returned in register B. In order to demonstrate local variables, as part of your subroutine, store `R1` times `R2` in an 16-bit local variable on the stack. Write a calling sequence that writes 100 into `R1` and `R2`, calls `PAR`, and moves the result to global variable `ANSWER`.

9. Write a shortest reentrant MC6805 `SWI` interrupt handler `PSHX` that will push X on the stack as the 6811 instruction `PSHX` works. Assume the location of `SWI` is at $281. This instruction must be reentrant.

10. Write a shortest 6805 subroutine `MOVE` that can move any number of words from any location to any other location in memory. The calling sequence will put the beginning address of the source in page-zero global variable `SRC`, the beginning address of the destination in `DST`, and the length in `LEN`. Use impure coding if necessary.

11. Write a position-independent reentrant 6805 subroutine `QUAD` that evaluates the quadratic function $ax^2 + bx + c$, where unsigned 8-bit arguments a, b, c, and x are passed as globals named `PARA`, `PARB`, `PARC`, and `PARX`, and the output is returned in register A. In order to demonstrate the absence of local variables, as part of your subroutine, store ax^2 in an 8-bit global variable `TEMP`. Write a calling sequence that writes 1, 2, 3, and 4 into `PARA`, `PARB`, `PARC`, and `PARX`, calls `QUAD`, and moves the result to global variable `ANSWER`.

12. Write a shortest position-independent reentrant 6805 subroutine `PAR` that computes the parallel resistance of two resistors `R1` and `R2`, where unsigned 8-bit arguments are passed as globals named `R1` and `R2`, and the output is returned in register A. In order to demonstrate the absence of local variables, as part of your subroutine, store `R1` times `R2` in a 16-bit global variable `TEMP`. Write a calling sequence that writes 100 into `R1` and `R2`, calls `PAR`, and moves the result to global variable `ANSWER`.

13. Write a shortest 6805 program segment that will jump to subroutines `L0` to `L7` depending on the value of X. If (X) = 0, jump to subroutine `L0`; if (X) = 1, jump to subroutine `L1`, and so on. Assume that there is a table `JTBL` as shown below:

```
JTBL DC.W L0, L1, L2, L3, L4, L5, L6, L7
```

Use self-modifying code if necessary.

14. Write a shortest 6805 subroutine to divide the unsigned number in X by the unsigned number in A, leaving the quotient in X and the remainder in A. Use only `TEMP1` and `TEMP2` to store variables needed by the subroutine.

15. Write a shortest 6805 subroutine to clear bit n of a 75-bit vector similar to the SET in Problem 8 in Chapter 3. The instruction `BCLR N,M` clears bit `N` of byte `M` and has opcode $ 11 + 2 * `N` followed by offset `M`. Use self-modifying code.

16. Write an 6805 subroutine to transmit the bits of the 75-bit vector set by `SET` described in Problem 8 in Chapter 3, bit 0 first, serially through the least-significant bit of output port A at location 0. Each time a bit is sent out, the second-least-significant bit of that output port is pulsed high and then low. The least-significant bit happens to be connected to a serial data input, and the second-least-significant bit is connected to a clock of a shift register that controls display lights.

17. Write a shortest 68300 series subroutine `CLRREG` to clear all the registers except A7. Assume that there is a block of 60 bytes of zeros, after LOC0, that is not in part of your program (i.e., use 32-bit direct addressing, and do not count these bytes when calculating the length of your subroutine). Be careful, because this one must be checked out, and the obvious solutions do not work.

18. Write a fastest 68300 series subroutine `MULT` that multiplies two 32-bit unsigned binary numbers in D0 and Dl, to produce a 64-bit product in D0:D1.

19. Write a position-independent, reentrant, fastest 68300 series subroutine `DOTPRD` that passes parameters on the stack, in the same manner as that subroutine in Figure 6.21.

20. Write a position-independent reentrant 68300 series subroutine `QUAD` that evaluates the quadratic function $ax^2 + bx + c$, where signed 16-bit arguments a, b, c, and x are passed on the stack from low to high addresses respectively, named `PARA`, `PARB`, `PARC`, and `PARX`, and the output is returned in register D0. In order to demonstrate local variables, as part of your subroutine, store ax^2 in a 16-bit local variable on the stack. Write a calling sequence that writes 1, 2, 3, and 4 into `PARA`, `PARB`, `PARC`, and `PARX`, calls `QUAD`, and moves the result to global variable `ANSWER`.

21. Write a shortest position-independent reentrant 68300 series subroutine PAR that computes the parallel resistance of two resistors `R1` and `R2`, where unsigned 16-bit arguments are passed on the stack and named `R1` and `R2`, and the output is returned in register D0. In order to demonstrate local variables, as part of your subroutine, store R1 times `R2` in a 16-bit local variable on the stack. Write a calling sequence that writes 100 into `R1` and `R2`, calls `PAR`, and moves the result to global variable `ANSWER`.

22. Write a fastest position-independent, reentrant, 68300 series subroutine `CAH` that converts a string of ASCII characters representing a hexadecimal number to an unsigned binary number in D0. The first character is pointed to by A0, and the length is in D0.

23. Write a position-independent reentrant 500 series subroutine `QUAD` that evaluates the quadratic function $ax^2 + bx + c$, where signed 16-bit arguments a, b, c, and x are passed in registers r27, r28, r29, and r30, respectively, and the output is returned in register r31. In order to demonstrate local variables, as part of your subroutine, store ax^2 in register r26. Write a calling sequence that writes 1, 2, 3, and 4 into the four registers holding a through x, resepectively, and moves the result to global variable `ANSWER`.

24. Write a shortest position-independent reentrant 500 series subroutine `PAR` that computes the parallel resistance of two resistors `R1` and `R2`, where unsigned 16-bit arguments are passed in registers r29 and r30, and the output is returned in register r31. In order to demonstrate local variables, as part of your subroutine, store `R1` times `R2` in a 16-bit local variable in register r28. Write a calling sequence that writes 100 into `R1` and `R2`, calls `PAR`, and moves the result to global variable `ANSWER`.

25. Select the most suitable microprocessor or microcomputer among the 6805, 6811, 6812, or for the following applications.

(a) A graphics terminal needing 250K bytes of programs and 100K bytes of data

(b) A motor controller, storing a 15K program, needing to quickly evaluate polynomials

(c) A text editor for a "smart terminal" needing 8K for programs and 40K for data storage

(d) A keyless entry system (combination lock for a door) requiring $D0 bytes of program memory and 2 parallel I/O ports.

This module from Axiom Manufacturing has an MMC2001 on a small plug-in board that can be purchased separately, and a mother board that has external flash memory, sockets for debug software, and LCD and keypad interfaces.

Appendix 1
Number Representations and Binary Arithmetic

This appendix contains material needed for the rest of the book that is usually found in an introductory course on logic design. The two topics are the representation of integers with different bases and binary arithmetic with unsigned and two's-complement numbers.

A1.1 Number Representations

If b and m are positive integers, and if N is a nonnegative integer less than bm, then N can be expressed uniquely with a finite series

$$N = c_{m-1} * b_{m-1} + c_{m-2} * b_{m-2} + + c_0 * b_0 \quad (1)$$

where $0 \le c_i < b$ for $0 \le i \le m - 1$. The integer b is called the *base* or *radix* and the sequence $c_{m-1} \ldots c_0$ is called a *base-b representation* of N. If b = 2, the digits $c_{m-1} \ldots c_0$ are called *bits*, and the sequence $c_{m-1} \ldots c_0$ is called an *m-bit binary representation* of N. Binary, octal (base 8) and hexadecimal (base 16) representations, as well as the ordinary decimal representation, are the ones used when discussing computers with hexadecimal being particularly useful with microcontrollers. When the hexadecimal representation is used, the numbers 10 through 15 are replaced by the letters A through F, respectively, so that hexadecimal sequences such as 112 will be unambiguous without the use of commas (e.g., without commas, 112 could be interpreted as 1,1,2, or 11,2, or 1,12, which are the decimal numbers 274, 178, or 28, respectively). Unless stated otherwise, all numbers will be given in decimal and, when confusion is possible, a binary sequence will be preceded by a % and a hexadecimal sequence by a $. For example, 110 denotes the integer one hundred and ten, %110 denotes the integer six, and $110 denotes the integer two hundred seventy-two.

To go from a base-b representation of N to its decimal representation, one has only to use (1). To go from decimal to base b, notice that

$$\begin{aligned} N &= c_0 + c_1 * b^1 + \ldots + c_{m-1} * b^{m-1} \\ &= c_0 + b * (c_1 + b * (c_2 + \ldots) \ldots) \end{aligned}$$

so that dividing N by b yields a remainder c_0. Dividing the quotient by b again yields a remainder equal to c_1 and so on. Although this is a fairly convenient method with a calculator, we shall see later that there is a more computationally efficient way to do it with an 8-bit microprocessor.

To go from binary to hexadecimal or octal, one only needs to generalize from the example:

%1101 0011 1011 = $D3B

Thus, to go from binary to hexadecimal, one first partitions the binary representation into groups of four 0s and 1s from right to left, adding leading 0s to get an exact multiple of four 0s and 1s, and then represents each group of four 0s and 1s by its hexadecimal equivalent. To go from hexadecimal to binary is just the reverse of this process.

A1.2 Binary Arithmetic

One can add the binary representations of unsigned numbers exactly like the addition of decimal representations, except that a carry is generated when 2 is obtained as a sum of particular bits. For example,

```
 1010        1100        1110
+0111       +0111       +0111
10001       10011       10101
```

Notice that when two 4-bit representations are added and a carry is produced from adding the last or most significant bits, five bits are needed to represent the sum.

Similarly, borrows are generated when 1 is subtracted from 0 for a particular bit. For example,

```
 1111        1011          1000
-0101       -0100         -1001
 1010        0111       (1)1111
```

In the last of the examples above, we had to borrow out of the most significant bit, effectively loaning the first number 24 to complete the subtraction. We have put a "(1)" before the 4-bit result to indicate this borrow. Of course, when a borrow occurs out of the most significant bit, the number being subtracted is larger than the number that we are subtracting it from.

When handling numbers in microprocessors, one usually has instructions that add and subtract m-bit numbers, yielding an m-bit result and a carry bit. Labeled **C**, the carry bit is put equal to the carry out of the most significant bit after an add instruction, while for subtraction, it is put equal to the borrow out of the most significant bit. The bit **C** thus indicates *unsigned overflow;* that is, it equals 1 when the addition of two positive m-bit numbers produces a result that cannot be expressed with m-bits, while with subtraction, it equals 1 when a positive number is subtracted from a smaller positive number so that the negative result cannot be expressed with equation (1).

We can picture the m-bit result of addition and subtraction of these nonnegative numbers (also called *unsigned* numbers) using Figure Al.l, where m is taken to be four bits. For example, to find the representation of **M** + **N**, one moves **N** positions clockwise from the representation of **M** while, for **M** - **N**, one moves **N** positions counterclockwise. Mathematically speaking, we are doing our addition and subtraction *modulo-16* when we truncate the result to four bits. In particular, we get all the usual answers as long as unsigned overflow does not occur, but with overflow, 9 + 8 is 1, 8–9 is 15, and so on.

We also want some way of representing negative numbers. If we restrict ourselves to m binary digits c_m, c_0, then we can clearly represent 2^m different integers. For example, with (1) we can represent all of the nonnegative integers in the range 0 to 2^m–1. Of course, only nonnegative integers are represented with (1), so that another representation is needed to assign negative integers to some of the m-bit sequences. With the usual decimal notation, plus and minus signs are used to distinguish between positive and negative numbers. Restricting ourselves to binary representations, the natural counterpart of this decimal convention is to use the first bit as a sign bit. For example, put c_m equal to 1 if **N** is negative, and put c_{m-1}, equal to 0 if **N** is positive or zero. The remaining binary digits $c_{m-2} \ldots, c_0$ are then used to represent the magnitude of N with (1).

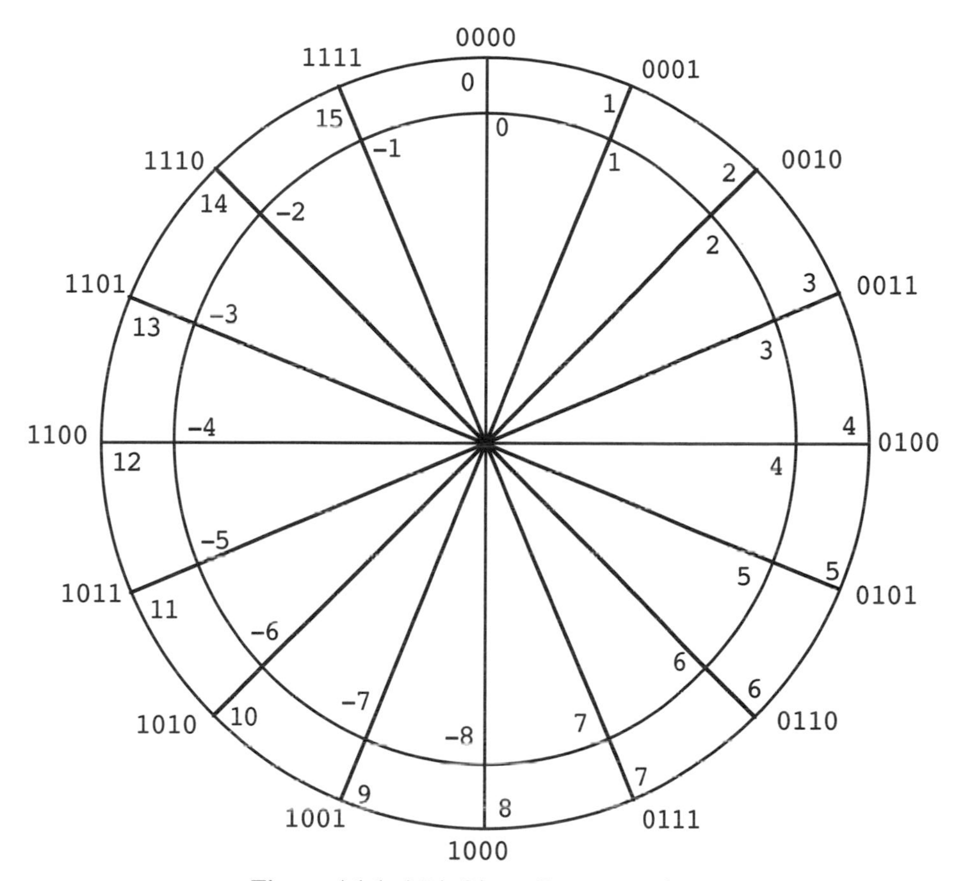

Figure A1.1. 4-Bit Binary Representations

This particular representation, called the *signed-magnitude* representation, has two problems. First, carrying out addition and subtraction is clumsy, particularly from a logic design point of view. Second, the number zero has two representations, 10 . . . 0 and 00 . . . 0, which, at best, has a perplexing feeling to it. The following representation, the *two's-complement* representation, essentially solves both of these problems.

Looking again at Figure A1.1, notice that when we subtract the representation of N from M, we get the same thing as adding the representation of M to that of 2^4–N because moving N positions counterclockwise is the same thing as moving clockwise 2^4–N positions. Thus, as far as modulo-16 addition is concerned, M–1 = M + (–1) is the same as M + 15, M–2 = M + (–2) is the same as M + 14, and so on. Letting –1 be represented with the sequence for 15, –2 with the sequence for 14, and so on, we get the two's-complement representation shown in Figure Al.l for m equal to four.

We choose to represent the negative integers –1 through –8 in Figure A1.1 because, with this choice, the leading bit is a sign bit, as with the signed magnitude representation (e.g., 1 for minus, 0 for plus). Additionally, the number zero, which is now considered positive, is represented by the single sequence 0000. However, the nicest feature is that the two's-complement representation of M + N is obtained by simply adding the two's-complement representations of M and N and truncating to four bits. This works, of course, as long as there is no *signed* overflow, that is, adding two m-bit two's-complement numbers whose sum cannot be represented with an m-bit two's-complement number. For addition, notice that signed overflow occurs when, and only when, the sign of the two representations added are the same but different from the sign of the result. A similar observation can be made for the two's-complement representation of M–N.

We can now summarize the facts for m-bit two's-complement representations. With m fixed and $-2^{m-1} \leq N < 2^{m-1}$, the m-bit two's-complement representation of N is given by:

1. The m-bit representation of N for $0 \leq N < 2^{m-1}$.
2. The m-bit representation of $2^m + N$ for $-2^{m-1} \leq N < 0$.

The first bit of the representation is a sign bit and, after a little thought, you should be able to see that if cm , c0 is the m-bit two's-complement representation of N, then

$$N = - c_{m-1} * 2^{m-1} + c_{m-2} * 2^{m-2} + ... + c_0 * 2^0 \qquad (2)$$

The difference between equations (2) and (1), of course, is that the first term in (2) is negative. Finally, the two's-complement representation of M + N is obtained by adding the two's-complement representations of M and N and truncating to m bits. The answer is correct except when signed overflow occurs or, equivalently, when the signs of the two representations are the same but different from that of the result.

If N is positive, its m-bit two's-complement representation y is just its ordinary m-bit representation. It is not difficult to see that the m-bit two's-complement representation of –N can be obtained by subtracting each bit of y from 1 and then adding 1 to the result. This procedure, sometimes called "taking the two's-complement" of y, works even if N is zero or negative, with two exceptions. If N is zero, the result needs to be truncated to m bits. If $N = -2^{m-1}$, one will just get back the two's-complement representation of -2^{m-1}. To see why this works for $-2^{m-1} < N < 2^{m-1}$, suppose that y = c_m, c_0 and let $d_j = 1-c_j$, $0 \leq j \leq m-1$. Then it is easy to see from (2) that $d_{m-1} \ldots d_0$ is the m-bit two's-complement representation of –N–1. That the procedure works now follows from the fact above for the addition of two's-complement representations.

One situation frequently encountered when two's-complement representations are used with microproccssors is that of finding the hexadecimal equivalent of the 8-bit two's-complement representation of a negative number. For example, for a –46 you could find the 8-bit representation of 46, use the technique just mentioned, and then find its hexadecimal equivalent. You could also use the two's-complement definition, finding the 8-bit representation of 24–46 and then converting this to hexadecimal. It would, however, usually be quicker just to convert 2r–46 = 210 to hexadecimal. Finally, one could also use a 16's-complement approach, that is, convert the number to hexadecimal, subtract each hexadecimal digit from 15, and then add 1 to get the result. For example,

46 = $2E and the 16's-complement of $2E is $D1 + 1 = $D2, the desired result. You should try to understand how this works. (See the problems at the end of this appendix.)

A1.3 Remarks

The material discussed here can be found in any introductory text on logic design. We recommend the book *Fundamentals of Logic Design, 4th ed.*, by C. H. Roth (PWS Publishing Co., Boston MA, 1995).

PROBLEMS

1. Find the hexadecimal equivalents of the 8-bit two's-complement representations of –44 and –121.

2. Explain why the 16's-complement technique works when used for calculations such as Problem 1.

3. Suppose that you were going to add a 16-bit two's-complement representation with an 8-bit one. How would you change the 8-bit representation so that the 16-bit result would be correct? This process is sometimes called *sign extension*.

4. Give a simple condition for signed overflow when two's-complement representations are subtracted.

5. One textbook reason for preferring the two's-complement representation of integers over the signed-magnitude representation is that the logic design of a device that adds and subtracts numbers is simpler. For example, suppose that M and N have m-bit two's-complement representations x and y. To subtract M from N, one can take the two's-complement of x and then add it to y, presumably simpler from the logic design viewpoint than dealing with signed-magnitudes. Does this always work? Try it with m = 8 N equal to –1, and M equal to –128. What is the condition for overflow? Does this work when N and M are interpreted as unsigned numbers? Interpret.

6. Suppose that we add two m-bit representations x and y, where x is the unsigned representation of M and y is the two's-complement representation of N. Will the answer, truncated to m bits, be correct in any sense? Explain.

Appendix 2
Using the HiWare CD-ROM

A2.1 Loading HIWARE Software

You can use the software on the enclosed CD-ROM to simulate your programs on a PC running Windows 95 or later, or Windows NT 5.0 or later, without using any extra hardware. You can also use this software with a real target microcontroller, so you can collect data from external hardware, and control external hardware. Open the CD-ROM, check "setpe.exe", and choose the Motorola HC12 target. If you have 60 megabytes of disk space, load all parts of the tool chain.

A2.2 Opening the HIWARE Toolbox

You can open the HIWARE Toolbox in at least two ways, depending on how HIWARE was installed. The best way is to click on the Start icon, to the Programs item, to the HIWARE item, to HC12 Projects item, to either ManualProject item or AssemblerProject item. This should open the toolbox for this book's experiments.

Another way to open the HIWARE Toolbox is to click on the Start icon, to the Programs item, to the HIWARE item, to the HIWARE TOOLS item. When the tool bar appears, click on its leftmost icon (the one with the picture of three pages and a pencil) to open a dialog box. In that dialog box, click on the Open.. icon. Click on folder and files names to find and select one of the folders described below. Then click on OK in the inner box, and the outer box. If a dialog box appears to confirm that you change the project, click on OK. This should open the toolbox for experiments in this book.

The folder ManualProgramFolder in the HIWARE folder provides examples of programs in the first three chapters of the book. These examples use just the HIWARE debugger applicaton. The folder AssemblyProgramFolder in the HIWARE folder provides examples of programs in the remaining chapters of the book. These examples use the NOTEPAD text editor, the HIWARE assembler and the HIWAVE debugger. You can also use other text editors such as WinEdit, in place of NOTEPAD, to generate source code for the assembler.

A2.3 Running Examples From the ManualProgramFolder

ManualProgramFolder has files which contain examples from this text book that you can load and run on the HIWAVE simulator, or download and run on the target machine. The file p.11.abs contains machine code for an example on page 11, which is Figure 1.5. Similarly, p.13a.abs is the example on the top of page 13, Figure 1.8. Click on the HIWAVE icon in the HIWARE toolbox. After HIWAVE starts, load an example by pulling the Simulator menu to the Load.. item. Select the file for the example, such as p.11.abs. First, double-click on the box next to SP in the Register window, and type a suitable stack pointer value such as hexadecimal A00 (just press the keys A, 0, and 0, and Enter). Then single-step though the program to learn how it works. You can single-step by clicking on the RUN menu and releasing on the Single Step item, or by using the F11 key, or by clicking on the single step icon, which is a U-shape with an arrow into the center of the U. Observe the changes in the registers as the program is executed. Note that by right-clicking on the Register window, you can change the register's format to binary or decimal representation. Use the format most suitable for the example.

To examine a part of memory, right-click on the Memory window, and select the Address... item. When the dialog box appears, type the address such as hexadecimal 800, and then type Enter. Alternatively, to display the memory around the address that is in a register such as the stack pointer, drag and drop from the register's box to the Memory window. The Memory window should now display the memory beginning with the address. The left part of the Memory window displays the memory data, and the right part displays the same data as ASCII characters (See Chapter 4). Note that by right-clicking on the Memory window, you can change the memory data format to binary or decimal representation, or change the memory display's word size to 16 bits or 32 bits.

You can change the data in memory after the program has been loaded to learn more about the topic covered by the example. Double-click on the display of a byte you wish to change in the memory window, and retype its hexadecimal value. After entering a hexadecimal number into the memory, press Enter or press another number key to enter a number into the next memory location. You can also double-click on the ASCII representation of the memory data to enter ASCII characters into memory.

You can run until an instruction is executed. Right-click on the Assembly window on the line displaying the instruction at which you wish to stop. Release on the Run To Cursor item. You can set breakpoints, which are instructions which you wish to stop at, and run the program to the first breakpoint it meets. Right-click on the Assembly window on the line displaying the instruction at which you set a breakpoint and release on the Set Breakpoint item. Then start execution by clicking on the RUN menu and releasing on the Start/Continue item, or by using the F5 key, or by clicking on the Start/Continue icon, which is a green arrow.

Finally, you can insert the instruction BRA *, whose opcode is $20FE, at the end of your program. Start the program as described above. After a moment, stop it by clicking on the Stop icon, which is a red tee.

To write a new program into memory, load the file blank.abs. This file merely has a value of zero written into location 800. Then type the program machine code into the Memory window. Note that you cannot type a program into the Assembly window. But after a program is manually entered into the Memory window, you can see the program in the Assembly window to verify that you got the correct machine code into memory.

A2.4 Running Examples From the AssemblyProgramFolder

AssemblyProgramFolder has files such as Ea4.txt, which contain examples from chapter 4 in this text book, and which you can copy and paste into an assembler source file such as Program.asm. To assemble Program.asm to generate an absolute file such as Program.abs, open the assembler by clicking on the assembler icon in the HIWARE toolbox. Type the source file's name, Program.asm, in the assembler's top center window. If the file name is already in the window, instead of retyping the file name, click on the icon immediately to the right of this window. The status of the assembly, including error messages, is output into the bottom window. When you have no errors, the Program.abs file is written. Program.abs can then be simulated as in §A2.3.

A2.5 Downloading to a 'B32 Board

You can use the HIWARE software to download and debug a program in a file, such as p.11.abs or program.abs, on the Motorola M68HC12B32EVB board as your target. Before you attempt to run your first program on a target, you should begin by simulating the program on the HIWAVE simulator, as in §A2.4. After you are comfortable with the simulator's operation, you should run it on actual hardware. First, make sure that W3 and W4 are in their 0 position to configure the board for EVB mode. Then connect the target's DB-9 connector to the personal computer COM1 port. Apply 5 volt power to the target. You should always apply the 5V power after all connections are made, and you should never change a connector while power is applied to the 'B32 board. Pull the Component menu to the Select_Target item, which brings up a dialog box. In that box, select the Asciimon target name. (If COM1 is not available, use another COM port, but change HIWAVE's communictations by pulling the Monitor menu to the Communication item, and change the port in the dialog box that appears).

HIWAVE behaves the same when used as a downloader/debugger as when used as a simulator. Follow §A2.4's procedures to run an example on the target microcontroller.

Having changed HIWAVE from a simulator to a downloader/debugger, you can pull the File menu to the Save Project item and release. This will save the current configuration so that when you restart HIWAVE it will start as a downloader/debugger, and you won't have to repeat the above procedure each time you use HIWAVE as a downloader/debugger. To change HIWAVE back to a simulator, pull the Components menu to the Set Target item, and click on the Sim entry.

A2.6 POD-Mode BDM Interface

You can run your program on a target, such as an Technological Arts Adapt-812 board, or an Axiom PB68HC12A4 board. This technique utilizes the state-of-the-art background debug module BDM in your target, providing a debugger that runs in the M68HC12B32EVB board (called the POD) that is isolated from the target.

Begin by running a program on the HIWAVE simulator, and then running it on the 'B32 board, as described in §A2.5. After you are comfortable with the simulator's and 'B32 board's operation, reconnect the W3 to its 0 position and W4 to its 1 position to configure the board for POD mode, and reset the POD. Plug the 6-wire cable into the POD's W11 connector and the target's BDM connector. Match pin 1 with the cable/connector pin 1. Restart HIWAVE. You should be able to duplicate what you did in the 'B32 board, running it on the Adapt-812 or PB68HC12A4 board as a target. You can use a more powerful Motorola BDI debugger, in place of the POD, and other target microcontrollers can be run using the POD or similar BDI debugger.

A2.7 Techniques for HiWare Tools

We have had some experiences with HiWare tools that might help you use them more efficiently. We add a note here on our suggestions to help you with this powerful software.

A problem with the current version is that when you change project files, the compiler/linker/HIWAVE debugger may read or write the wrong files, or fail to find the files it needs. We found that by shutting down all HiWare programs, and starting them up again, the problem goes away. But you do not have to restart the computer. If you have verified that the paths to the files are correct, but you are unable to access them through the compiler/linker/HIWAVE debugger, then try restarting all HiWare programs "from scratch," The same remedy is suggested when the HIWAVE simulator or debugger fails to execute single-step commands, or breakpoints, correctly.

When dealing with different environments such as your own PC running Windows 95, workstations running Windows NT, and a PC running Windows 98 in the laboratory, keep separate complete project folders for each environment, and copy the source code from one to another folder. That way, you will spend less time readjusting the paths to your programs and HiWare's applications when you switch platforms.

We hope that the CD-ROM supplied through HiWare makes your reading of this book much more profitable and enjoyable. We have found it to be most helpful in debugging our examples and problem solutions.

INDEX